AMUSEMENTS IN MATHEMATICS

by

HENRY ERNEST DUDENEY

In Mathematicks he was greater
Than Tycho Brahe or Erra Pater:
For he, by geometrick scale,
Could take the size of pots of ale;
Resolve, by sines and tangents, straight,
If bread or butter wanted weight;
And wisely tell what hour o' th' day
The clock does strike by algebra.
BUTLER'S *Hudibras.*

DOVER PUBLICATIONS, INC.

NEW YORK

This Dover edition, first published in 1958, is an unabridged and unaltered republication of the work originally published by Thomas Nelson & Sons, Limited, in 1917, and is published by special arrangement with Thomas Nelson & Sons, Limited.

This Dover edition also contains a Note on British Coins and Stamps and a Note on the Game of Cricket specially prepared for the American audience.

International Standard Book Number: 0-486-20473-1
Library of Congress Catalog Card Number: 58-13360

Manufactured in the United States of America

Dover Publications, Inc.
180 Varick Street
New York, N. Y. 10014

This edition is dedicated to the memory of

Henry E. Dudeney

1847 – 1930

with the hope that it will keep alive for many years his remarkable contributions to intellectual and mathematical recreations.

PREFACE

IN issuing this volume of my Mathematical Puzzles, of which some have appeared in periodicals and others are given here for the first time, I must acknowledge the encouragement that I have received from many unknown correspondents, at home and abroad, who have expressed a desire to have the problems in a collected form, with some of the solutions given at greater length than is possible in magazines and newspapers. Though I have included a few old puzzles that have interested the world for generations, where I felt that there was something new to be said about them, the problems are in the main original. It is true that some of these have become widely known through the press, and it is possible that the reader may be glad to know their source.

On the question of Mathematical Puzzles in general there is, perhaps, little more to be said than I have written elsewhere. The history of the subject entails nothing short of the actual story of the beginnings and development of exact thinking in man. The historian must start from the time when man first succeeded in counting his ten fingers and in dividing an apple into two approximately equal parts. Every puzzle that is worthy of consideration can be referred to mathematics and logic. Every man, woman, and child who tries to " reason out " the answer to the simplest puzzle is working, though not of necessity consciously, on mathematical lines. Even those puzzles that we have no way of attacking except by haphazard attempts can be brought under a method of what has been called " glorified trial "—a system of shortening our labours by avoiding or eliminating what our reason tells us is useless. It is, in fact, not easy to say sometimes where the " empirical " begins and where it ends.

When a man says, " I have never solved a puzzle in my life," it is difficult to know exactly what he means, for every intelligent individual is doing it every day. The unfortunate inmates of our lunatic asylums are sent there expressly because they cannot solve puzzles—because they have lost their powers of reason. If there were no puzzles to solve, there would be no questions to ask ; and if there were no questions to be asked, what a world it would be ! We should all be equally omniscient, and conversation would be useless and idle.

It is possible that some few exceedingly sober-minded mathematicians, who are impatient of any terminology in their favourite science but the academic, and who object to the elusive x and y appearing under any other names, will have wished that various problems had been presented in a less popular dress and introduced with a less flippant phraseology. I can only refer them to the first word of my title and remind them that we are primarily out to be amused—not, it is true, without some hope of picking up morsels of knowledge by the way. If the manner is light, I can only say, in the words of Touchstone, that it is " an ill-favoured thing, sir, but my own ; a poor humour of mine, sir."

As for the question of difficulty, some of the puzzles, especially in the Arithmetical and Algebraical category, are quite easy. Yet some of those examples that look the simplest should not be passed over without a little consideration, for now and again it will be found that there is some more or less subtle pitfall or trap into which the reader may be apt to fall. It is good exercise to cultivate the habit of being very wary over the exact wording of a puzzle. It teaches exactitude and caution. But some of the problems are very hard nuts indeed, and not unworthy of the attention of the advanced mathematician. Readers will doubtless select according to their individual tastes.

In many cases only the mere answers are given. This leaves the beginner something to do on his own behalf in working out the method of solution, and saves space that would be wasted from the point of view of the advanced student. On the other hand, in particular cases where it seemed likely to interest, I have given rather extensive solutions and treated problems in a general manner. It will often be found that the notes on one problem will serve to elucidate a good many others in the book ; so that the reader's difficulties will sometimes be found cleared up as he advances. Where it is possible to say a thing in a manner that may be " understanded of the people " generally, I prefer to use this simple phraseology, and so engage the attention and interest of a larger public. The mathematician will in such cases have no difficulty in expressing the matter under consideration in terms of his familiar symbols.

I have taken the greatest care in reading the proofs, and trust that any errors that may have crept in are very few. If any such should occur, I can only plead, in the words of Horace, that " good Homer sometimes nods," or, as the bishop put it, " Not even the youngest curate in my diocese is infallible."

I have to express my thanks in particular to the proprietors of *The Strand Magazine, Cassell's Magazine, The Queen, Tit-Bits,* and *The Weekly Dispatch* for their courtesy in allowing me to reprint some of the puzzles that have appeared in their pages.

THE AUTHORS' CLUB
March 25, 1917

CONTENTS

A Note on British Coins and Stamps

SEVERAL of Dudeney's puzzles are based upon British money, which might not be familiar to the American reader. The basic unit is the pound, which contains twenty shillings; the shilling in turn contains twelve pence (pennies). The symbol for pound is £, which is placed before the number like our dollar sign. The symbols for shilling and pence are, respectively, s. and d., which are placed, like our cent sign, after the number. In place notation these symbols are sometimes omitted. Common ways of writing British prices are:

£2–6–6	Two pounds, six shillings, six pence
2 6 6	Two pounds, six shillings, six pence
7/6 or 7–6	or 7 s. 6 d., seven shillings and six pence.

Just as in American money one does not write down 101¢, but $1.01, in British notation one writes down twenty-five shillings as £1 5 s.; eighteen pence as 1 s. 6 d.; and twelve pence as 1 s.

At the time that Dudeney wrote his books British coinage was not exactly the same as it is today. These coins could have been available to his readers:

Name	Notation	Value
Farthing	¼ d.	One quarter of a penny
Half penny	½ d.	One half of a penny
Penny	1 d.	
Two pence	2 d.	Two pence (1/6th of a shilling)
Three pence	3 d.	Three pence (1/4th of a shilling)
Four pence	4 d.	Four pence (1/3rd of a shilling)
Six pence	6 d.	Six pence (1/2 of a shilling)

Name	Notation	Value
Shilling	1 s.	Twelve pence
Florin	2 s.	Two shillings
Half crown	2 s. 6 d.	Two shillings and six pence
Double florin	4 s.	Four shillings
Crown	5 s.	Five shillings
Half sovereign	10 s.	Ten shillings (half a pound)
Sovereign	£1	Twenty shillings or one pound
Guinea	£1 1 s.	Twenty-one shillings

Many of these coins are no longer in circulation. The guinea, even though the coin itself is no longer in circulation, is still used as a value term for £1 1 s., and its exact multiples. £5 5 s., for example, is still spoken of as five guineas.

The postal stamps in circulation in Dudeney's time were of the following denominations: ½ d., 1 d., 1½ d., 2 d., 2½ d., 3 d., 4 d., 5 d., 6 d., 9 d., 10 d., 1 s., 2 s. 6 d., 5 s., 10 s., £1, £5.

A Note on the Game of Cricket

A few comments may make Puzzles 387 and 388 easier for Americans who do not understand cricket.

CRICKET is played on a field at the centre of which two "wickets" are set 22 yards apart. Each wicket consists of three "stumps," wooden sticks 28″ high and 9″ apart, on top of which are lightly balanced two smaller sticks called "bails." A line is marked on either side of the wicket (the "bowling crease") and another (the "popping crease") 4′ in front.

The game is played between two teams of 11 men each. The defending or "fielding" side are in the field all at once; two "batsmen" from the opposite ("batting") side must also be in the field. One of the latter (the "striker") stands with one foot inside the popping crease at one end of the pitch, and at the other a member of the fielding side (the "bowler") delivers the ball, with a straight-armed action (bowls it), having one foot behind the bowling crease at his own end. His aim is to put the batsman out by knocking down the wicket, or by having the ball caught by a member of the fielding side after the batsman has hit it, or by having one of the batsmen trapped in the zone between the popping creases when a fieldsman knocks down the nearer wicket. (There are other ways of putting a batsman out, which are not relevant.)

The batsman endeavours to score "runs," primarily by hitting the ball far enough away from any fieldsman to enable him and his partner to change ends. To score, the two batsmen *must* step over, or put their bats over, the popping creases at the ends opposite to where they were when the ball was bowled.

A change of ends constitutes "one run." If, however, one or other batsman omits to comply with the rule about crossing the popping crease, the run cannot be scored. This is why the umpires in puzzle

387 call "three short"—in their haste, Messrs. Dumkins and Podder both failed to observe the rule. Similarly, in the second innings, they call "two short."

The question arising, therefore, is simply: which, if either, pair of batsmen succeeded in accomplishing a legal change of ends.

For puzzle 388, the American reader should remember that there must always be *two* batsmen in play; therefore, when ten men of the 11 in the side have been put out, the surviving player cannot continue by himself and is spoken of as "not out" or as having "carried his bat."

"Drawing stumps" is the conclusion of the day's play. The "duck's egg" (more commonly just "duck") is, of course, zero.

AMUSEMENTS IN MATHEMATICS.

ARITHMETICAL AND ALGEBRAICAL PROBLEMS.

"And what was he?
Forsooth, a great arithmetician."
Othello, I. i.

THE puzzles in this department are roughly thrown together in classes for the convenience of the reader. Some are very easy, others quite difficult. But they are not arranged in any order of difficulty—and this is intentional, for it is well that the solver should not be warned that a puzzle is just what it seems to be. It may, therefore, prove to be quite as simple as it looks, or it may contain some pitfall into which, through want of care or over-confidence, we may stumble.

Also, the arithmetical and algebraical puzzles are not separated in the manner adopted by some authors, who arbitrarily require certain problems to be solved by one method or the other. The reader is left to make his own choice and determine which puzzles are capable of being solved by him on purely arithmetical lines.

MONEY PUZZLES.

"Put not your trust in money, but put your money in trust."
OLIVER WENDELL HOLMES.

1.—A POST-OFFICE PERPLEXITY.

IN every business of life we are occasionally perplexed by some chance question that for the moment staggers us. I quite pitied a young lady in a branch post-office when a gentleman entered and deposited a crown on the counter with this request : " Please give me some twopenny stamps, six times as many penny stamps, and make up the rest of the money in twopence-halfpenny stamps." For a moment she seemed bewildered, then her brain cleared, and with a smile she handed over stamps in exact fulfilment of the order. How long would it have taken you to think it out ?

2.—YOUTHFUL PRECOCITY.

THE precocity of some youths is surprising. One is disposed to say on occasion, " That boy of yours is a genius, and he is certain to do great things when he grows up ; " but past experience has taught us that he invariably becomes quite an ordinary citizen. It is so often the case, on the contrary, that the dull boy becomes a great man. You never can tell. Nature loves to present to us these queer paradoxes. It is well known that those wonderful " lightning calculators," who now and again surprise the world by their feats, lose all their mysterious powers directly they are taught the elementary rules of arithmetic.

A boy who was demolishing a choice banana was approached by a young friend, who, regarding him with envious eyes, asked, "How much did you pay for that banana, Fred ? " The prompt answer was quite remarkable in its way : " The man what I bought it of receives just half as many sixpences for sixteen dozen dozen bananas as he gives bananas for a fiver."

Now, how long will it take the reader to say correctly just how much Fred paid for his rare and refreshing fruit ?

3.—AT A CATTLE MARKET.

THREE countrymen met at a cattle market. " Look here," said Hodge to Jakes, " I'll give you six of my pigs for one of your horses, and then you'll have twice as many animals here as I've got." " If that's your way of doing business," said Durrant to Hodge, " I'll give you fourteen of my sheep for a horse, and then you'll have three times as many animals as I." " Well, I'll go better than that," said Jakes to Durrant ; " I'll give you four cows for a horse, and then you'll have six times as many animals as I've got here."

No doubt this was a very primitive way of bartering animals, but it is an interesting little puzzle to discover just how many animals Jakes, Hodge, and Durrant must have taken to the cattle market.

4.—THE BEANFEAST PUZZLE.

A NUMBER of men went out together on a beanfeast. There were four parties invited—namely, 25 cobblers, 20 tailors, 18 hatters, and 12 glovers. They spent altogether £6, 13s. It was found that five cobblers spent as much as four tailors ; that twelve tailors spent as much as nine hatters; and that six hatters spent as much as eight glovers. The puzzle is to find out how much each of the four parties spent.

5.—A QUEER COINCIDENCE.

SEVEN men, whose names were Adams, Baker, Carter, Dobson, Edwards, Francis, and Gudgeon, were recently engaged in play. The name of the particular game is of no consequence. They had agreed that whenever a player won a game he should double the money of each of the other players—that is, he was to give the players just as much money as they had already in their pockets. They played seven games, and, strange to say, each won a game in turn, in the order in which their names are given. But a more curious coincidence is this—that when they had finished play each of the seven men had exactly the same amount —two shillings and eightpence—in his pocket. The puzzle is to find out how much money each man had with him before he sat down to play.

6.—A CHARITABLE BEQUEST.

A MAN left instructions to his executors to distribute once a year exactly fifty-five shillings among the poor of his parish; but they were only to continue the gift so long as they could make it in different ways, always giving eighteenpence each to a number of women and half a crown each to men. During how many years could the charity be administered ? Of course, by " different ways " is meant a different number of men and women every time.

7.—THE WIDOW'S LEGACY.

A GENTLEMAN who recently died left the sum of £8,000 to be divided among his widow, five sons, and four daughters. He directed that every son should receive three times as much as a daughter, and that every daughter should have twice as much as their mother. What was the widow's share ?

8.—INDISCRIMINATE CHARITY.

A CHARITABLE gentleman, on his way home one night, was appealed to by three needy persons in succession for assistance. To the first person he gave one penny more than half the money he had in his pocket ; to the second person he gave twopence more than half the money he then had in his pocket ; and to the third person he handed over threepence more than half of what he had left. On entering his house he had only one penny in his pocket. Now, can you say exactly how much money that gentleman had on him when he started for home ?

9.—THE TWO AEROPLANES.

A MAN recently bought two aeroplanes, but afterwards found that they would not answer the purpose for which he wanted them. So he sold them for £600 each, making a loss of 20 per cent. on one machine and a profit of 20 per cent. on the other. Did he make a profit on the whole transaction, or a loss ? And how much ?

10.—BUYING PRESENTS.

" WHOM do you think I met in town last week, Brother William ? " said Uncle Benjamin. " That old skinflint Jorkins. His family had been taking him around buying Christmas presents. He said to me, ' Why cannot the government abolish Christmas, and make the giving of presents punishable by law ? I came out this morning with a certain amount of money in my pocket, and I find I have spent just half of it. In fact, if you will believe me, I take home just as many shillings as I had pounds, and half as many pounds as I had shillings. It is monstrous!'" Can you say exactly how much money Jorkins had spent on those presents ?

11.—THE CYCLISTS' FEAST.

'TWAS last Bank Holiday, so I've been told,
  Some cyclists rode abroad in glorious weather.
Resting at noon within a tavern old,
  They all agreed to have a feast together.
" Put it all in one bill, mine host," they said,
  " For every man an equal share will pay."
The bill was promptly on the table laid,
  And four pounds was the reckoning that day.
But, sad to state, when they prepared to square,
  'Twas found that two had sneaked outside and fled.
So, for two shillings more than his due share
  Each honest man who had remained was bled.
They settled later with those rogues, no doubt.
How many were they when they first set out ?

12.—A QUEER THING IN MONEY.

IT will be found that £66, 6s. 6d. equals 15,918 pence. Now, the four 6's added together make 24, and the figures in 15,918 also add to 24. It is a curious fact that there is only one other sum of money, in pounds, shillings, and pence (all similarly repetitions of one figure), of which the digits shall add up the same as the digits of the amount in pence. What is the other sum of money ?

13.—A NEW MONEY PUZZLE.

THE largest sum of money that can be written in pounds, shillings, pence, and farthings, using each of the nine digits once and only once, is

£98,765, 4s. 3½d. Now, try to discover the smallest sum of money that can be written down under precisely the same conditions. There must be some value given for each denomination—pounds, shillings, pence, and farthings—and the nought may not be used. It requires just a little judgment and thought.

14.—SQUARE MONEY.

"This is queer," said McCrank to his friend. "Twopence added to twopence is fourpence, and twopence multiplied by twopence is also fourpence." Of course, he was wrong in thinking you can multiply money by money. The multiplier must be regarded as an abstract number. It is true that two feet multiplied by two feet will make four square feet. Similarly, two pence multiplied by two pence will produce four square pence! And it will perplex the reader to say what a " square penny " is. But we will assume for the purposes of our puzzle that twopence multiplied by twopence is fourpence. Now, what two amounts of money will produce the next smallest possible result, the same in both cases, when added or multiplied in this manner? The two amounts need not be alike, but they must be those that can be paid in current coins of the realm.

15.—POCKET MONEY.

What is the largest sum of money—all in current silver coins and no four-shilling piece—that I could have in my pocket without being able to give change for a half-sovereign?

16.—THE MILLIONAIRE'S PERPLEXITY.

Mr. Morgan G. Bloomgarten, the millionaire, known in the States as the Clam King, had, for his sins, more money than he knew what to do with. It bored him. So he determined to persecute some of his poor but happy friends with it. They had never done him any harm, but he resolved to inoculate them with the " source of all evil." He therefore proposed to distribute a million dollars among them and watch them go rapidly to the bad. But he was a man of strange fancies and superstitions, and it was an inviolable rule with him never to make a gift that was not either one dollar or some power of seven—such as 7, 49, 343, 2,401, which numbers of dollars are produced by simply multiplying sevens together. Another rule of his was that he would never give more than six persons exactly the same sum. Now, how was he to distribute the 1,000,000 dollars? You may distribute the money among as many people as you like, under the conditions given.

17.—THE PUZZLING MONEY-BOXES.

Four brothers—named John, William, Charles, and Thomas—had each a money-box. The boxes were all given to them on the same day, and they at once put what money they had into them ; only, as the boxes were not very large, they first changed the money into as few coins as possible. After they had done this, they told one another how much money they had saved, and it was found that if John had had 2s. more in his box than at present, if William had had 2s. less, if Charles had had twice as much, and if Thomas had had half as much, they would all have had exactly the same amount.

Now, when I add that all four boxes together contained 45s., and that there were only six coins in all in them, it becomes an entertaining puzzle to discover just what coins were in each box.

18.—THE MARKET WOMEN.

A number of market women sold their various products at a certain price per pound (different in every case), and each received the same amount—2s. 2¼d. What is the greatest number of women there could have been? The price per pound in every case must be such as could be paid in current money.

19.—THE NEW YEAR'S EVE SUPPERS.

The proprietor of a small London café has given me some interesting figures. He says that the ladies who come alone to his place for refreshment spend each on an average eighteenpence, that the unaccompanied men spend half a crown each, and that when a gentleman brings in a lady he spends half a guinea. On New Year's Eve he supplied suppers to twenty-five persons, and took five pounds in all. Now, assuming his averages to have held good in every case, how was his company made up on that occasion? Of course, only single gentlemen, single ladies, and pairs (a lady and gentleman) can be supposed to have been present, as we are not considering larger parties.

20.—BEEF AND SAUSAGES.

" A neighbour of mine," said Aunt Jane, " bought a certain quantity of beef at two shillings a pound, and the same quantity of sausages at eighteenpence a pound. I pointed out to her that if she had divided the same money equally between beef and sausages she would have gained two pounds in the total weight. Can you tell me exactly how much she spent ? "

" Of course, it is no business of mine," said Mrs. Sunniborne ; " but a lady who could pay such prices must be somewhat inexperienced in domestic economy."

" I quite agree, my dear," Aunt Jane replied, " but you see that is not the precise point under discussion, any more than the name and morals of the tradesman."

21.—A DEAL IN APPLES.

I paid a man a shilling for some apples, but they were so small that I made him throw in two extra apples. I find that made them cost just a penny a dozen less than the first price he asked. How many apples did I get for my shilling ?

22.—A DEAL IN EGGS.

A man went recently into a dairyman's shop to buy eggs. He wanted them of various qualities.

The salesman had new-laid eggs at the high price of fivepence each, fresh eggs at one penny each, eggs at a halfpenny each, and eggs for election-eering purposes at a greatly reduced figure, but as there was no election on at the time the buyer had no use for the last. However, he bought some of each of the three other kinds and obtained exactly one hundred eggs for eight and fourpence. Now, as he brought away exactly the same number of eggs of two of the three qualities, it is an interesting puzzle to determine just how many he bought at each price.

23.—THE CHRISTMAS-BOXES.

SOME years ago a man told me he had spent one hundred English silver coins in Christmas-boxes, giving every person the same amount, and it cost him exactly £1, 10s. 1d. Can you tell just how many persons received the present, and how he could have managed the distribution? That odd penny looks queer, but it is all right.

24.—A SHOPPING PERPLEXITY.

Two ladies went into a shop where, through some curious eccentricity, no change was given, and made purchases amounting together to less than five shillings. "Do you know," said one lady, "I find I shall require no fewer than six current coins of the realm to pay for what I have bought." The other lady considered a moment, and then exclaimed: "By a peculiar coincidence, I am exactly in the same dilemma." "Then we will pay the two bills together." But, to their astonishment, they still required six coins. What is the smallest possible amount of their purchases—both different?

25.—CHINESE MONEY.

THE Chinese are a curious people, and have strange inverted ways of doing things. It is said that they use a saw with an upward pressure instead of a downward one, that they plane a deal board by pulling the tool toward them instead of pushing it, and that in building a house they first construct the roof and, having raised that into position, proceed to work downwards. In money the currency of the country consists of taels of fluctuating value. The tael became thinner and thinner until 2,000 of them piled together made less than three inches in height. The common cash consists of brass coins of varying thicknesses, with a round, square, or triangular hole in the centre, as in our illustration.

These are strung on wires like buttons. Supposing that eleven coins with round holes are worth fifteen ching-changs, that eleven with square holes are worth sixteen ching-changs, and that eleven with triangular holes are worth seventeen ching-changs, how can a Chinaman give me change for half a crown, using no coins other than the three mentioned? A ching-chang is worth exactly twopence and four-fifteenths of a ching-chang.

26.—THE JUNIOR CLERK'S PUZZLE.

Two youths, bearing the pleasant names of Moggs and Snoggs, were employed as junior clerks by a merchant in Mincing Lane. They were both engaged at the same salary—that is, commencing at the rate of £50 a year, payable half-yearly. Moggs had a yearly rise of £10, and Snoggs was offered the same, only he asked, for reasons that do not concern our puzzle, that he might take his rise at £2, 10s. half-yearly, to which his employer (not, perhaps, unnaturally!) had no objection.

Now we come to the real point of the puzzle. Moggs put regularly into the Post Office Savings Bank a certain proportion of his salary, while Snoggs saved twice as great a proportion of his, and at the end of five years they had together saved £268, 15s. How much had each saved? The question of interest can be ignored.

27.—GIVING CHANGE.

EVERY one is familiar with the difficulties that frequently arise over the giving of change, and how the assistance of a third person with a few coins in his pocket will sometimes help us to set the matter right. Here is an example. An Englishman went into a shop in New York and bought goods at a cost of thirty-four cents. The only money he had was a dollar, a three-cent piece, and a two-cent piece. The tradesman had only a half-dollar and a quarter-dollar. But another customer happened to be present, and when asked to help produced two dimes, a five-cent piece, a two-cent piece, and a one-cent piece. How did the tradesman manage to give change? For the benefit of those readers who are not familiar with the American coinage, it is only necessary to say that a dollar is a hundred cents and a dime ten cents. A puzzle of this kind should rarely cause any difficulty if attacked in a proper manner.

28.—DEFECTIVE OBSERVATION.

OUR observation of little things is frequently defective, and our memories very liable to lapse. A certain judge recently remarked in a case that he had no recollection whatever of putting the wedding-ring on his wife's finger. Can you correctly answer these questions without having the coins in sight? On which side of a penny is the date given? Some people are so unobservant that, although they are handling the coin nearly every day of their lives, they are at a loss to answer this simple question. If I lay a penny flat on the table, how many other pennies can I place around it, every one also lying flat on the table, so that they all touch the first one? The geometrician will, of course, give the answer at once, and not need to make any experiment.

He will also know that, since all circles are similar, the same answer will necessarily apply to any coin. The next question is a most interesting one to ask a company, each person writing down his answer on a slip of paper, so that no one shall be helped by the answers of others. What is the greatest number of three-penny-pieces that may be laid flat on the surface of a half-crown, so that no piece lies on another or overlaps the surface of the half-crown ? It is amazing what a variety of different answers one gets to this question. Very few people will be found to give the correct number. Of course the answer must be given without looking at the coins.

29.—THE BROKEN COINS.

A man had three coins—a sovereign, a shilling, and a penny—and he found that exactly the same fraction of each coin had been broken away. Now, assuming that the original intrinsic value of these coins was the same as their nominal value—that is, that the sovereign was worth a pound, the shilling worth a shilling, and the penny worth a penny—what proportion of each coin has been lost if the value of the three remaining fragments is exactly one pound ?

30.—TWO QUESTIONS IN PROBABILITIES.

There is perhaps no class of puzzle over which people so frequently blunder as that which involves what is called the theory of probabilities. I will give two simple examples of the sort of puzzle I mean. They are really quite easy, and yet many persons are tripped up by them. A friend recently produced five pennies and said to me : " In throwing these five pennies at the same time, what are the chances that at least four of the coins will turn up either all heads or all tails ? " His own solution was quite wrong, but the correct answer ought not to be hard to discover. Another person got a wrong answer to the following little puzzle which I heard him propound : " A man placed three sovereigns and one shilling in a bag. How much should be paid for permission to draw one coin from it ? " It is, of course, understood that you are as likely to draw any one of the four coins as another.

31.—DOMESTIC ECONOMY.

Young Mrs. Perkins, of Putney, writes to me as follows : " I should be very glad if you could give me the answer to a little sum that has been worrying me a good deal lately. Here it is : We have only been married a short time, and now, at the end of two years from the time when we set up housekeeping, my husband tells me that he finds we have spent a third of his yearly income in rent, rates, and taxes, one-half in domestic expenses, and one-ninth in other ways. He has a balance of £190 remaining in the bank. I know this last, because he accidentally left out his pass-book the other day, and I peeped into it. Don't you think that a husband ought to give his wife his entire confidence in

his money matters ? Well, I do ; and—will you believe it ?—he has never told me what his income really is, and I want, very naturally, to find out. Can you tell me what it is from the figures I have given you ? "

Yes ; the answer can certainly be given from the figures contained in Mrs. Perkins's letter. And my readers, if not warned, will be practically unanimous in declaring the income to be something absurdly in excess of the correct answer !

32.—THE EXCURSION TICKET PUZZLE.

When the big flaming placards were exhibited at the little provincial railway station, announcing that the Great —— Company would run cheap excursion trains to London for the Christmas holidays, the inhabitants of Mudley-cum-Turmits were in quite a flutter of excitement. Half an hour before the train came in the little booking office was crowded with country passengers, all bent on visiting their friends in the great Metropolis. The booking clerk was unaccustomed to dealing with crowds of such a dimension, and he told me afterwards, while wiping his manly brow, that what caused him so much trouble was the fact that these rustics paid their fares in such a lot of small money.

He said that he had enough farthings to supply a West End draper with change for a week, and a sufficient number of threepenny pieces for the congregations of three parish churches. " That excursion fare," said he, " is nineteen shillings and ninepence, and I should like to know in just how many different ways it is possible for such an amount to be paid in the current coin of this realm."

Here, then, is a puzzle : In how many different ways may nineteen shillings and ninepence be paid in our current coin ? Remember that the fourpenny-piece is not now current.

33.—A PUZZLE IN REVERSALS.

Most people know that if you take any sum of money in pounds, shillings, and pence, in which the number of pounds (less than £12) exceeds that of the pence, reverse it (calling the pounds pence and the pence pounds), find the difference, then reverse and add this difference, the result is always £12, 18s. 11d. But if we omit the condition, " less than £12," and allow nought to represent shillings or pence—(1) What is the lowest amount to which the rule will not apply ? (2) What is the highest amount to which it will apply ? Of course, when reversing such a sum as £14, 15s. 3d. it may be written £3, 16s. 2d., which is the same as £3, 15s. 14d.

34.—THE GROCER AND DRAPER.

A country " grocer and draper " had two rival assistants, who prided themselves on their rapidity in serving customers. The young man on the grocery side could weigh up two one-pound parcels of sugar per minute, while the drapery assistant could cut three one-yard lengths of cloth in the same time. Their employer, one slack day, set them a race, giving the grocer a barrel of sugar and telling him to weigh up forty-eight one-pound parcels of sugar while the draper divided a roll of forty-eight yards of cloth into yard pieces. The two men were interrupted together by customers for nine minutes, but the draper was disturbed seventeen times as long as the grocer. What was the result of the race ?

35.—JUDKINS'S CATTLE.

HIRAM B. JUDKINS, a cattle-dealer of Texas, had five droves of animals, consisting of oxen, pigs, and sheep, with the same number of animals in each drove. One morning he sold all that he had to eight dealers. Each dealer bought the same number of animals, paying seventeen dollars for each ox, four dollars for each pig, and two dollars for each sheep ; and Hiram received in all three hundred and one dollars. What is the greatest number of animals he could have had ? And how many would there be of each kind ?

36.—BUYING APPLES.

As the purchase of apples in small quantities has always presented considerable difficulties, I think it well to offer a few remarks on this subject. We all know the story of the smart boy who, on being told by the old woman that she was selling her apples at four for threepence, said : " Let me see ! Four for threepence ; that's three for twopence, two for a penny, one for nothing—I'll take one ! "

There are similar cases of perplexity. For example, a boy once picked up a penny apple from a stall, but when he learnt that the woman's pears were the same price he exchanged it, and was about to walk off. " Stop ! " said the woman. " You haven't paid me for the pear ! " " No," said the boy, " of course not. I gave you the apple for it." " But you didn't pay for the apple ! " " Bless the woman ! You don't expect me to pay for the apple and the pear too ! " And before the poor creature could get out of the tangle the boy had disappeared.

Then, again, we have the case of the man who gave a boy sixpence and promised to repeat the gift as soon as the youngster had made it into ninepence. Five minutes later the boy returned. " I have made it into ninepence," he said, at the same time handing his benefactor threepence. " How do you make that out ? " he was asked. " I bought threepennyworth of apples." " But that does not make it into ninepence ! " " I should rather think it did," was the boy's reply. " The apple woman has threepence, hasn't she ? Very well, I have threepennyworth of apples, and I have just given you the other threepence. What's that but ninepence ? "

I cite these cases just to show that the small boy really stands in need of a little instruction in the art of buying apples. So I will give a simple poser dealing with this branch of commerce.

An old woman had apples of three sizes for sale—one a penny, two a penny, and three a penny. Of course two of the second size and three of the third size were respectively equal to one apple of the largest size. Now, a gentleman who had an equal number of boys and girls gave his children sevenpence to be spent amongst them all on these apples. The puzzle is to give each child an equal distribution of apples. How was the sevenpence spent, and how many children were there ?

37.—BUYING CHESTNUTS.

THOUGH the following little puzzle deals with the purchase of chestnuts, it is not itself of the " chestnut " type. It is quite new. At first sight it has certainly the appearance of being of the " nonsense puzzle " character, but it is all right when properly considered.

A man went to a shop to buy chestnuts. He said he wanted a pennyworth, and was given five chestnuts. " It is not enough ; I ought to have a sixth," he remarked. " But if I give you one chestnut more." the shopman replied, " you will have five too many." Now, strange to say, they were both right. How many chestnuts should the buyer receive for half a crown ?

38.—THE BICYCLE THIEF.

HERE is a little tangle that is perpetually cropping up in various guises. A cyclist bought a bicycle for £15 and gave in payment a cheque for £25. The seller went to a neighbouring shopkeeper and got him to change the cheque for him, and the cyclist, having received his £10 change, mounted the machine and disappeared. The cheque proved to be valueless, and the salesman was requested by his neighbour to refund the amount he had received. To do this, he was compelled to borrow the £25 from a friend, as the cyclist forgot to leave his address, and could not be found. Now, as the bicycle cost the salesman £11, how much money did he lose altogether ?

39.—THE COSTERMONGER'S PUZZLE.

" How much did yer pay for them oranges, Bill ? "

" I ain't a-goin' to tell yer, Jim. But I beat the old cove down fourpence a hundred."

" What good did that do yer ? "

" Well, it meant five more oranges on every ten shillin's-worth."

Now, what price did Bill actually pay for the oranges ? There is only one rate that will fit in with his statements.

AGE AND KINSHIP PUZZLES.

" The days of our years are threescore years and ten."—*Psalm* xc. 10.

FOR centuries it has been a favourite method of propounding arithmetical puzzles to pose them in the form of questions as to the age of an individual. They generally lend themselves to very easy solution by the use of algebra, though often the difficulty lies in stating them correctly. They may be made very complex and may demand considerable ingenuity, but no general laws can well be laid down for their solution. The solver must use his own sagacity.

As for puzzles in relationship or kinship, it is quite curious how bewildering many people find these things. Even in ordinary conversation, some statement as to relationship, which is quite clear in the mind of the speaker, will immediately tie the brains of other people into knots. Such expressions as " He is my uncle's son-in-law's sister " convey absolutely nothing to some people without a detailed and laboured explanation. In such cases the best course is to sketch a brief genealogical table, when the eye comes immediately to the assistance of the brain. In these days, when we have a growing lack of respect for pedigrees, most people have got out of the habit of rapidly drawing such tables, which is to be regretted, as they would save a lot of time and brain racking on occasions.

40.—MAMMA'S AGE.

TOMMY : " How old are you, mamma ? "
Mamma : " Let me think, Tommy. Well, our three ages add up to exactly seventy years."
Tommy : " That's a lot, isn't it ? And how old are you, papa ? "
Papa : " Just six times as old as you, my son."
Tommy : " Shall I ever be half as old as you, papa ? "
Papa : " Yes, Tommy ; and when that happens our three ages will add up to exactly twice as much as to-day."
Tommy : " And supposing I was born before you, papa ; and supposing mamma had forgot all about it, and hadn't been at home when I came ; and supposing——"
Mamma : " Supposing, Tommy, we talk about bed. Come along, darling. You'll have a headache."

Now, if Tommy had been some years older he might have calculated the exact ages of his parents from the information they had given him. Can you find out the exact age of mamma ?

41.—THEIR AGES.

" My husband's age," remarked a lady the other day, " is represented by the figures of my own age reversed. He is my senior, and the difference between our ages is one-eleventh of their sum."

42.—THE FAMILY AGES.

WHEN the Smileys recently received a visit from the favourite uncle, the fond parents had all the five children brought into his presence. First came Billie and little Gertrude, and the uncle was informed that the boy was exactly twice as old as the girl. Then Henrietta arrived, and it was pointed out that the combined ages of herself and Gertrude equalled twice the age of Billie. Then Charlie came running in, and somebody remarked that now the combined ages of the two boys were exactly twice the combined ages of the two girls. The uncle was ex-

pressing his astonishment at these coincidences when Janet came in. " Ah ! uncle," she exclaimed, " you have actually arrived on my twenty-first birthday ! " To this Mr. Smiley added the final staggerer : " Yes, and now the combined ages of the three girls are exactly equal to twice the combined ages of the two boys." Can you give the age of each child ?

43.—MRS. TIMPKINS'S AGE.

EDWIN : " Do you know, when the Timpkinses married eighteen years ago Timpkins was three times as old as his wife, and to-day he is just twice as old as she ? "
Angelina : " Then how old was Mrs. Timpkins on the wedding day ? "
Can you answer Angelina's question ?

44.—A CENSUS PUZZLE.

MR. and Mrs. Jorkins have fifteen children, all born at intervals of one year and a half. Miss Ada Jorkins, the eldest, had an objection to state her age to the census man, but she admitted that she was just seven times older than little Johnnie, the youngest of all. What was Ada's age ? Do not too hastily assume that you have solved this little poser. You may find that you have made a bad blunder !

45.—MOTHER AND DAUGHTER.

" MOTHER, I wish you would give me a bicycle," said a girl of twelve the other day.
" I do not think you are old enough yet, my dear," was the reply. " When I am only three times as old as you are you shall have one."
Now, the mother's age is forty-five years. When may the young lady expect to receive her present ?

46.—MARY AND MARMADUKE.

MARMADUKE : " Do you know, dear, that in seven years' time our combined ages will be sixty-three years ? "
Mary : " Is that really so ? And yet it is a fact that when you were my present age you were twice as old as I was then. I worked it out last night."
Now, what are the ages of Mary and Marmaduke ?

47.—ROVER'S AGE.

" Now, then, Tommy, how old is Rover ? " Mildred's young man asked her brother.
" Well, five years ago," was the youngster's reply, " sister was four times older than the dog, but now she is only three times as old."
Can you tell Rover's age ?

48.—CONCERNING TOMMY'S AGE.

TOMMY SMART was recently sent to a new school. On the first day of his arrival the teacher asked him his age, and this was his curious reply : " Well, you see, it is like this. At the time I was born—I forget the year—my only sister, Ann, happened to be just one-quarter the age of mother, and she is now one-third the age of father." "That's all very well," said the teacher, "but what I want is not the age of your sister Ann, but your own age." "I was just coming to that," Tommy answered; "I am just a quarter of mother's present age, and in four years' time I shall be a quarter the age of father. Isn't that funny?"

This was all the information that the teacher could get out of Tommy Smart. Could you have told, from these facts, what was his precise age? It is certainly a little puzzling.

49.—NEXT-DOOR NEIGHBOURS.

THERE were two families living next door to one another at Tooting Bec—the Jupps and the Simkins. The united ages of the four Jupps amounted to one hundred years, and the united ages of the Simkins also amounted to the same. It was found in the case of each family that the sum obtained by adding the squares of each of the children's ages to the square of the mother's age equalled the square of the father's age. In the case of the Jupps, however, Julia was one year older than her brother Joe, whereas Sophy Simkin was two years older than her brother Sammy. What was the age of each of the eight individuals?

50.—THE BAG OF NUTS.

THREE boys were given a bag of nuts as a Christmas present, and it was agreed that they should be divided in proportion to their ages, which together amounted to $17\frac{1}{2}$ years. Now the bag contained 770 nuts, and as often as Herbert took four Robert took three, and as often as Herbert took six Christopher took seven. The puzzle is to find out how many nuts each had, and what were the boys' respective ages.

51.—HOW OLD WAS MARY?

HERE is a funny little age problem, by the late Sam Loyd, which has been very popular in the United States. Can you unravel the mystery?

The combined ages of Mary and Ann are forty-four years, and Mary is twice as old as Ann was when Mary was half as old as Ann will be when Ann is three times as old as Mary was when Mary was three times as old as Ann. How old is Mary? That is all, but can you work it out? If not, ask your friends to help you, and watch the shadow of bewilderment creep over their faces as they attempt to grip the intricacies of the question.

52.—QUEER RELATIONSHIPS.

"SPEAKING of relationships," said the Parson, at a certain dinner-party, "our legislators are getting the marriage law into a frightful tangle. Here, for example, is a puzzling case that has come under my notice. Two brothers married two sisters. One man died and the other man's wife also died. Then the survivors married."

"The man married his deceased wife's sister, under the recent Act?" put in the Lawyer.

"Exactly. And therefore, under the civil law, he is legally married and his child is legitimate. But, you see, the man is the woman's deceased husband's brother, and therefore, also under the civil law, she is not married to him and her child is illegitimate."

"He is married to her and she is not married to him!" said the Doctor.

"Quite so. And the child is the legitimate son of his father, but the illegitimate son of his mother."

"Undoubtedly 'the law is a hass,'" the Artist exclaimed, "if I may be permitted to say so," he added, with a bow to the Lawyer.

"Certainly," was the reply. "We lawyers try our best to break in the beast to the service of man. Our legislators are responsible for the breed."

"And this reminds me," went on the Parson, "of a man in my parish who married the sister of his widow. This man——"

"Stop a moment, sir," said the Professor. "Married the sister of his widow? Do you marry dead men in your parish?"

"No; but I will explain that later. Well, this man has a sister of his own. Their names are Stephen Brown and Jane Brown. Last week a young fellow turned up whom Stephen introduced to me as his nephew. Naturally, I spoke of Jane as his aunt, but, to my astonishment, the youth corrected me, assuring me that, though he was the nephew of Stephen, he was not the nephew of Jane, the sister of Stephen. This perplexed me a good deal, but it is quite correct."

The Lawyer was the first to get at the heart of the mystery. What was his solution?

53.—hEARD ON THE TUBE RAILWAY.

FIRST LADY: "And was he related to you, dear?"

Second Lady: "Oh, yes. You see, that gentleman's mother was my mother's mother-in-law, but he is not on speaking terms with my papa."

First Lady: "Oh, indeed!" (But you could see that she was not much wiser.)

How was the gentleman related to the Second Lady?

54.—A FAMILY PARTY.

A CERTAIN family party consisted of 1 grandfather, 1 grandmother, 2 fathers, 2 mothers, 4 children, 3 grandchildren, 1 brother, 2 sisters, 2 sons, 2 daughters, 1 father-in-law, 1 mother-in-law, and 1 daughter-in-law. Twenty-three people, you will say. No; there were only seven persons present. Can you show how this might be?

55.—A MIXED PEDIGREE.

JOSEPH BLOGGS: "I can't follow it, my dear boy. It makes me dizzy!"

John Snoggs: "It's very simple. Listen again! You happen to be my father's brother-in-law, my brother's father-in-law, and also my father-in-law's brother. You see, my father was——"

But Mr. Bloggs refused to hear any more. Can the reader show how this extraordinary triple relationship might have come about?

56.—WILSON'S POSER.

" SPEAKING of perplexities——" said Mr. Wilson, throwing down a magazine on the table in the commercial room of the Railway Hotel.

" Who was speaking of perplexities? " inquired Mr. Stubbs.

" Well, then, reading about them, if you want to be exact—it just occurred to me that perhaps you three men may be interested in a little matter connected with myself."

It was Christmas Eve, and the four commercial travellers were spending the holiday at Grassminster. Probably each suspected that the others had no homes, and perhaps each was conscious of the fact that he was in that predicament himself. In any case they seemed to be perfectly comfortable, and as they drew round the cheerful fire the conversation became general.

" What is the difficulty? " asked Mr. Packhurst.

" There's no difficulty in the matter, when you rightly understand it. It is like this. A man named Parker had a flying-machine that would carry two. He was a venturesome sort of chap—reckless, I should call him—and he had some bother in finding a man willing to risk his life in making an ascent with him. However, an uncle of mine thought he would chance it, and one fine morning he took his seat in the machine and she started off well. When they were up about a thousand feet, my nephew suddenly——"

" Here, stop, Wilson! What was your nephew doing there? You said your uncle," interrupted Mr. Stubbs.

" Did I? Well, it does not matter. My nephew suddenly turned to Parker and said that the engine wasn't running well, so Parker called out to my uncle——"

" Look here," broke in Mr. Waterson, " we are getting mixed. Was it your uncle or your nephew? Let's have it one way or the other."

" What I said is quite right. Parker called out to my uncle to do something or other, when my nephew——"

" There you are again, Wilson," cried Mr. Stubbs; " once for all, are we to understand that both your uncle and your nephew were on the machine? "

" Certainly. I thought I made that clear. Where was I? Well, my nephew shouted back to Parker——"

" Phew! I'm sorry to interrupt you again, Wilson, but we can't get on like this. Is it true that the machine would only carry two? "

" Of course. I said at the start that it only carried two."

" Then what in the name of aerostation do you mean by saying that there were three persons on board? " shouted Mr. Stubbs.

" Who said there were three? "

" You have told us that Parker, your uncle, and your nephew went up on this blessed flying-machine."

" That's right."

" And the thing would only carry two! "

" Right again."

" Wilson, I have known you for some time as a truthful man and a temperate man," said Mr. Stubbs, solemnly. " But I am afraid since you took up that new line of goods you have overworked yourself."

" Half a minute, Stubbs," interposed Mr. Waterson. " I see clearly where we all slipped a cog. Of course, Wilson, you meant us to understand that Parker is either your uncle or your nephew. Now we shall be all right if you will just tell us whether Parker is your uncle or nephew."

" He is no relation to me whatever."

The three men sighed and looked anxiously at one another. Mr. Stubbs got up from his chair to reach the matches, Mr. Packhurst proceeded to wind up his watch, and Mr. Waterson took up the poker to attend to the fire. It was an awkward moment, for at the season of goodwill nobody wished to tell Mr. Wilson exactly what was in his mind.

" It's curious," said Mr. Wilson, very deliberately, " and it's rather sad, how thickheaded some people are. You don't seem to grip the facts. It never seems to have occurred to either of you that your uncle and my nephew are one and the same man."

" What! " exclaimed all three together.

" Yes; David George Linklater is my uncle, and he is also my nephew. Consequently, I am both his uncle and nephew. Queer, isn't it? I'll explain how it comes about."

Mr. Wilson put the case so very simply that the three men saw how it might happen without any marriage within the prohibited degrees. Perhaps the reader can work it out for himself.

CLOCK PUZZLES.

" Look at the clock! "
Ingoldsby Legends.

IN considering a few puzzles concerning clocks and watches, and the times recorded by their hands under given conditions, it is well that a particular convention should always be kept in mind. It is frequently the case that a solution requires the assumption that the hands can actually record a time involving a minute fraction of a second. Such a time, of course, cannot be really indicated. Is the puzzle, therefore, impossible of solution? The conclusion deduced from a logical syllogism depends for its truth on the two premises assumed, and it is the same in mathematics. Certain things are antecedently assumed, and the answer depends entirely on the truth of those assumptions.

" If two horses," says Lagrange, " can pull a load of a certain weight, it is natural to suppose that four horses could pull a load of double that weight, six horses a load of three times that weight. Yet, strictly speaking, such is not the case. For the inference is based on the assumption that the four horses pull alike in amount and direction, which in practice can scarcely ever be the case. It so happens that we are frequently led in our reckonings to results which diverge widely from reality ; But the fault is not the fault of mathematics ; for mathematics always gives back to us exactly what we have put into it. The ratio was constant according to that supposition. The result is founded upon that supposition. If the supposition is false the result is necessarily false."

If one man can reap a field in six days, we say two men will reap it in three days, and three men will do the work in two days. We here assume, as in the case of Lagrange's horses, that all the men are exactly equally capable of work. But we assume even more than this. For when three men get together they may waste time in gossip or play ; or, on the other hand, a spirit of rivalry may spur them on to greater diligence. We may assume any conditions we like in a problem, provided they be clearly expressed and understood, and the answer will be in accordance with those conditions.

57.—WHAT WAS THE TIME ?

" I SAY, Rackbrane, what is the time ? " an acquaintance asked our friend the professor the other day. The answer was certainly curious.

" If you add one quarter of the time from noon till now to half the time from now till noon to-morrow, you will get the time exactly."

What was the time of day when the professor spoke ?

58.—A TIME PUZZLE.

How many minutes is it until six o'clock if fifty minutes ago it was four times as many minutes past three o'clock ?

59.—A PUZZLING WATCH.

A FRIEND pulled out his watch and said, " This watch of mine does not keep perfect time ; I must have it seen to. I have noticed that the minute hand and the hour hand are exactly together every sixty-five minutes." Does that watch gain or lose, and how much per hour ?

60.—THE WAPSHAW'S WHARF MYSTERY.

THERE was a great commotion in Lower Thames Street on the morning of January 12, 1887. When the early members of the staff arrived at Wapshaw's Wharf they found that the safe had been broken open, a considerable sum of money removed, and the offices left in great disorder. The night watchman was nowhere to be found, but nobody who had been acquainted with him for one moment suspected him to be guilty of the robbery. In this belief the proprietors were confirmed when, later in the day, they were informed that the poor fellow's body had been picked up by the River Police. Certain marks of violence pointed to the fact that he had been brutally attacked and thrown into the river. A watch found in his pocket had stopped, as is invariably the case in such circumstances, and this was a valuable clue to the time of the outrage. But a very stupid officer (and we invariably find one or two stupid individuals in the most intelligent bodies of men) had actually amused himself by turning the hands round and round, trying to set the watch going again. After he had been severely reprimanded for this serious indiscretion, he was asked whether he could remember the time that was indicated by the watch when found. He replied that he could not, but he recollected that the hour hand and minute hand were exactly together, one above the other, and the second hand had just passed the forty-ninth second. More than this he could not remember.

What was the exact time at which the watchman's watch stopped ? The watch is, of course, assumed to have been an accurate one.

61.—CHANGING PLACES.

THE above clock face indicates a little before 42 minutes past 4. The hands will again point at exactly the same spots a little after 23 minutes past 8. In fact, the hands will have changed places. How many times do the hands of a clock change places between three o'clock p.m. and midnight ? And out of all the pairs of times indicated by these changes, what is the exact time when the minute hand will be nearest to the point IX ?

62.—THE CLUB CLOCK.

ONE of the big clocks in the Cogitators' Club was found the other night to have stopped just when, as will be seen in the illustration, the second hand was exactly midway between the other two hands. One of the members proposed to some of his friends that they should tell him the exact time when (if the clock had not

stopped) the second hand would next again have been midway between the minute hand and the hour hand. Can you find the correct time that it would happen ?

63.—THE STOP-WATCH.

WE have here a stop-watch with three hands. The second hand, which travels once round the face in a minute, is the one with the little ring at its end near the centre. Our dial indicates the exact time when its owner stopped the watch. You will notice that the three hands are nearly equidistant. The hour and minute hands point to spots that are exactly a third of the circumference apart, but the second hand is a little too advanced. An exact equidistance for the three hands is not possible. Now, we want to know what the time will be when the three hands are next at exactly the same distances as shown from one another. Can you state the time ?

64.—THE THREE CLOCKS.

ON Friday, April 1, 1898, three new clocks were all set going precisely at the same time—twelve noon. At noon on the following day it was found that clock A had kept perfect time, that clock B had gained exactly one minute, and that clock C had lost exactly one minute. Now, supposing that the clocks B and C had not been regulated, but all three allowed to go on as they had begun, and that they maintained the same rates of progress without stopping, on what date and at what time of day would all three pairs of hands again point at the same moment at twelve o'clock ?

65.—THE RAILWAY STATION CLOCK.

A CLOCK hangs on the wall of a railway station, 71 ft. 9 in. long and 10 ft. 4 in. high. Those are the dimensions of the wall, not of the clock ! While waiting for a train we noticed that the hands of the clock were pointing in opposite directions, and were parallel to one of the diagonals of the wall. What was the exact time ?

66.—THE VILLAGE SIMPLETON.

A FACETIOUS individual who was taking a long walk in the country came upon a yokel sitting on a stile. As the gentleman was not quite sure of his road, he thought he would make inquiries of the local inhabitant; but at the first glance he jumped too hastily to the conclusion that he had dropped on the village idiot. He therefore decided to test the fellow's intelligence by first putting to him the simplest question he could think of, which was, " What day of the week is this, my good man ? " The following is the smart answer that he received :—

" When the day after to-morrow is yesterday, to-day will be as far from Sunday as to-day was from Sunday when the day before yesterday was to-morrow."

Can the reader say what day of the week it was ? It is pretty evident that the countryman was not such a fool as he looked. The gentleman went on his road a puzzled but a wiser man.

LOCOMOTION AND SPEED PUZZLES.

" The race is not to the swift."—*Ecclesiastes* ix. 11.

67.—AVERAGE SPEED.

IN a recent motor ride it was found that we had gone at the rate of ten miles an hour, but we did the return journey over the same route, owing to the roads being more clear of traffic, at fifteen miles an hour. What was our average speed ? Do not be too hasty in your answer to this simple little question, or it is pretty certain that you will be wrong.

68.—THE TWO TRAINS.

I PUT this little question to a stationmaster, and his correct answer was so prompt that I am convinced there is no necessity to seek talented railway officials in America or elsewhere.

Two trains start at the same time, one from London to Liverpool, the other from Liverpool to London. If they arrive at their destinations one hour and four hours respectively after passing one another, how much faster is one train running than the other?

69.—THE THREE VILLAGES.

I SET out the other day to ride in a motor-car from Acrefield to Butterford, but by mistake I took the road going *via* Cheesebury, which is nearer Acrefield than Butterford, and is twelve miles to the left of the direct road I should have travelled. After arriving at Butterford I found that I had gone thirty-five miles. What are the three distances between these villages, each being a whole number of miles? I may mention that the three roads are quite straight.

half miles an hour, so that it takes her just six hours to make the double journey. Can any of you tell me how far it is from the bottom of the hill to the top?"

71.—SIR EDWYN DE TUDOR.

IN the illustration we have a sketch of Sir Edwyn de Tudor going to rescue his lady-love, the fair Isabella, who was held a captive by a neighbouring wicked baron. Sir Edwyn calculated that if he rode fifteen miles an hour he would arrive at the castle an hour too soon, while if he rode ten miles an hour he would get there just an hour too late. Now, it was of the first importance that he should arrive at the exact time appointed, in order that the rescue that he had planned should be a success, and the time of the tryst was five o'clock, when the captive lady would be taking her afternoon tea. The puzzle is to discover exactly how far Sir Edwyn de Tudor had to ride.

70.—DRAWING HER PENSION.

"SPEAKING of odd figures," said a gentleman who occupies some post in a Government office, "one of the queerest characters I know is an old lame widow who climbs up a hill every week to draw her pension at the village post office. She crawls up at the rate of a mile and a half an hour and comes down at the rate of four and a

72.—THE HYDROPLANE QUESTION.

THE inhabitants of Slocomb-on-Sea were greatly excited over the visit of a certain flying man. All the town turned out to see the flight of the wonderful hydroplane, and, of course, Dobson and his family were there. Master Tommy was in good form, and informed his father that Englishmen made better airmen than Scotsmen and Irishmen because they are not so heavy. "How do you make that out?" asked Mr. Dobson. "Well, you see," Tommy replied, "it is true that in Ireland there are men of Cork and in Scotland men of Ayr, which is better still, but in England there are lightermen." Unfortunately it had to be explained to Mrs. Dobson, and this took the edge off the thing. The hydroplane flight was from Slocomb to the neighbouring watering-place Poodleville—five miles distant. But there was a strong wind, which so helped the airman that he made the outward journey in the short time of ten minutes, though it took him an hour to get back to the starting point at Slocomb, with the wind dead against him. Now, how long would the ten miles have taken him if there had been a perfect calm? Of course, the hydroplane's engine worked uniformly throughout.

73.—DONKEY RIDING.

DURING a visit to the seaside Tommy and Evangeline insisted on having a donkey race over the mile course on the sands. Mr. Dobson and some of his friends whom he had met on the beach acted as judges, but, as the donkeys were familiar acquaintances and declined to part company the whole way, a dead heat was unavoidable. However, the judges, being stationed at different points on the course, which was marked off in quarter-miles, noted the following results :—The first three-quarters were run in six and three-quarter minutes, the first half-mile took the same time as the second half, and the third quarter was run in exactly the same time as the last quarter. From these results Mr. Dobson amused himself in discovering just how long it took those two donkeys to run the whole mile. Can you give the answer?

74.—THE BASKET OF POTATOES.

A MAN had a basket containing fifty potatoes. He proposed to his son, as a little recreation, that he should place these potatoes on the ground in a straight line. The distance between the first and second potatoes was to be one yard, between the second and third three yards, between the third and fourth five yards, between the fourth and fifth seven yards, and so on—an increase of two yards for every successive potato laid down. Then the boy was to pick them up and put them in the basket one at a time, the basket being placed beside the first potato. How far would the boy have to travel to accomplish the feat of picking them all up? We will not consider the journey involved in placing the potatoes, so that he starts from the basket with them all laid out.

75.—THE PASSENGER'S FARE.

AT first sight you would hardly think there was matter for dispute in the question involved in the following little incident, yet it took the two persons concerned some little time to come to an agreement. Mr. Smithers hired a motor-car to take him from Addleford to Clinkerville and back again for £3. At Bakenham, just midway, he picked up an acquaintance, Mr. Tompkins, and agreed to take him on to Clinkerville and bring him back to Bakenham on the return journey. How much should he have charged the passenger? That is the question. What was a reasonable fare for Mr. Tompkins?

DIGITAL PUZZLES.

"Nine worthies were they called."
DRYDEN: *The Flower and the Leaf.*

I GIVE these puzzles, dealing with the nine digits, a class to themselves, because I have always thought that they deserve more consideration than they usually receive. Beyond the mere trick of "casting out nines," very little seems to be generally known of the laws involved in these problems, and yet an acquaintance with the properties of the digits often supplies, among other uses, a certain number of arithmetical checks that are of real value in the saving of labour. Let me give just one example—the first that occurs to me.

If the reader were required to determine whether or not $15,763,530,163,289$ is a square number, how would he proceed? If the number had ended with a 2, 3, 7, or 8 in the digits place, of course he would know that it could not be a square, but there is nothing in its apparent form to prevent its being one. I suspect that in such a case he would set to work, with a sigh or a groan, at the laborious task of extracting the square root. Yet if he had given a little attention to the study of the digital properties of numbers, he would settle the question in this simple way. The sum of the digits is 59, the sum of which is 14, the sum of which is 5 (which I call the "digital root"), and therefore I know that the number cannot be a square, and for this reason. The digital root of successive square numbers from 1 upwards is always 1, 4, 7, or 9, and can never be anything else. In fact, the series, 1, 4, 9, 7, 7, 9, 4, 1, 9, is repeated into infinity. The analogous series for triangular numbers is 1, 3, 6, 1, 6, 3, 1, 9, 9. So here we have a similar negative check, for a number cannot be triangular (that is, $\frac{n^2+n}{2}$) if its digital root be 2, 4, 5, 7, or 8.

76.—THE BARREL OF BEER.

A MAN bought an odd lot of wine in barrels and one barrel containing beer. These are shown in the illustration, marked with the number of gallons that each barrel contained. He sold a quantity of the wine to one man and twice the quantity to another, but kept the beer to himself. The puzzle is to point out which barrel contains beer. Can you say which one it is? Of course, the man sold the barrels just as he

bought them, without manipulating in any way the contents.

77.—DIGITS AND SQUARES.

IT will be seen in the diagram that we have so arranged the nine digits in a square that the number in the second row is twice that in the

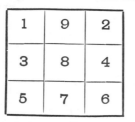

first row, and the number in the bottom row three times that in the top row. There are three other ways of arranging the digits so as to produce the same result. Can you find them?

78.—ODD AND EVEN DIGITS.

THE odd digits, 1, 3, 5, 7, and 9, add up 25, while the even figures, 2, 4, 6, and 8, only add up 20. Arrange these figures so that the odd ones and the even ones add up alike. Complex and improper fractions and recurring decimals are not allowed.

79.—THE LOCKERS PUZZLE.

A MAN had in his office three cupboards, each containing nine lockers, as shown in the dia-gram. He told his clerk to place a different one-figure number on each locker of cupboard A, and to do the same in the case of B, and of C. As we are here allowed to call nought a digit, and he was not prohibited from using nought as a number, he clearly had the option of omitting any one of ten digits from each cupboard.

Now, the employer did not say the lockers were to be numbered in any numerical order, and he was surprised to find, when the work was done, that the figures had apparently been mixed up indiscriminately. Calling upon his clerk for an explanation, the eccentric lad stated that the notion had occurred to him so to arrange the figures that in each case they formed a simple addition sum, the two upper rows of figures producing the sum in the lowest row. But the most surprising point was this : that he had so arranged them that the addition in A gave the smallest possible sum, that the addition in C gave the largest possible sum, and that all the nine digits in the three totals were different. The puzzle is to show how this could be done. No decimals are allowed and the nought may not appear in the hundreds place.

80.—THE THREE GROUPS.

THERE appeared in " Nouvelles Annales de Mathématiques " the following puzzle as a modification of one of my " Canterbury Puzzles." Arrange the nine digits in three groups of two, three, and four digits, so that the first two numbers when multiplied together make the third. Thus, $12 \times 483 = 5,796$. I now also propose to include the cases where there are one, four, and four digits, such as $4 \times 1,738 = 6,952$. Can you find all the possible solutions in both cases ?

81.—THE NINE COUNTERS.

I HAVE nine counters, each bearing one of the nine digits, 1, 2, 3, 4, 5, 6, 7, 8 and 9. I arranged them on the table in two groups, as shown in the illustration, so as to form two multiplication sums, and found that both sums gave the same product. You will find that 158 multiplied by 23 is 3,634, and that 79 multiplied by 46 is also 3,634. Now, the puzzle I propose is to rearrange the counters so as to get as large a product as possible. What is the best way of placing them? Remember both groups must multiply to the same amount, and there must be three counters multiplied by two in one case, and two multiplied by two counters in the other, just as at present.

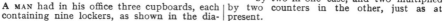

82.—THE TEN COUNTERS.

In this case we use the nought in addition to the 1, 2, 3, 4, 5, 6, 7, 8, 9. The puzzle is, as in the last case, so to arrange the ten counters that the products of the two multiplications shall be the same, and you may here have one or more figures in the multiplier, as you choose. The above is a very easy feat ; but it is also required to find the two arrangements giving pairs of the highest and lowest products possible. Of course every counter must be used, and the cipher may not be placed to the left of a row of figures where it would have no effect. Vulgar fractions or decimals are not allowed.

83.—DIGITAL MULTIPLICATION.

Here is another entertaining problem with the nine digits, the nought being excluded. Using each figure once, and only once, we can form two multiplication sums that have the same product, and this may be done in many ways. For example, 7 × 658 and 14 × 329 contain all the digits once, and the product in each case is the same—4,606. Now, it will be seen that the sum of the digits in the product is 16, which is neither the highest nor the lowest sum so obtainable. Can you find the solution of the problem that gives the lowest possible sum of digits in the common product ? Also that which gives the highest possible sum ?

84.—THE PIERROT'S PUZZLE.

Pierrot as in the example shown, or to place a single figure on one side and three figures on the other. If we only used three digits instead of four, the only possible ways are these : 3 multiplied by 51 equals 153, and 6 multiplied by 21 equals 126.

85.—THE CAB NUMBERS.

A London policeman one night saw two cabs drive off in opposite directions under suspicious circumstances. This officer was a particularly careful and wide-awake man, and he took out his pocket-book to make an entry of the numbers of the cabs, but discovered that he had lost his pencil. Luckily, however, he found a small piece of chalk, with which he marked the two numbers on the gateway of a wharf close by. When he returned to the same spot on his beat he stood and looked again at the numbers, and noticed this peculiarity, that all the nine digits (no nought) were used and that no figure was repeated, but that if he multiplied the two numbers together they again produced the nine digits, all once, and once only. When one of the clerks arrived at the wharf in the early morning, he observed the chalk marks and carefully rubbed them out. As the policeman could not remember them, certain mathematicians were then consulted as to whether there was any known method for discovering all the pairs of numbers that have the peculiarity that the officer had noticed ; but they knew of none. The investigation, however, was

The Pierrot in the illustration is standing in a posture that represents the sign of multiplication. He is indicating the peculiar fact that 15 multiplied by 93 produces exactly the same figures (1,395), differently arranged. The puzzle is to take any four digits you like (all different) and similarly arrange them so that the number formed on one side of the Pierrot when multiplied by the number on the other side shall produce the same figures. There are very few ways of doing it, and I shall give all the cases possible. Can you find them all ? You are allowed to put two figures on each side of the

interesting, and the following question out of many was proposed : What two numbers, containing together all the nine digits, will, when multiplied together, produce another number (the *highest possible*) containing also all the nine digits ? The nought is not allowed anywhere.

86.—QUEER MULTIPLICATION.

If I multiply 51,249,876 by 3 (thus using all the nine digits once, and once only), I get 153,749,628 (which again contains all the nine digits once). Similarly, if I multiply 16,583,742 by 9 the result is 149,253,678, where in each case all the nine digits are used. Now, take 6 as your multiplier and try to arrange the remaining eight digits so as to produce by multiplication a number containing all nine once, and once only. You will find it far from easy, but it can be done.

87.—THE NUMBER-CHECKS PUZZLE.

WHERE a large number of workmen are employed on a building it is customary to provide every man with a little disc bearing his number. These are hung on a board by the men as they arrive, and serve as a check on punctuality. Now, I once noticed a foreman remove a number of these checks from his board and place them on a split-ring which he carried in his pocket. This at once gave me the idea for a good puzzle. In fact, I will confide to my readers that this is just how ideas for puzzles arise. You cannot really create an idea: it happens—and you have to be on the alert to seize it when it does so happen.

It will be seen from the illustration that there are ten of these checks on a ring, numbered 1 to 9 and 0. The puzzle is to divide them into three groups without taking any off the ring, so that the first group multiplied by the second makes the third group. For example, we can divide them into the three groups, 2—8 9 0 7—1 5 4 6 3, by bringing the 6 and the 3 round to the 4, but unfortunately the first two when multiplied together do not make the third. Can you separate them correctly ? Of course you may have as many of the checks as you like in any group. The puzzle calls for some ingenuity, unless you have the luck to hit on the answer by chance.

88.—DIGITAL DIVISION.

IT is another good puzzle so to arrange the nine digits (the nought excluded) into two groups so that one group when divided by the other produces a given number without remainder. For example, 1 3 4 5 8 divided by 6 7 2 9 gives 2. Can the reader find similar arrangements producing 3, 4, 5, 6, 7, 8, and 9 respectively ? Also,

can he find the pairs of smallest possible numbers in each case ? Thus, 1 4 6 5 8 divided by 7 3 2 9 is just as correct for 2 as the other example we have given, but the numbers are higher.

89.—ADDING THE DIGITS.

IF I write the sum of money, £987, 5s. 4½d., and add up the digits, they sum to 36. No digit has thus been used a second time in the amount or addition. This is the largest amount possible under the conditions. Now find the smallest possible amount, pounds, shillings, pence, and farthings being all represented. You need not use more of the nine digits than you choose, but no digit may be repeated throughout. The nought is not allowed.

90.—THE CENTURY PUZZLE.

CAN you write 100 in the form of a mixed number, using all the nine digits once, and only once ? The late distinguished French mathematician, Edouard Lucas, found seven different ways of doing it, and expressed his doubts as to there being any other ways. As a matter of fact there are just eleven ways and no more. Here is one of them, $91\frac{5742}{638}$. Nine of the other ways have similarly two figures in the integral part of the number, but the eleventh expression has only one figure there. Can the reader find this last form ?

91.—MORE MIXED FRACTIONS.

WHEN I first published my solution to the last puzzle, I was led to attempt the expression of all numbers in turn up to 100 by a mixed fraction containing all the nine digits. Here are twelve numbers for the reader to try his hand at : 13, 14, 15, 16, 18, 20, 27, 36, 40, 69, 72, 94. Use every one of the nine digits once, and only once, in every case.

92.—DIGITAL SQUARE NUMBERS.

HERE are the nine digits so arranged that they form four square numbers : 9, 81, 324, 576. Now, can you put them all together so as to form a single square number—(I) the smallest possible, and (II) the largest possible ?

93.—THE MYSTIC ELEVEN.

CAN you find the largest possible number containing any nine of the ten digits (calling nought a digit) that can be divided by 11 without a remainder ? Can you also find the smallest possible number produced in the same way that is divisible by 11 ? Here is an example, where the digit 5 has been omitted : 896743012. This number contains nine of the digits and is divisible by 11, but it is neither the largest nor the smallest number that will work.

94.—THE DIGITAL CENTURY.

1   2   3   4   5   6   7   8   9 = 100.

IT is required to place arithmetical signs between the nine figures so that they shall equal

100. Of course, you must not alter the present numerical arrangement of the figures. Can you give a correct solution that employs (1) the fewest possible signs, and (2) the fewest possible separate strokes or dots of the pen? That is, it is necessary to use as few signs as possible, and those signs should be of the simplest form. The signs of addition and multiplication (+ and ×) will thus count as two strokes, the sign of subtraction (−) as one stroke, the sign of division (÷) as three, and so on.

95.—THE FOUR SEVENS.

In the illustration Professor Rackbrane is seen demonstrating one of the little posers with which he is accustomed to entertain his class. He believes that by taking his pupils off the beaten tracks he is the better able to secure their attention, and to induce original and ingenious methods of thought. He has, it will be seen, just shown how four 5's may be written with simple arithmetical signs so as to represent 100. Every juvenile reader will see at a glance that his example is quite correct. Now, what he wants you to do is this: Arrange four 7's (neither more nor less) with arithmetical signs so that they shall represent 100. If he had said we were to use four 9's we might at once have written 99⅜, but the four 7's call for rather more ingenuity. Can you discover the little trick?

96.—THE DICE NUMBERS.

I HAVE a set of four dice, not marked with spots in the ordinary way, but with Arabic figures, as shown in the illustration. Each die, of course, bears the numbers 1 to 6. When put together

they will form a good many different numbers. As represented they make the number 1246. Now, if I make all the different four-figure

numbers that are possible with these dice (never putting the same figure more than once in any number), what will they all add up to? You are allowed to turn the 6 upside down, so as to represent a 9. I do not ask, or expect, the reader to go to all the labour of writing out the full list of numbers and then adding them up. Life is not long enough for such wasted energy. Can you get at the answer in any other way?

VARIOUS ARITHMETICAL AND ALGEBRAICAL PROBLEMS.

"Variety's the very spice of life,
That gives it all its flavour."
COWPER: *The Task.*

97.—THE SPOT ON THE TABLE.

A BOY, recently home from school, wished to give his father an exhibition of his precocity. He pushed a large circular table into the corner of the room, as shown in the illustration, so that it touched both walls, and he then pointed to a spot of ink on the extreme edge.

"Here is a little puzzle for you, pater," said the youth. "That spot is exactly eight inches from one wall and nine inches from the other. Can you tell me the diameter of the table without measuring it?"

The boy was overheard to tell a friend, "It

fairly beat the guv'nor;" but his father is known to have remarked to a City acquaintance that he solved the thing in his head in a minute. I often wonder which spoke the truth.

98.—ACADEMIC COURTESIES.

IN a certain mixed school, where a special feature was made of the inculcation of good manners, they had a curious rule on assembling every morning. There were twice as many girls as boys. Every girl made a bow to every other girl, to every boy, and to the teacher. Every boy made a bow to every other boy, to every girl, and to the teacher. In all there were nine hundred bows made in that model academy every morning. Now, can you say exactly how many boys there were in the school? If you are not very careful, you are likely to get a good deal out in your calculation.

99.—THE THIRTY-THREE PEARLS.

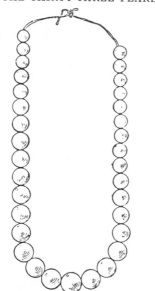

"A MAN I know," said Teddy Nicholson at a certain family party, "possesses a string of thirty-three pearls. The middle pearl is the largest and best of all, and the others are so selected and arranged that, starting from one end, each successive pearl is worth £100 more than the preceding one, right up to the big pearl. From the other end the pearls increase in value by £150 up to the large pearl. The whole string is worth £65,000. What is the value of that large pearl?"

"Pearls and other articles of clothing," said Uncle Walter, when the price of the precious gem had been discovered, "remind me of Adam and Eve. Authorities, you may not know, differ as to the number of apples that were eaten by Adam and Eve. It is the opinion of some that Eve 8 (ate) and Adam 2 (too), a total of 10 only. But certain mathematicians have figured it out differently, and hold that Eve 8 and Adam 8, a total of 16. Yet the most recent investigators think the above figures entirely wrong, for if Eve 8 and Adam 8 2, the total must be 90."

"Well," said Harry, "it seems to me that if there were giants in those days, probably Eve 8 1 and Adam 8 2, which would give a total of 163."

"I am not at all satisfied," said Maud. "It seems to me that if Eve 8 1 and Adam 8 1 2, they together consumed 893."

"I am sure you are all wrong," insisted Mr. Wilson, "for I consider that Eve 8 1 4 Adam, and Adam 8 1 2 4 Eve, so we get a total of 8,938."

"But, look here," broke in Herbert. "If Eve 8 1 4 Adam and Adam 8 1 2 4 2 oblige Eve, surely the total must have been 82,056!"

At this point Uncle Walter suggested that they might let the matter rest. He declared it to be clearly what mathematicians call an indeterminate problem.

100.—THE LABOURER'S PUZZLE.

PROFESSOR RACKBRANE, during one of his rambles, chanced to come upon a man digging a deep hole.

"Good morning," he said. "How deep is that hole?"

"Guess," replied the labourer. "My height is exactly five feet ten inches."

"How much deeper are you going?" said the professor.

"I am going twice as deep," was the answer, "and then my head will be twice as far below ground as it is now above ground."

Rackbrane now asks if you could tell how deep that hole would be when finished.

101.—THE TRUSSES OF HAY.

FARMER TOMPKINS had five trusses of hay, which he told his man Hodge to weigh before delivering them to a customer. The stupid fellow weighed them two at a time in all possible ways, and informed his master that the weights in pounds were 110, 112, 113, 114, 115, 116, 117, 118, 120, and 121. Now, how was Farmer Tompkins to find out from these figures how much every one of the five trusses weighed singly? The reader may at first think that he ought to be told "which pair is which pair," or something of that sort, but it is quite unnecessary. Can you give the five correct weights?

102.—MR. GUBBINS IN A FOG.

MR. GUBBINS, a diligent man of business, was much inconvenienced by a London fog. The electric light happened to be out of order and he had to manage as best he could with two candles. His clerk assured him that though both were of the same length one candle would burn for four hours and the other for five hours. After he had been working some time he put the candles out as the fog had lifted, and he then noticed that what remained of one candle was exactly four times the length of what was left of the other.

When he got home that night Mr. Gubbins, who liked a good puzzle, said to himself, " Of course it is possible to work out just how long those two candles were burning to-day. I'll have a shot at it." But he soon found himself in a worse fog than the atmospheric one. Could you have assisted him in his dilemma ? How long were the candles burning ?

103.—PAINTING THE LAMP-POSTS.

TIM MURPHY and Pat Donovan were engaged by the local authorities to paint the lamp-posts in a certain street. Tim, who was an early riser, arrived first on the job, and had painted three on the south side when Pat turned up and pointed out that Tim's contract was for the north side. So Tim started afresh on the north side and Pat continued on the south. When Pat had finished his side he went across the street and painted six posts for Tim, and then the job was finished. As there was an equal number of lamp-posts on each side of the street, the simple question is : Which man painted the more lamp-posts, and just how many more ?

104.—CATCHING THE THIEF.

" Now, constable," said the defendant's counsel in cross-examination, " you say that the prisoner was exactly twenty-seven steps ahead of you when you started to run after him ? "

" Yes, sir."

" And you swear that he takes eight steps to your five ? "

" That is so."

" Then I ask you, constable, as an intelligent man, to explain how you ever caught him, if that is the case ? "

" Well, you see, I have got a longer stride. In fact, two of my steps are equal in length to five of the prisoner's. If you work it out, you will find that the number of steps I required would bring me exactly to the spot where I captured him."

Here the foreman of the jury asked for a few minutes to figure out the number of steps the constable must have taken. Can you also say how many steps the officer needed to catch the thief ?

105.—THE PARISH COUNCIL ELECTION.

HERE is an easy problem for the novice. At the last election of the parish council of Tittlebury-in-the-Marsh there were twenty-three candidates for nine seats. Each voter was qualified to vote for nine of these candidates or for any less number. One of the electors wants to know in just how many different ways it was possible for him to vote.

106.—THE MUDDLETOWN ELECTION.

AT the last Parliamentary election at Muddletown 5,473 votes were polled. The Liberal was elected by a majority of 18 over the Conservative, by 146 over the Independent, and by 575 over the Socialist. Can you give a simple rule for figuring out how many votes were polled for each candidate ?

107.—THE SUFFRAGISTS' MEETING.

AT a recent secret meeting of Suffragists a serious difference of opinion arose. This led to a split, and a certain number left the meeting. " I had half a mind to go myself," said the chairwoman, "and if I had done so, two-thirds of us would have retired." " True," said another member ; " but if I had persuaded my friends Mrs. Wild and Christine Armstrong to remain we should only have lost half our number." Can you tell how many were present at the meeting at the start ?

108.—THE LEAP-YEAR LADIES.

LAST leap-year ladies lost no time in exercising the privilege of making proposals of marriage. If the figures that reached me from an occult source are correct, the following represents the state of affairs in this country.

A number of women proposed once each, of whom one-eighth were widows. In consequence, a number of men were to be married of whom one-eleventh were widowers. Of the proposals made to widowers, one-fifth were declined. All the widows were accepted. Thirty-five forty-fourths of the widows married bachelors. One thousand two hundred and twenty-one spinsters were declined by bachelors. The number of spinsters accepted by bachelors was seven times the number of widows accepted by bachelors. Those are all the particulars that I was able to obtain. Now, how many women proposed ?

109.—THE GREAT SCRAMBLE.

AFTER dinner, the five boys of a household happened to find a parcel of sugar-plums. It was quite unexpected loot, and an exciting scramble ensued, the full details of which I will recount with accuracy, as it forms an interesting puzzle.

You see, Andrew managed to get possession of just two-thirds of the parcel of sugar-plums. Bob at once grabbed three-eighths of these, and Charlie managed to seize three-tenths also. Then young David swooped upon all that Andrew had left, except one-seventh, which Edgar artfully secured for himself by a cunning trick. Now the fun began in real earnest, for Andrew and Charlie jointly set upon Bob, who stumbled against the fender and dropped half of all that he had, which were equally picked up by David and Edgar, who had crawled under a table and were waiting. Next, Bob sprang on Charlie from a chair, and upset all the latter's collection on to the floor. Of this prize Andrew got just a quarter, Bob gathered up one-third, David got two-sevenths, while Charlie and Edgar divided equally what was left of that stock.

They were just thinking the fray was over

when David suddenly struck out in two directions at once, upsetting three-quarters of what Bob and Andrew had last acquired. The two latter, with the greatest difficulty, recovered five-eighths of it in equal shares, but the three others each carried off one-fifth of the same. Every sugar-plum was now accounted for, and they called a truce, and divided equally amongst them the remainder of the parcel. What is the smallest number of sugar-plums there could have been at the start, and what proportion did each boy obtain?

110.—THE ABBOT'S PUZZLE.

THE first English puzzlist whose name has come down to us was a Yorkshireman—no other than Alcuin, Abbot of Canterbury (A.D. 735–804). Here is a little puzzle from his works, which is at least interesting on account of its antiquity. " If 100 bushels of corn were distributed among 100 people in such a manner that each man received three bushels, each woman two, and each child half a bushel, how many men, women, and children were there ? "

Now, there are six different correct answers, if we exclude a case where there would be no women. But let us say that there were just five times as many women as men, then what is the correct solution ?

111.—REAPING THE CORN.

A FARMER had a square cornfield. The corn was all ripe for reaping, and, as he was short of men, it was arranged that he and his son should share the work between them. The farmer first cut one rod wide all round the square, thus leaving a smaller square of standing corn in the middle

of the field. " Now," he said to his son, " I have cut my half of the field, and you can do your share." The son was not quite satisfied as to the proposed division of labour, and as the village schoolmaster happened to be passing, he appealed to that person to decide the matter. He found the farmer was quite correct, provided there was no dispute as to the size of the field, and on this point they were agreed. Can you tell the area of the field, as that ingenious schoolmaster succeeded in doing ?

112.—A PUZZLING LEGACY.

A MAN left a hundred acres of land to be divided among his three sons—Alfred, Benjamin, and Charles—in the proportion of one-third, one-fourth, and one-fifth respectively. But Charles died. How was the land to be divided fairly between Alfred and Benjamin ?

113.—THE TORN NUMBER.

I HAD the other day in my possession a label bearing the number 3 0 2 5 in large figures. This got accidentally torn in half, so that 3 0 was on one piece and 2 5 on the other, as shown on the illustration. On looking at these pieces I began

to make a calculation, scarcely conscious of what I was doing, when I discovered this little peculiarity. If we add the 3 0 and the 2 5 together and square the sum we get as the result the complete original number on the label ! Thus, 3 0 added to 2 5 is 5 5, and 5 5 multiplied by 5 5 is 3 0 2 5. Curious, is it not ? Now, the puzzle is to find another number, composed of four figures, all different, which may be divided in the middle and produce the same result.

114.—CURIOUS NUMBERS.

THE number 48 has this peculiarity, that if you add 1 to it the result is a square number (49, the square of 7), and if you add 1 to its half, you also get a square number (25, the square of 5). Now, there is no limit to the numbers that have this peculiarity, and it is an interesting puzzle to find three more of them—the smallest possible numbers. What are they ?

115.—A PRINTER'S ERROR.

IN a certain article a printer had to set up the figures 5⁴.2³, which, of course, means that the fourth power of 5 (625) is to be multiplied by the cube of 2 (8), the product of which is 5,000. But he printed 5⁴.2³ as 5 4 2 3, which is not correct. Can you place four digits in the manner shown, so that it will be equally correct if the printer sets it up aright or makes the same blunder ?

116.—THE CONVERTED MISER.

MR. JASPER BULLYON was one of the very few misers who have ever been converted to a sense of their duty towards their less fortunate fellowmen. One eventful night he counted out his accumulated wealth, and resolved to distribute it amongst the deserving poor.

He found that if he gave away the same number of pounds every day in the year, he could exactly spread it over a twelvemonth without there being anything left over ; but if he rested on the Sundays, and only gave away a fixed number of pounds every week-day, there would be one sovereign left over on New Year's Eve. Now, putting it at the lowest possible, what was the exact number of pounds that he had to distribute ?

Could any question be simpler ? A sum of pounds divided by one number of days leaves no remainder, but divided by another number of days leaves a sovereign over. That is all ; and yet, when you come to tackle this little question, you will be surprised that it can become so puzzling.

117.—A FENCE PROBLEM.

THE practical usefulness of puzzles is a point that we are liable to overlook. Yet, as a matter of fact, I have from time to time received quite a large number of letters from individuals who have found that the mastering of some little principle upon which a puzzle was built has proved of considerable value to them in a most unexpected way. Indeed, it may be accepted as a good maxim that a puzzle is of little real value unless, as well as being amusing and perplexing, it conceals some instructive and possibly useful feature. It is, however, very curious how these little bits of acquired knowledge dovetail into the occasional requirements of everyday life, and equally curious to what strange and mysterious uses some of our readers seem to apply them. What, for example, can be the object of Mr. Wm. Oxley, who writes to me all the way from Iowa, in wishing to ascertain the dimensions of a field that he proposes to enclose, containing just as many acres as there shall be rails in the fence ?

The man wishes to fence in a perfectly square field which is to contain just as many acres as there are rails in the required fence. Each hurdle, or portion of fence, is seven rails high, and two lengths would extend one pole (16½ ft.) : that is to say, there are fourteen rails to the pole, lineal measure. Now, what must be the size of the field ?

118.—CIRCLING THE SQUARES.

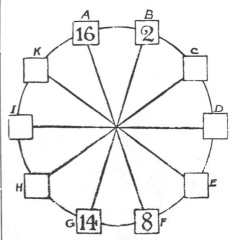

THE puzzle is to place a different number in each of the ten squares so that the sum of the squares of any two adjacent numbers shall be equal to the sum of the squares of the two numbers diametrically opposite to them. The four numbers placed, as examples, must stand as they are. The square of 16 is 256, and the square of 2 is 4. Add these together, and the result is 260. Also —the square of 14 is 196, and the square of 8 is 64. These together also make 260. Now, in precisely the same way, B and C should be equal to G and H (the sum will not necessarily be 260), A and K to F and E, H and I to C and D, and so on, with any two adjoining squares in the circle.

All you have to do is to fill in the remaining six numbers. Fractions are not allowed, and I shall show that no number need contain more than two figures.

119.—RACKBRANE'S LITTLE LOSS.

PROFESSOR RACKBRANE was spending an evening with his old friends, Mr. and Mrs. Potts, and they engaged in some game (he does not say what game) of cards. The professor lost the first game, which resulted in doubling the money that both Mr. and Mrs. Potts had laid on the table. The second game was lost by Mrs. Potts, which doubled the money then held by her husband and the professor. Curiously enough, the third game was lost by Mr. Potts, and had the effect of doubling the money then held by his wife and the professor. It was then found that each person had exactly the same money, but the professor had lost five shillings in the course of play. Now, the professor asks, what was the sum of money with which he sat down at the table ? Can you tell him ?

120.—THE FARMER AND HIS SHEEP.

FARMER LONGMORE had a curious aptitude for arithmetic, and was known in his district as the "mathematical farmer." The new vicar was not aware of this fact when, meeting his worthy parishioner one day in the lane, he asked him

will know exactly how many sheep Farmer Longmore owned.

121.—HEADS OR TAILS.

CROOKS, an inveterate gambler, at Goodwood recently said to a friend, " I'll bet you half the money in my pocket on the toss of a coin—heads I win, tails I lose." The coin was tossed and the money handed over. He repeated the offer again and again, each time betting half the money then in his possession. We are not told how long the game went on, or how many times the coin was tossed, but this we know, that the number of times that Crooks lost was exactly

in the course of a short conversation, " Now, how many sheep have you altogether ? " He was therefore rather surprised at Longmore's answer, which was as follows : " You can divide my sheep into two different parts, so that the difference between the two numbers is the same as the difference between their squares. Maybe, Mr. Parson, you will like to work out the little sum for yourself."

Can the reader say just how many sheep the farmer had ? Supposing he had possessed only twenty sheep, and he divided them into the two parts 12 and 8. Now, the difference between 12 and 8 is 4, but the difference between their squares, 144 and 64, is 80. So that will not do, for 4 and 80 are certainly not the same. If you can find numbers that work out correctly, you

equal to the number of times that he won. Now, did he gain or lose by this little venture ?

122.—THE SEE-SAW PUZZLE.

NECESSITY is, indeed, the mother of invention. I was amused the other day in watching a boy who wanted to play see-saw and, in his failure to find another child to share the sport with him, had been driven back upon the ingenious resort of tying a number of bricks to one end of the plank to balance his weight at the other.

As a matter of fact, he just balanced against sixteen bricks, when these were fixed to the short end of plank, but if he fixed them to the long end of plank he only needed eleven as balance.

Now, what was that boy's weight, if a brick

weighs equal to a three-quarter brick and three-quarters of a pound ?

123.—A LEGAL DIFFICULTY.

" A CLIENT of mine," said a lawyer, " was on the point of death when his wife was about to present him with a child. I drew up his will, in which he settled two-thirds of his estate upon his son (if it should happen to be a boy) and one-third on the mother. But if the child should be a girl, then two-thirds of the estate should go to the mother and one-third to the daughter. As a matter of fact, after his death twins were born—a boy and a girl. A very nice point then arose. How was the estate to be equitably divided among the three in the closest possible accordance with the spirit of the dead man's will ? "

124.—A QUESTION OF DEFINITION.

" MY property is exactly a mile square," said one landowner to another.
" Curiously enough, mine is a square mile," was the reply.
" Then there is no difference ? "
Is this last statement correct ?

125.—THE MINERS' HOLIDAY.

SEVEN coal-miners took a holiday at the sea-side during a big strike. Six of the party spent exactly half a sovereign each, but Bill Harris was more extravagant. Bill spent three shillings more than the average of the party. What was the actual amount of Bill's expenditure ?

126.—SIMPLE MULTIPLICATION.

IF we number six cards 1, 2, 4, 5, 7, and 8, and arrange them on the table in this order :—

    1   4   2   8   5   7

we can demonstrate that in order to multiply by 3 all that is necessary is to remove the 1 to the other end of the row, and the thing is done. The answer is 428571. Can you find a number that, when multiplied by 3 and divided by 2, the answer will be the same as if we removed the first card (which in this case is to be a 3) from the beginning of the row to the end ?

127.—SIMPLE DIVISION.

SOMETIMES a very simple question in elementary arithmetic will cause a good deal of perplexity. For example, I want to divide the four numbers, 701, 1,059, 1,417, and 2,312, by the largest number possible that will leave the same remainder in every case. How am I to set to work ? Of course, by a laborious system of trial one can in time discover the answer, but there is quite a simple method of doing it if you can only find it.

128.—A PROBLEM IN SQUARES.

WE possess three square boards. The surface of the first contains five square feet more than the second, and the second contains five square feet more than the third. Can you give exact measurements for the sides of the boards ? If you can solve this little puzzle, then try to find three squares in arithmetical progression, with a common difference of 7 and also of 13.

129.—THE BATTLE OF HASTINGS.

ALL historians know that there is a great deal of mystery and uncertainty concerning the details of the ever-memorable battle on that fatal day, October 14, 1066. My puzzle deals with a curious passage in an ancient monkish chronicle that may never receive the attention that it deserves, and if I am unable to vouch for the authenticity of the document it will none the less serve to furnish us with a problem that can hardly fail to interest those of my readers who have arithmetical predilections. Here is the passage in question.

" The men of Harold stood well together, as their wont was, and formed sixty and one squares, with a like number of men in every square thereof, and woe to the hardy Norman who ventured to enter their redoubts ; for a single blow of a Saxon war-hatchet would break his lance and cut through his coat of mail. . . . When Harold threw himself into the fray the Saxons were one mighty square of men, shouting the battle-cries, ' Ut ! ' ' Olicrosse ! ' ' Godemitè ! ' "

Now, I find that all the contemporary authorities agree that the Saxons did actually fight in this solid order. For example, in the " Carmen de Bello Hastingensi," a poem attributed to Guy, Bishop of Amiens, living at the time of the battle, we are told that " the Saxons stood fixed in a dense mass," and Henry of Huntingdon records that " they were like unto a castle, impenetrable to the Normans ; " while Robert Wace, a century after, tells us the same thing. So in this respect my newly-discovered chronicle may not be greatly in error. But I have reason to believe that there is something wrong with the actual figures. Let the reader see what he can make of them.

The number of men would be sixty-one times a square number ; but when Harold himself joined in the fray they were then able to form one large square. What is the smallest possible number of men there could have been ?

In order to make clear to the reader the simplicity of the question, I will give the lowest solutions in the case of 60 and 62, the numbers immediately preceding and following 61. They are $60 \times 4^2 + 1 = 31^2$, and $62 \times 8^2 + 1 = 63^2$. That is, 60 squares of 16 men each would be 960 men, and when Harold joined them they would be 961 in number, and so form a square with 31 men on every side. Similarly in the case of the figures I have given for 62. Now, find the lowest answer for 61.

130.—THE SCULPTOR'S PROBLEM.

AN ancient sculptor was commissioned to supply two statues, each on a cubical pedestal. It is with these pedestals that we are concerned. They were of unequal sizes, as will be seen in the illustration, and when the time arrived for payment a dispute arose as to whether the agreement was based on lineal or cubical measurement. But as soon as they came to measure the two pedestals the matter was at once settled, because, curiously enough, the number of lineal feet was exactly the same as the number of

cubical feet. The puzzle is to find the dimensions for two pedestals having this peculiarity, in the smallest possible figures. You see, if the two pedestals, for example, measure respectively 3 ft. and 1 ft. on every side, then the lineal measurement would be 4 ft. and the cubical contents 28 ft., which are not the same, so these measurements will not do.

131.—THE SPANISH MISER.

THERE once lived in a small town in New Castile a noted miser named Don Manuel Rodriguez. His love of money was only equalled by a strong passion for arithmetical problems. These puzzles usually dealt in some way or other with his accumulated treasure, and were propounded by him solely in order that he might have the pleasure of solving them himself. Unfortunately very few of them have survived, and when travelling through Spain, collecting material for a proposed work on "The Spanish Onion as a Cause of National Decadence," I only discovered a very few. One of these concerns the three boxes that appear in the accompanying authentic portrait.

Each box contained a different number of golden doubloons. The difference between the number of doubloons in the upper box and the number in the middle box was the same as the difference between the number in the middle box and the number in the bottom box. And if the contents of any two of the boxes were united they would form a square number. What is the smallest number of doubloons that there could have been in any one of the boxes?

132.—THE NINE TREASURE BOXES.

THE following puzzle will illustrate the importance on occasions of being able to fix the minimum and maximum limits of a required number. This can very frequently be done. For example, it has not yet been ascertained in how many different ways the knight's tour can be performed on the chess board; but we know that it is fewer than the number of combinations of 168 things taken 63 at a time and is greater than 31,054,144—for the latter is the number of routes of a particular type. Or, to take a more familiar case, if you ask a man how many coins he has in his pocket, he may tell you that he has not the slightest idea. But on further questioning you will get out of him some such statement as the following: "Yes, I am positive that I have more than three coins, and equally certain that there are not so many as twenty-five." Now, the knowledge that a certain number lies between 2 and 12 in my puzzle will enable the solver to find the exact answer; without that information there would be an infinite number of answers, from which it would be impossible to select the correct one.

This is another puzzle received from my friend Don Manuel Rodriguez, the cranky miser of New Castile. On New Year's Eve in 1879 he showed me nine treasure boxes, and after informing me that every box contained a square number of golden doubloons, and that the difference between the contents of A and B was the same as between B and C, D and E, E and F, G and H, or H and I, he requested me to tell him the number of coins in every one of the boxes. At first I thought this was impossible, as there would be an infinite number of different answers, but on consideration I found that this

was not the case. I discovered that while every box contained coins, the contents of A, B, C increased in weight in alphabetical order ; so did D, E, F ; and so did G, H, I ; but D or E need not be heavier than C, nor G or H heavier than F. It was also perfectly certain that box A could not contain more than a dozen coins at the outside ; there might not be half that number, but I was positive that there were not more than twelve. With this knowledge I was able to arrive at the correct answer.

In short, we have to discover nine square numbers such that A, B, C ; and D, E, F ; and G, H, I are three groups in arithmetical progression, the common difference being the same in each group, and A being less than 12. How many doubloons were there in every one of the nine boxes ?

133.—THE FIVE BRIGANDS.

THE five Spanish brigands, Alfonso, Benito, Carlos, Diego, and Esteban, were counting their spoils after a raid, when it was found that they had captured altogether exactly 200 doubloons. One of the band pointed out that if Alfonso had twelve times as much, Benito three times as much, Carlos the same amount, Diego half as much, and Esteban one-third as much, they would still have altogether just 200 doubloons. How many doubloons had each ?

There are a good many equally correct answers to this question. Here is one of them :

A	.	.	.	6 × 12	=	72
B	.	.	.	12 × 3	=	36
C	.	.	.	17 × 1	=	17
D	.	.	.	120 × ½	=	60
E	.	.	.	45 × ⅓	=	15
				200		200

The puzzle is to discover exactly how many different answers there are, it being understood that every man had something and that there is to be no fractional money—only doubloons in every case.

This problem, worded somewhat differently, was propounded by Tartaglia (died 1559), and he flattered himself that he had found one solution ; but a French mathematician of note (M. A. Labosne), in a recent work, says that his readers will be astonished when he assures them that there are 6,639 different correct answers to the question. Is this so ? How many answers are there ?

134.—THE BANKER'S PUZZLE.

A BANKER had a sporting customer who was always anxious to wager on anything. Hoping to cure him of his bad habit, he proposed as a wager that the customer would not be able to divide up the contents of a box containing only sixpences into an exact number of equal piles of sixpences. The banker was first to put in one or more sixpences (as many as he liked) ; then the customer was to put in one or more (but in his case not more than a pound in value), neither knowing what the other put in. Lastly, the customer was to transfer from the banker's counter to the box as many sixpences as the banker desired him to put in. The puzzle is to find how many sixpences the banker should first put in and how many he should ask the customer to transfer, so that he may have the best chance of winning.

135.—THE STONEMASON'S PROBLEM.

A STONEMASON once had a large number of cubic blocks of stone in his yard, all of exactly the same size. He had some very fanciful little ways, and one of his queer notions was to keep these blocks piled in cubical heaps, no two heaps containing the same number of blocks. He had discovered for himself (a fact that is well known to mathematicians) that if he took all the blocks contained in any number of heaps in regular order, beginning with the single cube, he could always arrange those on the ground so as to form a perfect square. This will be clear to the reader, because one block is a square, $1 + 8 = 9$ is a square, $1 + 8 + 27 = 36$ is a square, $1 + 8 + 27 + 64 = 100$ is a square, and so on. In fact, the sum of any number of consecutive cubes, beginning always with 1, is in every case a square number.

One day a gentleman entered the mason's yard and offered him a certain price if he would supply him with a consecutive number of these cubical heaps which should contain altogether a number of blocks that could be laid out to form a square, but the buyer insisted on more than three heaps and *declined to take the single block* because it contained a flaw. What was the smallest possible number of blocks of stone that the mason had to supply ?

136.—THE SULTAN'S ARMY.

A CERTAIN Sultan wished to send into battle an army that could be formed into two perfect squares in twelve different ways. What is the smallest number of men of which that army could be composed ? To make it clear to the novice, I will explain that if there were 130 men, they could be formed into two squares in only two different ways—81 and 49, or 121 and 9. Of course, all the men must be used on every occasion.

137.—A STUDY IN THRIFT.

CERTAIN numbers are called triangular, because if they are taken to represent counters or coins they may be laid out on the table so as to form triangles. The number 1 is always regarded as triangular, just as 1 is a square and a cube number. Place one counter on the table—that is, the first triangular number. Now place two more counters beneath it, and you have a triangle of three counters ; therefore 3 is triangular. Next place a row of three more counters, and you have a triangle of six counters ; therefore 6 is triangular. We see that every row of counters that we add, containing just one more counter than the row above it, makes a larger triangle.

Now, half the sum of any number and its square is always a triangular number. Thus half of $2 + 2^2 = 3$; half of $3 + 3^2 = 6$; half of $4 + 4^2 = 10$; half of $5 + 5^2 = 15$; and so on. So if we want to form a triangle with 8 counters on each side we shall require half of $8 + 8^2$, or 36 counters. This is a pretty little property of numbers. Before going further, I will here say that if the reader refers to the "Stonemason's Problem" (No. 135) he will remember that the sum of any number of consecutive cubes beginning with 1 is always a square, and these form the series 1^2, 3^2, 6^2, 10^2, etc. It will now be understood when I say that one of the keys to the puzzle was the fact that these are always the squares of triangular numbers—that is, the squares of 1, 3, 6, 10, 15, 21, 28, etc., any of which numbers we have seen will form a triangle.

Every whole number is either triangular, or the sum of two triangular numbers or the sum of three triangular numbers. That is, if we take any number we choose we can always form one, two, or three triangles with them. The number 1 will obviously, and uniquely, only form one triangle; some numbers will only form two triangles (as 2, 4, 11, etc.); some numbers will only form three triangles (as 5, 8, 14, etc.). Then, again, some numbers will form both one and two triangles (as 6), others both one and three triangles (as 3 and 10), others both two and three triangles (as 7 and 9), while some numbers (like 21) will form one, two, or three triangles, as we desire. Now for a little puzzle in triangular numbers.

Sandy McAllister, of Aberdeen, practised strict domestic economy, and was anxious to train his good wife in his own habits of thrift. He told her last New Year's Eve that when she had saved so many sovereigns that she could lay them all out on the table so as to form a perfect square, or a perfect triangle, or two triangles, or three triangles, just as he might choose to ask, he would add five pounds to her treasure. Soon she went to her husband with a little bag of £36 in sovereigns and claimed her reward. It will be found that the thirty-six coins will form a square (with side 6), that they will form a single triangle (with side 8), that they will form two triangles (with sides 5 and 6), and that they will form three triangles (with sides 3, 5, and 5). In each of the four cases all the thirty-six coins are used, as required, and Sandy therefore made his wife the promised present like an honest man.

The Scotsman then undertook to extend his promise for five more years, so that if next year the increased number of sovereigns that she has saved can be laid out in the same four different ways she will receive a second present; if she succeeds in the following year she will get a third present, and so on until she has earned six presents in all. Now, how many sovereigns must she put together before she can win the sixth present?

What you have to do is to find five numbers, the smallest possible, higher than 36, that can be displayed in the four ways—to form a square, to form a triangle, to form two triangles, and to form three triangles. The highest of your five numbers will be your answer.

138.—THE ARTILLERYMEN'S DILEMMA.

"ALL cannon-balls are to be piled in square pyramids," was the order issued to the regiment. This was done. Then came the further order, "All pyramids are to contain a square number of balls." Whereupon the trouble arose. "It can't be done," said the major. "Look at this pyramid, for example; there are sixteen balls at the base, then nine, then four, then one at the top, making thirty balls in all. But there must be six more balls, or five fewer, to make a square number." "It *must* be done," insisted the general. "All you have to do is to put the right number of balls in your pyramids." "I've got it!" said a lieutenant, the mathematical genius of the regiment. "Lay the balls out singly." "Bosh!" exclaimed the general. "You can't *pile* one ball into a pyramid!" Is it really possible to obey both orders?

139.—THE DUTCHMEN'S WIVES.

I WONDER how many of my readers are acquainted with the puzzle of the "Dutchmen's Wives"—in which you have to determine the names of three men's wives, or, rather, which wife belongs to each husband. Some thirty years ago it was "going the rounds," as something quite new, but I recently discovered it in the *Ladies' Diary* for 1739-40, so it was clearly familiar to the fair sex over one hundred and seventy years ago. How many of our mothers, wives, sisters, daughters, and aunts could solve the puzzle to-day? A far greater proportion than then, let us hope.

Three Dutchmen, named Hendrick, Elas, and Cornelius, and their wives, Gurtrün, Katrün, and Anna, purchase hogs. Each buys as many as (he or she) gives shillings for one. Each husband pays altogether three guineas more than his wife. Hendrick buys twenty-three more hogs than Katrün, and Elas eleven more

daughters, bought cloth at the same shop. Each of the ten paid as many farthings per foot as she bought feet, and each mother spent 8s. 5½d. more than her daughter. Mrs. Robinson spent 6s. more than Mrs. Evans, who spent about a quarter as much as Mrs. Jones. Mrs. Smith spent most of all. Mrs. Brown bought 21 yards more than Bessie — one of the girls. Annie bought 16 yards more than Mary and spent £3, 0s. 8d. more than Emily. The Christian name of the other girl was Ada. Now, what was her surname ?

141.—SATURDAY MARKETING.

HERE is an amusing little case of marketing which, although it deals with a good many items of money, leads up to a question of a totally different character. Four married couples went into their village on a recent Saturday night to do a little marketing. They had to be very economical, for among them they only possessed forty shilling coins. The fact is, Ann spent 1s., Mary spent 2s., Jane spent 3s., and Kate spent 4s. The men were rather more extravagant than their wives, for Ned Smith spent as much as his wife, Tom Brown twice as much as his wife, Bill Jones three times as much as his wife, and Jack Robinson four times as much as his wife. On the way home somebody suggested that they should divide what coin they had left equally among them. This was done, and the puzzling question is simply this : What was the surname of each woman ? Can you pair off the four couples ?

than Gurtrün. Now, what was the name of each man's wife ?

140.—FIND ADA'S SURNAME.

THIS puzzle closely resembles the last one, my remarks on the solution of which the reader may like to apply in another case. It was recently submitted to a Sydney evening newspaper that indulges in " intellect sharpeners," but was rejected with the remark that it is childish and that they only published problems capable of solution ! Five ladies, accompanied by their

GEOMETRICAL PROBLEMS.

> " God geometrizes continually."
> PLATO.

" THERE is no study," said Augustus de Morgan, " which presents so simple a beginning as that of geometry ; there is none in which difficulties grow more rapidly as we proceed." This will be found when the reader comes to consider the following puzzles, though they are not arranged in strict order of difficulty. And the fact that they have interested and given pleasure to man for untold ages is no doubt due in some measure to the appeal they make to the eye as well as to the brain. Sometimes an algebraical formula or theorem seems to give pleasure to the mathematician's eye, but it is probably only an intellectual pleasure. But there can be no doubt that in the case of certain geometrical problems, notably dissection or superposition puzzles, the æsthetic faculty in man contributes to the delight. For example, there are probably few readers who will examine the various cuttings of the Greek cross in the following pages without being in some degree stirred by a sense of beauty. Law and order in Nature are always pleasing to contemplate, but when they come under the very eye they seem to make a specially strong appeal. Even the person with no geometrical knowledge whatever is induced after the inspection of such things to exclaim, " How very pretty!" In fact, I have known more than one person led on to a study of geometry by the fascination of cutting-out puzzles. I have, therefore, thought it well to keep these dissection puzzles distinct from the geometrical problems on more general lines.

DISSECTION PUZZLES.

> " Take him and cut him out in little stars."
> Romeo and Juliet, iii. 2.

PUZZLES have infinite variety, but perhaps there is no class more ancient than dissection, cutting-out, or superposition puzzles. They were certainly known to the Chinese several thousand years before the Christian era. And they are just as fascinating to-day as they can have been at any period of their history. It is supposed by those who have investigated the matter that the ancient Chinese philosophers used these

puzzles as a sort of kindergarten method of im-
parting the principles of geometry. Whether
this was so or not, it is certain that all good
dissection puzzles (for the nursery type of jig-
saw puzzle, which merely consists in cutting
up a picture into pieces to be put together
again, is not worthy of serious consideration)
are really based on geometrical laws. This
statement need not, however, frighten off the
novice, for it means little more than this, that
geometry will give us the " reason why," if we
are interested in knowing it, though the solutions may often be discovered by any intelligent
person after the exercise of patience, ingenuity,
and common sagacity.

If we want to cut one plane figure into parts
that by readjustment will form another figure,
the first thing is to find a way of doing it at all,
and then to discover how to do it in the fewest
possible pieces. Often a dissection problem is
quite easy apart from this limitation of pieces.
At the time of the publication in the *Weekly
Dispatch*, in 1902, of a method of cutting an
equilateral triangle into four parts that will
form a square (see No. 26, " Canterbury
Puzzles "), no geometrician would have had any
difficulty in doing what is required in five
pieces : the whole point of the discovery lay
in performing the little feat in four pieces only.

Mere approximations in the case of these
problems are valueless ; the solution must be
geometrically exact, or it is not a solution at
all. Fallacies are cropping up now and again,
and I shall have occasion to refer to one or two
of these. They are interesting merely as fallacies. But I want to say something on two
little points that are always arising in cutting-
out puzzles—the questions of " hanging by a
thread " and " turning over." These points
can best be illustrated by a puzzle that is frequently to be found in the old books, but invariably with a false solution. The puzzle is
to cut the figure shown in Fig. 1 into three

there will be four pieces—or six pieces in all.
It is, therefore, not a solution in three pieces.

If, however, the reader will look at the solution in Figs. 3 and 4, he will see that no such

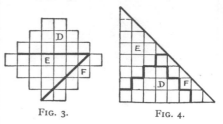

FIG. 3.   FIG. 4.

fault can be found with it. There is no question
whatever that there are three pieces, and the
solution is in this respect quite satisfactory.
But another question arises. It will be found
on inspection that the piece marked F, in Fig. 3,
is turned over in Fig. 4—that is to say, a different
side has necessarily to be presented. If the
puzzle were merely to be cut out of cardboard
or wood, there might be no objection to this
reversal, but it is quite possible that the material would not admit of being reversed. There
might be a pattern, a polish, a difference of
texture, that prevents it. But it is generally
understood that in dissection puzzles you are
allowed to turn pieces over unless it is distinctly stated that you may not do so. And
very often a puzzle is greatly improved by the
added condition, " no piece may be turned
over." I have often made puzzles, too, in which
the diagram has a small repeated pattern, and
the pieces have then so to be cut that not only
is there no turning over, but the pattern has to
be matched, which cannot be done if the pieces
are turned round, even with the proper side
uppermost.

Before presenting a varied series of cutting-
out puzzles, some very easy and others difficult,
I propose to consider one family alone—those
problems involving what is known as the Greek
cross with the square. This will exhibit a
great variety of curious transpositions, and, by
having the solutions as we go along, the reader
will be saved the trouble of perpetually turning
to another part of the book, and will have everything under his eye. It is hoped that in this
way the article may prove somewhat instructive
to the novice and interesting to others.

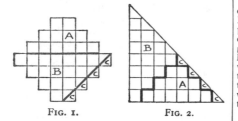

FIG. 1.   FIG. 2.

pieces that will fit together and form a half-
square triangle. The answer that is invariably
given is that shown in Figs. 1 and 2. Now, it
is claimed that the four pieces marked C are
really only one piece, because they may be so
cut that they are left " hanging together by a
mere thread." But no serious puzzle lover will
ever admit this. If the cut is made so as to
leave the four pieces joined in one, then it cannot result in a perfectly exact solution. If, on
the other hand, the solution is to be exact, then

GREEK CROSS PUZZLES.

" To fret thy soul with crosses."
                                    SPENSER.
" But, for my part, it was Greek to me."
                            Julius Cæsar, i. 2.

MANY people are accustomed to consider the
cross as a wholly Christian symbol. This is
erroneous : it is of very great antiquity. The
ancient Egyptians employed it as a sacred

symbol, and on Greek sculptures we find representations of a cake (the supposed real origin of our hot cross buns) bearing a cross. Two such cakes were discovered at Herculaneum. Cecrops offered to Jupiter Olympus a sacred cake or *boun* of this kind. The cross and ball, so frequently found on Egyptian figures, is a circle and the *tau* cross. The circle signified the eternal preserver of the world, and the T, named from the Greek letter *tau*, is the monogram of Thoth, the Egyptian Mercury, meaning wisdom. This *tau* cross is also called by Christians the cross of St. Anthony, and is borne on a badge in the bishop's palace at Exeter. As for the Greek or mundane cross, the cross with four equal arms, we are told by competent antiquaries that it was regarded by ancient occultists for thousands of years as a sign of the dual forces of Nature—the male and female spirit of everything that was everlasting.

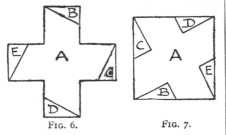

FIG. 5.

The Greek cross, as shown in Fig. 5, is formed by the assembling together of five equal squares. We will start with what is known as the Hindu problem, supposed to be upwards of three thousand years old. It appears in the seal of Harvard College, and is often given in old works as symbolical of mathematical science and ex-

FIG. 6.   FIG. 7.

actitude. Cut the cross into five pieces to form a square. Figs. 6 and 7 show how this

is done. It was not until the middle of the nineteenth century that we found that the cross might be transformed into a square in only four pieces. Figs. 8 and 9 will show how to do

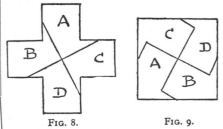

FIG. 8.   FIG. 9.

it, if we further require the four pieces to be all of the same size and shape. This Fig. 9 is remarkable because, according to Dr. Le Plongeon and others, as expounded in a work by Professor Wilson of the Smithsonian Institute, here we have the great Swastika, or sign, of " good luck to you "—the most ancient symbol of the human race of which there is any record. Professor Wilson's work gives some four hundred illustrations of this curious sign as found in the Aztec mounds of Mexico, the pyramids of Egypt, the ruins of Troy, and the ancient lore of India and China. One might almost say there is a curious affinity between the Greek cross and Swastika! If, however, we require that the four pieces shall be produced by only two clips of the scissors (assuming the puzzle is in paper form), then we must cut as in Fig. 10 to form Fig. 11, the first clip of the scissors being from *a* to *b*. Of course folding the paper, or holding the pieces together after the first cut, would not in this case be allowed. But there is an infinite number of different ways of making the cuts to solve the puzzle in four pieces. To this point I propose to return.

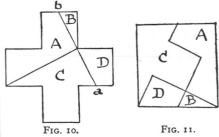

FIG. 10.   FIG. 11.

It will be seen that every one of these puzzles has its reverse puzzle—to cut a square into pieces to form a Greek cross. But as a square has not so many angles as the cross, it is not always equally easy to discover the true directions of the cuts. Yet in the case of the examples given, I will leave the reader to determine their direction for himself, as they are rather obvious from the diagrams.

Cut a square into five pieces that will form two separate Greek crosses of *different sizes.* This is quite an easy puzzle. As will be seen in Fig. 12, we have only to divide our square into 25 little squares and then cut as shown. The cross A is cut out entire, and the pieces B, C, D, and E form the larger cross in Fig. 13. The reader may here like to cut the single piece, B, into four pieces all similar in shape to itself, and form a cross with them in the manner shown in Fig. 13. I hardly need give the solution.

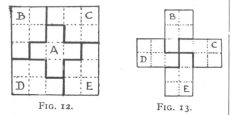

FIG. 12.   FIG. 13.

Cut a square into five pieces that will form two separate Greek crosses of exactly the *same size.* This is more difficult. We make the cuts as in Fig. 14, where the cross A comes out entire and the other four pieces form the cross in Fig. 15. The direction of the cuts is pretty obvious. It will be seen that the sides of the square in Fig. 14 are marked off into six equal parts. The sides of the cross are found by ruling lines from certain of these points to others.

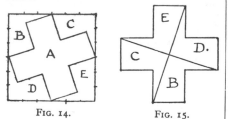

FIG. 14.   FIG. 15.

I will now explain, as I promised, why a Greek cross may be cut into four pieces in an infinite number of different ways to make a square. Draw a cross, as in Fig. 16. Then draw on transparent paper the square shown in Fig. 17, taking care that the distance c to d is exactly the same as the distance a to b in the cross. Now place the transparent paper over the cross and slide it about into different positions, only be very careful always to keep the square at the same angle to the cross as shown, where a b is parallel to c d. If you place the point c exactly over a the lines will indicate the solution (Figs. 10 and 11). If you place c in the very centre of the dotted square, it will give the solution in Figs. 8 and 9. You will now see that by sliding the square about so that the point c is always within the dotted square you may get as many different solutions as you like;

because, since an infinite number of different points may theoretically be placed within this square, there must be an infinite number of

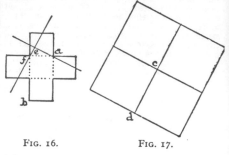

FIG. 16.   FIG. 17.

different solutions. But the point c need not necessarily be placed within the dotted square. It may be placed, for example, at point e to give a solution in four pieces. Here the joins at a and f may be as slender as you like. Yet if you once get over the edge at a or f you no longer have a solution in four pieces. This proof will be found both entertaining and instructive. If you do not happen to have any transparent paper at hand, any thin paper will of course do if you hold the two sheets against a pane of glass in the window.

It may have been noticed from the solutions of the puzzles that I have given that the side of the square formed from the cross is always equal to the distance a to b in Fig. 16. This must necessarily be so, and I will presently try to make the point quite clear.

We will now go one step further. I have already said that the ideal solution to a cutting-out puzzle is always that which requires the fewest possible pieces. We have just seen that two crosses of the same size may be cut out of a square in five pieces. The reader who succeeded in solving this perhaps asked himself: " Can it be done in fewer pieces ? " This is just the sort of question that the true puzzle lover is always asking, and it is the right attitude for him to adopt. The answer to the question is that the puzzle may be solved in four pieces—the fewest possible. This, then, is a new puzzle. Cut a square into four pieces that will form two Greek crosses of the same size.

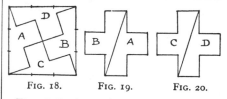

FIG. 18.   FIG. 19.   FIG. 20.

The solution is very beautiful. If you divide by points the sides of the square into three equal parts, the directions of the lines in Fig. 18 will be quite obvious. If you cut along these lines, the pieces A and B will form the cross in Fig. 19 and the pieces C and D the similar cross in Fig. 20. In this square we have another form of Swastika.

The reader will here appreciate the truth of my remark to the effect that it is easier to find the directions of the cuts when transforming a cross to a square than when converting a square into a cross. Thus, in Figs. 6, 8, and 10 the directions of the cuts are more obvious than in Fig. 14, where we had first to divide the sides of the square into six equal parts, and in Fig. 18, where we divide them into three equal parts. Then, supposing you were required to cut two equal Greek crosses, each into two pieces, to form a square, a glance at Figs. 19 and 20 will show how absurdly more easy this is than the reverse puzzle of cutting the square to make two crosses.

Referring to my remarks on " fallacies," I will now give a little example of these " solutions " that are not solutions. Some years ago a young correspondent sent me what he evidently thought was a brilliant new discovery—the transforming of a square into a Greek cross in four pieces by cuts all parallel to the sides of the square. I give his attempt in Figs. 21 and 22, where it will be seen that the four pieces do not form a symmetrical Greek cross, because the four arms are not really squares but oblongs. To make it a true Greek cross we should require the additions that I have indicated with dotted lines. Of course his solution produces a cross, but it is not the symmetrical Greek variety required by the conditions of the puzzle. My young friend thought his attempt was " near enough " to be correct ; but if he bought a penny apple with a sixpence he probably would not have thought it " near enough " if he had been given only fourpence change. As the reader advances he will realize the importance of this question of exactitude.

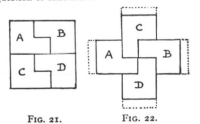

FIG. 21.   FIG. 22.

In these cutting-out puzzles it is necessary not only to get the directions of the cutting lines as correct as possible, but to remember that these lines have no width. If after cutting up one of the crosses in a manner indicated in these articles you find that the pieces do not exactly fit to form a square, you may be certain that the fault is entirely your own. Either your cross was not exactly drawn, or your cuts were not made quite in the right directions, or (if you used wood and a fretsaw) your saw was not sufficiently fine. If you cut out the puzzles in paper with scissors, or in cardboard with a penknife, no material is lost ; but with a saw, however fine, there is a certain loss. In the case of most puzzles this slight loss is not sufficient to be appreciable, if the puzzle is cut out on a large scale, but there have been instances where I have found it desirable to draw and cut out each part separately—not from one diagram—in order to produce a perfect result.

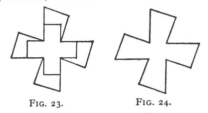

FIG. 23.   FIG. 24.

Now for another puzzle. If you have cut out the five pieces indicated in Fig. 14, you will find that these can be put together so as to form the curious cross shown in Fig. 23. So if I asked you to cut Fig. 24 into five pieces to form either a square or two equal Greek crosses you would know how to do it. You would make the cuts as in Fig. 23, and place them together as in Figs. 14 and 15. But I want something better than that, and it is this. Cut Fig. 24 into only four pieces that will fit together and form a square.

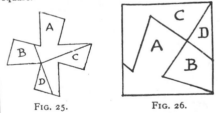

FIG. 25.   FIG. 26.

The solution to the puzzle is shown in Figs. 25 and 26. The direction of the cut dividing A and C in the first diagram is very obvious, and the second cut is made at right angles to it. That the four pieces should fit together and form a square will surprise the novice, who will do well to study the puzzle with some care, as it is most instructive.

I will now explain the beautiful rule by which we determine the size of a square that shall have the same area as a Greek cross, for it is applicable, and necessary, to the solution of almost every dissection puzzle that we meet with. It was first discovered by the philosopher Pythagoras, who died 500 B.C., and is the 47th proposition of Euclid. The young reader who knows nothing of the elements of geometry will get some idea of the fascinating character of that science. The triangle A B C in Fig. 27 is what we call a right-angled triangle, because the side B C is at right angles to the side A B. Now if we build up a square on each side of the triangle, the squares on A B and B C will together be exactly equal to the square on the long side A C, which we call the hypotenuse. This is proved in the case I have given by subdividing the three squares into cells of equal dimensions.

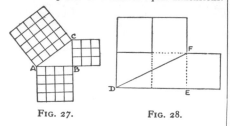

FIG. 27.   FIG. 28.

It will be seen that 9 added to 16 equals 25, the number of cells in the large square. If you make triangles with the sides 5, 12 and 13, or with 8, 15 and 17, you will get similar arithmetical proofs, for these are all "rational" right-angled triangles, but the law is equally true for all cases. Supposing we cut off the lower arm of a Greek cross and place it to the left of the upper arm, as in Fig. 28, then the square on E F added to the square on D E exactly equals a square on D F. Therefore we know that the square of D F will contain the same area as the cross. This fact we have proved practically by the solutions of the earlier puzzles of this series. But whatever length we give to D E and E F, we can never give the exact length of D F in numbers, because the triangle is not a "rational" one. But the law is none the less geometrically true.

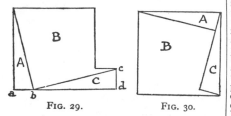

FIG. 29.   FIG. 30.

Now look at Fig. 29, and you will see an elegant method for cutting a piece of wood of the shape of two squares (of any relative dimensions) into three pieces that will fit together and form a single square. If you mark off the distance a b equal to the side c d the directions of the cuts are very evident. From what we have just been considering, you will at once see why b c must be the length of the side of the new square. Make the experiment as often as you like, taking different relative proportions for the two squares, and you will find the rule always come true. If you make the two squares of exactly the same size, you will see that the diagonal of any square is always the side of a square that is twice the size. All this, which is so simple that anybody can understand it, is

very essential to the solving of cutting-out puzzles. It is in fact the key to most of them. And it is all so beautiful that it seems a pity that it should not be familiar to everybody.

We will now go one step further and deal with the half-square. Take a square and cut it in half diagonally. Now try to discover how to cut this triangle into four pieces that will form a Greek cross. The solution is shown in Figs. 31 and 32. In this case it will be seen that we divide two of the sides of the triangle into three equal parts and the long side into four equal parts. Then the direction of the cuts will be easily found. It is a pretty puzzle, and a little more difficult than some of the others that I have given. It should be noted again that it would have been much easier to locate the cuts in the reverse puzzle of cutting the cross to form a half-square triangle.

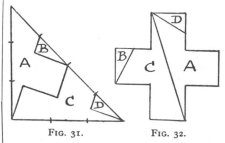

FIG. 31.   FIG. 32.

Another ideal that the puzzle maker always keeps in mind is to contrive that there shall, if possible, be only one correct solution. Thus, in the case of the first puzzle, if we only require that a Greek cross shall be cut into four pieces to form a square, there is, as I have shown, an infinite number of different solutions. It makes a better puzzle to add the condition that all the four pieces shall be of the same size and shape, because it can then be solved in only one way, as in Figs. 8 and 9. In this way, too, a puzzle that is too easy to be interesting may be improved by such an addition. Let us take an example. We have seen in Fig. 28 that Fig. 33 can be cut into two pieces to form a Greek cross. I suppose an intelligent child would do it in five

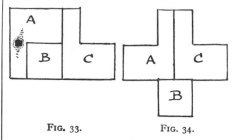

FIG. 33.   FIG. 34.

minutes. But suppose we say that the puzzle has to be solved with a piece of wood that has a bad knot in the position shown in Fig. 33—a knot that we must not attempt to cut through —then a solution in two pieces is barred out, and it becomes a more interesting puzzle to solve it in three pieces. I have shown in Figs. 33 and 34 one way of doing this, and it will be found entertaining to discover other ways of doing it. Of course I could bar out all these other ways by introducing more knots, and so reduce the puzzle to a single solution, but it would then be overloaded with conditions.

And this brings us to another point in seeking the ideal. Do not overload your conditions, or you will make your puzzle too complex to be interesting. The simpler the conditions of a puzzle are, the better. The solution may be as complex and difficult as you like, or as happens, but the conditions ought to be easily understood, or people will not attempt a solution.

If the reader were now asked " to cut a half-square into as few pieces as possible to form a Greek cross," he would probably produce our solution, Figs. 31–32, and confidently claim that he had solved the puzzle correctly. In this way he would be wrong, because it is not now stated that the square is to be divided diagonally. Although we should always observe the exact conditions of a puzzle we must not read into it conditions that are not there. Many puzzles are based entirely on the tendency that people have to do this.

The very first essential in solving a puzzle is to be sure that you understand the exact conditions. Now, if you divided your square in half so as to produce Fig. 35 it is possible to cut it into as few as three pieces to form a Greek cross. We thus save a piece.

I give another puzzle in Fig. 36. The dotted lines are added merely to show the correct proportions of the figure—a square of 25 cells with the four corner cells cut out. The puzzle is to cut this figure into five pieces that will form a Greek cross (entire) and a square.

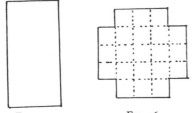

FIG. 35.   FIG. 36.

The solution to the first of the two puzzles last given—to cut a rectangle of the shape of a half-square into three pieces that will form a Greek cross—is shown in Figs. 37 and 38. It will be seen that we divide the long sides of the oblong into six equal parts and the short sides into three equal parts, in order to get the points that will indicate the direction of the cuts. The reader should compare this solution with some of the previous illustrations. He will see, for example, that if we continue the cut that divides B and C in the cross, we get Fig. 15.

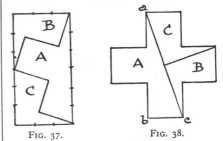

FIG. 37.   FIG. 38.

The other puzzle, like the one illustrated in Figs. 12 and 13, will show how useful a little arithmetic may sometimes prove to be in the solution of dissection puzzles. There are twenty-one of those little square cells into which our figure is subdivided, from which we have to form both a square and a Greek cross. Now, as the cross is built up of five squares, and 5 from 21 leaves 16—a square number—we ought easily to be led to the solution shown in Fig. 39. It will be

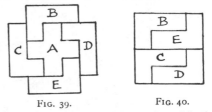

FIG. 39.   FIG. 40.

seen that the cross is cut out entire, while the four remaining pieces form the square in Fig. 40.

Of course a half-square rectangle is the same as a double square, or two equal squares joined together. Therefore, if you want to solve the puzzle of cutting a Greek cross into four pieces to form two separate squares of the same size, all you have to do is to continue the short cut in Fig. 38 right across the cross, and you will have four pieces of the same size and shape. Now divide Fig. 37 into two equal squares by a hori-

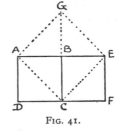

FIG. 41.

zontal cut midway and you will see the four pieces forming the two squares.

Cut a Greek cross into five pieces that will form two separate squares, one of which shall contain half the area of one of the arms of the cross. In further illustration of what I have

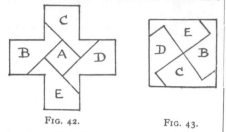

FIG. 42. FIG. 43.

already written, if the two squares of the same size A B C D and B C F E, in Fig. 41, are cut in the manner indicated by the dotted lines, the four pieces will form the large square A G E C. We thus see that the diagonal A C is the side of a square twice the size of A B C D. It is also clear that half the diagonal of any square is equal

pieces B, C, D, and E. After what I have written, the reader will have no difficulty in seeing that the square A is half the size of one of the arms of the cross, because the length of the diagonal of the former is clearly the same as the side of the latter. The thing is now self-evident. I have thus tried to show that some of these puzzles that many people are apt to regard as quite wonderful and bewildering, are really not difficult if only we use a little thought and judgment.

In conclusion of this particular subject I will give four Greek cross puzzles, with detached solutions.

142.—THE SILK PATCHWORK.

THE lady members of the Wilkinson family had made a simple patchwork quilt, as a small Christmas present, all composed of square pieces of the same size, as shown in the illustration. It only lacked the four corner pieces to make it complete. Somebody pointed out to them that if you unpicked the Greek cross in the middle and then cut the stitches along the dark joins, the four pieces all of the same size and shape would fit together and form a square. This the reader knows, from the solution in Fig. **39,**

to the side of a square of half the area. Therefore, if the large square in the diagram is one of the arms of your cross, the small square is the size of one of the squares required in the puzzle. The solution is shown in Figs. 42 and 43. It will be seen that the small square is cut out whole and the large square composed of the four

is quite easily done. But George Wilkinson suddenly suggested to them this poser. He said, "Instead of picking out the cross entire, and forming the square from four equal pieces, can you cut out a square entire and four equal pieces that will form a perfect Greek cross?" The puzzle is, of course, now quite easy.

143.—TWO CROSSES FROM ONE.

CUT a Greek cross into five pieces that will form two such crosses, both of the same size. The solution of this puzzle is very beautiful.

144.—THE CROSS AND THE TRIANGLE.

CUT a Greek cross into six pieces that will form an equilateral triangle. This is another hard problem, and I will state here that a solution is practically impossible without a previous knowledge of my method of transforming an equilateral triangle into a square (see No. 26, "Canterbury Puzzles").

145.—THE FOLDED CROSS.

CUT out of paper a Greek cross; then so fold it that with a single straight cut of the scissors the four pieces produced will form a square.

VARIOUS DISSECTION PUZZLES.

WE will now consider a small miscellaneous selection of cutting-out puzzles, varying in degrees of difficulty.

146.—AN EASY DISSECTION PUZZLE.

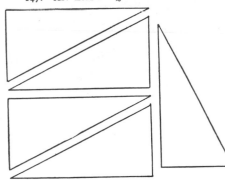

FIRST, cut out a piece of paper or cardboard of the shape shown in the illustration. It will be seen at once that the proportions are simply those of a square attached to half of another similar square, divided diagonally. The puzzle is to cut it into four pieces all of precisely the same size and shape.

147.—AN EASY SQUARE PUZZLE.

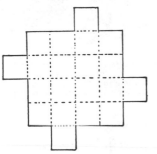

IF you take a rectangular piece of cardboard, twice as long as it is broad, and cut it in half diagonally, you will get two of the pieces shown in the illustration. The puzzle is with five such pieces of equal size to form a square. One of the pieces may be cut in two, but the others must be used intact.

148.—THE BUN PUZZLE.

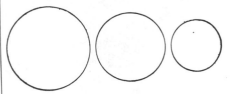

THE three circles represent three buns, and it is simply required to show how these may be equally divided among four boys. The buns must be regarded as of equal thickness throughout and of equal thickness to each other. Of course, they must be cut into as few pieces as possible. To simplify it I will state the rather surprising fact that only five pieces are necessary, from which it will be seen that one boy gets his share in two pieces and the other three receive theirs in a single piece. I am aware that this statement "gives away" the puzzle, but it should not destroy its interest to those who like to discover the "reason why."

149.—THE CHOCOLATE SQUARES.

HERE is a slab of chocolate, indented at the dotted lines so that the twenty squares can be easily separated. Make a copy of the slab in paper or cardboard and then try to cut it into nine pieces so that they will form four perfect squares all of exactly the same size.

150.—DISSECTING A MITRE.

THE figure that is perplexing the carpenter in the illustration represents a mitre. It will be seen that its proportions are those of a square with one quarter removed. The puzzle is to cut it into five pieces that will fit together and form a perfect square. I show an attempt, published in America, to perform the feat in

four pieces, based on what is known as the "step principle," but it is a fallacy.

We are told first to cut off the pieces 1 and 2 and pack them into the triangular space marked off by the dotted line, and so form a rectangle.

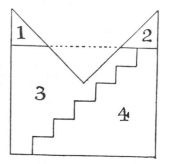

So far, so good. Now, we are directed to apply the old step principle, as shown, and, by moving down the piece 4 one step, form the required square. But, unfortunately, it does *not* produce a square : only an oblong. Call the three long sides of the mitre 84 in. each. Then, before cutting the steps, our rectangle in three pieces will be 84 × 63. The steps must be 10½ in. in height and 12 in. in breadth. Therefore, by moving down a step we reduce by 12 in. the side 84 in. and increase by 10½ in. the side 63 in. Hence our final rectangle must be 72 in. × 73½ in., which certainly is not a square ! The fact is, the step principle can only be applied to rectangles with sides of particular relative lengths. For example, if the shorter side in this case were 61⁴⁄₇ (instead of 63), then the step method would apply. For the steps would then be 10⁴⁄₇ in. in height and 12 in. in breadth. Note that 61⁴⁄₇ × 84 = the square of 72. At present no solution has been found in four pieces, and I do not believe one possible.

151.—THE JOINER'S PROBLEM.

I HAVE often had occasion to remark on the practical utility of puzzles, arising out of an application to the ordinary affairs of life of the little tricks and "wrinkles" that we learn while solving recreation problems.

The joiner, in the illustration, wants to cut the piece of wood into as few pieces as possible to form a square table-top, without any waste

by cutting it into five pieces, the parts fit together and form a square, as shown in the illustration. Now, it is quite an interesting puzzle

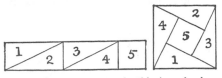

to discover how we can do this in only four pieces.

154.—MRS. HOBSON'S HEARTHRUG.

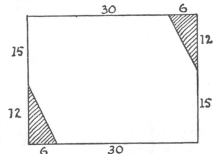

of material. How should he go to work? How many pieces would you require?

152.—ANOTHER JOINER'S PROBLEM.

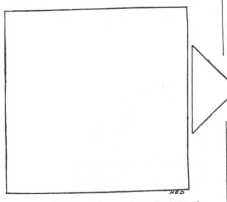

A JOINER had two pieces of wood of the shapes and relative proportions shown in the diagram. He wished to cut them into as few pieces as possible so that they could be fitted together, without waste, to form a perfectly square tabletop. How should he have done it? There is no necessity to give measurements, for if the smaller piece (which is half a square) be made a little too large or a little too small it will not affect the method of solution.

153.—A CUTTING-OUT PUZZLE.

HERE is a little cutting-out poser. I take a strip of paper, measuring five inches by one inch, and,

MRS. HOBSON's boy had an accident when playing with the fire, and burnt two of the corners of a pretty hearthrug. The damaged corners have been cut away, and it now has the appearance and proportions shown in my diagram. How is Mrs. Hobson to cut the rug into the fewest possible pieces that will fit together and form a perfectly square rug?

It will be seen that the rug is in the proportions 36 × 27 (it does not matter whether we say inches or yards), and each piece cut away measured 12 and 6 on the outside.

155.—THE PENTAGON AND SQUARE.

I WONDER how many of my readers, amongst those who have not given any close attention to the elements of geometry, could draw a regular pentagon, or five-sided figure, if they suddenly required to do so. A regular hexagon, or six-sided figure, is easy enough, for everybody knows that all you have to do is to describe a circle and then, taking the radius as the length of one of the sides, mark off the six points round the circumference. But a pentagon is quite another matter. So, as my puzzle has to do with the cutting up of a regular pentagon, it will perhaps be well if I first show my less experienced readers how this figure is to be correctly drawn. Describe a circle and draw the two lines H B and D G, in the diagram, through the centre at right angles. Now find the point A, midway between C and B. Next place the point of your compasses at A and with the distance A D describe the arc. cutting H B at E. Then place the point of your compasses at D and with the distance D E describe the arc cutting the circumference at F. Now, D F

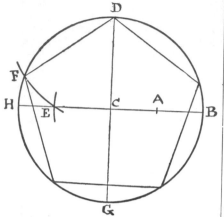

is one of the sides of your pentagon, and you have simply to mark off the other sides round the circle. Quite simple when you know how, but otherwise somewhat of a poser.

Having formed your pentagon, the puzzle is to cut it into the fewest possible pieces that will fit together and form a perfect square.

has simply cut out of paper an equilateral triangle—that is, a triangle with all its three sides of the same length. He proposes that it shall be cut into five pieces in such a way that they will fit together and form either two or three smaller equilateral triangles, using all the material in each case. Can you discover how the cuts should be made ?

Remember that when you have made your five pieces, you must be able, as desired, to put them together to form either the single original triangle or to form two triangles or to form three triangles—all equilateral.

157.—THE TABLE-TOP AND STOOLS.

I HAVE frequently had occasion to show that the published answers to a great many of the oldest and most widely known puzzles are either quite incorrect or capable of improvement. I propose to consider the old poser of the table-top and stools that most of my readers have probably seen in some form or another in books compiled for the recreation of childhood.

The story is told that an economical and ingenious schoolmaster once wished to convert a circular table-top, for which he had no use, into seats for two oval stools, each with a hand-hole in the centre. He instructed the carpenter to make the cuts as in the illustration and then join the eight pieces together in the manner shown. So impressed was he with the ingenuity of his performance that he set the puzzle to his geometry class as a little study in dissection. But the remainder of the story has

156.—THE DISSECTED TRIANGLE.

A GOOD puzzle is that which the gentleman in the illustration is showing to his friends. He

never been published, because, so it is said, it was a characteristic of the principals of academies that they would never admit that they could err. I get my information from a descendant of the original boy who had most reason to be interested in the matter.

The clever youth suggested modestly to the master that the hand-holes were too big, and that a small boy might perhaps fall through them. He therefore proposed another way of making the cuts that would get over this objection. For his impertinence he received such severe chastisement that he became convinced that the larger the hand-hole in the stools the more comfortable might they be.

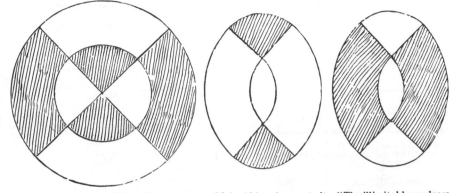

Now what was the method the boy proposed ? Can you show how the circular table-top may be cut into eight pieces that will fit together and form two oval seats for stools (each of exactly the same size and shape) and each having similar hand-holes of smaller dimensions than in the case shown above ? Of course, all the wood must be used.

158.—THE GREAT MONAD.

HERE is a symbol of tremendous antiquity which is worthy of notice. It is borne on the Korean ensign and merchant flag, and has been adopted as a trade sign by the Northern Pacific Railroad Company, though probably few are aware that it is the Great Monad, as shown in the sketch below. This sign is to the Chinaman what the cross is to the Christian. It is the sign of Deity and eternity, while the two parts into which the circle is divided are called the Yin and the Yan—the male and female forces of nature. A writer on the subject more than three thousand years ago is reported to have said in reference to it : "The illimitable produces the great extreme. The great extreme produces the two principles. The two principles produce the four quarters, and from the four quarters we develop the quadrature of the eight diagrams of Feuh-hi." I hope readers will not ask me to explain this, for I have not the slightest idea what it means. Yet I am persuaded that for ages the symbol has had occult and probably mathematical meanings for the esoteric student.

I will introduce the Monad in its elementary form. Here are three easy questions respecting this great symbol :—

(I.) Which has the greater area, the inner circle containing the Yin and the Yan, or the outer ring ?

(II.) Divide the Yin and the Yan into four pieces of the same size and shape by one cut.

(III.) Divide the Yin and the Yan into four pieces of the same size, but different shape, by one straight cut.

159.—THE SQUARE OF VENEER.

THE following represents a piece of wood in my possession, 5 in. square. By markings on the surface it is divided into twenty-five square inches. I want to discover a way of cutting this piece of wood into the fewest possible pieces that will fit together and form two perfect squares of different sizes and of known dimensions. But, unfortunately, at every one of the sixteen intersections of the cross lines a small nail has been driven in at some time or other, and my fret-saw will be injured if it comes in contact with any of these. I have therefore to find a method of doing the work that will not necessitate my cutting through any of those

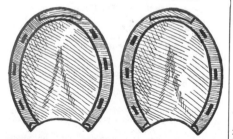

sixteen points. How is it to be done? Remember, the exact dimensions of the two squares must be given.

160.—THE TWO HORSESHOES.

WHY horseshoes should be considered " lucky " is one of those things which no man can understand. It is a very old superstition, and John Aubrey (1626–1700) says, " Most houses at the West End of London have a horseshoe on the threshold." In Monmouth Street there were seventeen in 1813 and seven so late as 1855. Even Lord Nelson had one nailed to the mast of the ship *Victory*. To-day we find it more conducive to " good luck " to see that they are securely nailed on the feet of the horse we are about to drive.

Nevertheless, so far as the horseshoe, like the Swastika and other emblems that I have had occasion at times to deal with, has served to symbolize health, prosperity, and goodwill towards men, we may well treat it with a certain amount of respectful interest. May there not, moreover, be some esoteric or lost mathematical mystery concealed in the form of a horseshoe? I have been looking into this matter, and I wish to draw my readers' attention to the very remarkable fact that the pair of horseshoes shown in my illustration are related in a striking and beautiful manner to the circle, which is the symbol of eternity. I present this fact in the form of a simple problem, so that it may be seen how subtly this relation has been concealed for ages and ages. My readers will, I know, be pleased when they find the key to the mystery.

Cut out the two horseshoes carefully round the outline and then cut them into four pieces, all different in shape, that will fit together and form a perfect circle. Each shoe must be cut into two pieces and all the part of the horse's hoof contained within the outline is to be used and regarded as part of the area.

161.—THE BETSY ROSS PUZZLE.

A CORRESPONDENT asked me to supply him with the solution to an old puzzle that is attributed to a certain Betsy Ross, of Philadelphia, who showed it to George Washington. It consists in so folding a piece of paper that with one clip of the scissors a five-pointed star of Freedom may be produced. Whether the story of the puzzle's origin is a true one or not I cannot say, but I have a print of the old house in Philadelphia where the lady is said to have lived, and I believe it still stands there. But my readers will doubtless be interested in the little poser.

Take a circular piece of paper and so fold it that with one cut of the scissors you can produce a perfect five-pointed star.

162.—THE CARDBOARD CHAIN.

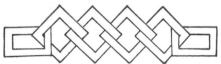

CAN you cut this chain out of a piece of cardboard without any join whatever? Every link is solid, without its having been split and afterwards joined at any place. It is an interesting old puzzle that I learnt as a child, but I have no knowledge as to its inventor.

163.—THE PAPER BOX.

IT may be interesting to introduce here, though it is not strictly a puzzle, an ingenious method for making a paper box.

Take a square of stout paper and by successive foldings make all the creases indicated by the dotted lines in the illustration. Then cut away the eight little triangular pieces that are shaded, and cut through the paper along the dark lines. The second illustration shows the box half folded up, and the reader will have no difficulty in effecting its completion. Before folding up, the reader might cut out the circular piece indicated in the diagram, for a purpose I will now explain.

This box will be found to serve excellently for the production of vortex rings. These rings, which were discussed by Von Helmholtz in 1858, are most interesting, and the box (with the hole cut out) will produce them to perfection. Fill the box with tobacco smoke by blowing it

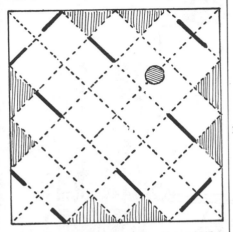

gently through the hole. Now, if you hold it horizontally, and softly tap the side that is opposite to the hole, an immense number of perfect rings can be produced from one mouthful of smoke. It is best that there should be no currents of air in the room. People often do not realise that these rings are formed in the air when no smoke is used. The smoke only makes them visible. Now, one of these rings, if properly directed on its course, will travel across the room and put out the flame of a candle, and this feat is much more striking if you can manage to do it without the smoke. Of course, with a little practice, the rings may be blown from the mouth, but the box produces them in much greater perfection, and no skill whatever is required. Lord Kelvin propounded the theory that matter may consist of vortex

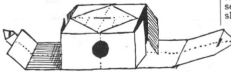

rings in a fluid that fills all space, and by a development of the hypothesis he was able to explain chemical combination.

164.—THE POTATO PUZZLE.

Take a circular slice of potato, place it on the table, and see into how large a number of pieces you can divide it with six cuts of a knife. Of course you must not readjust the pieces or pile them after a cut. What is the greatest number of pieces you can make?

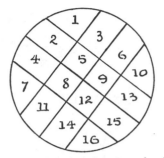

The illustration shows how to make sixteen pieces. This can, of course, be easily beaten.

165.—THE SEVEN PIGS.

Here is a little puzzle that was put to one of the sons of Erin the other day and perplexed him unduly, for it is really quite easy. It will be seen from the illustration that he was shown a sketch of a square pen containing seven pigs. He was asked how he would intersect the pen with three straight fences so as to enclose every pig in a separate sty. In other words, all you have to do is to take your pencil and, with three straight strokes across the square, enclose each pig separately. Nothing could be simpler.

The Irishman complained that the pigs would not keep still while he was putting up the fences. He said that they would all flock together, or one obstinate beast would go into a corner and flock all by himself. It was pointed out to him that for the purposes of the puzzle the pigs were stationary. He answered that Irish pigs are not stationery—they are pork. Being persuaded to make the attempt, he drew three lines, one of which cut through a pig. When it was explained that this is not allowed, he protested

that a pig was no use until you cut its throat. "Begorra, if it's bacon ye want without cutting your pig, it will be all gammon." We will not do the Irishman the injustice of suggesting that the miserable pun was intentional. However, he failed to solve the puzzle. Can you do it?

166.—THE LANDOWNER'S FENCES.

THE landowner in the illustration is consulting with his bailiff over a rather puzzling little question. He has a large plan of one of his fields, in which there are eleven trees. Now, he wants to divide the field into just eleven enclosures by means of straight fences, so that every enclosure shall contain one tree as a shelter for his cattle. How is he to do it with as few fences as possible? Take your pencil and draw straight lines across the field until you have marked off the eleven enclosures (and no more), and then see how many fences you require. Of course the fences may cross one another.

167.—THE WIZARD'S CATS.

A WIZARD placed ten cats inside a magic circle as shown in our illustration, and hypnotized them so that they should remain stationary

during his pleasure. He then proposed to draw three circles inside the large one, so that no cat could approach another cat without crossing a

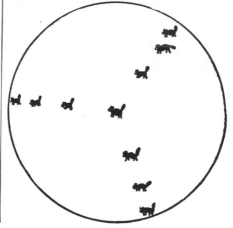

magic circle. Try to draw the three circles so that every cat has its own enclosure and cannot reach another cat without crossing a line.

168.—THE CHRISTMAS PUDDING.

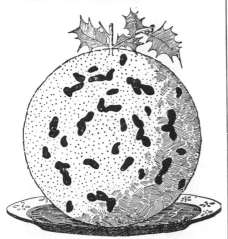

"Speaking of Christmas puddings," said the host, as he glanced at the imposing delicacy at the other end of the table. "I am reminded of the fact that a friend gave me a new puzzle the other day respecting one. Here it is," he added, diving into his breast pocket.

"'Problem : To find the contents,' I suppose," said the Eton boy.

"No ; the proof of that is in the eating. I will read you the conditions."

"'Cut the pudding into two parts, each of exactly the same size and shape, without touching any of the plums. The pudding is to be regarded as a flat disc, not as a sphere.'"

"Why should you regard a Christmas pudding as a disc ? And why should any reasonable person ever wish to make such an accurate division ? " asked the cynic.

"It is just a puzzle—a problem in dissection."

All in turn had a look at the puzzle, but nobody succeeded in solving it. It is a little difficult unless you are acquainted with the principle involved in the making of such puddings, but easy enough when you know how it is done.

169.—A TANGRAM PARADOX.

MANY pastimes of great antiquity, such as chess, have so developed and changed down the centuries that their original inventors would scarcely recognize them. This is not the case with Tangrams, a recreation that appears to be at least four thousand years old, that has apparently never been dormant, and that has not been altered or "improved upon" since the legendary Chinaman Tan first cut out the seven

pieces shown in Diagram 1. If you mark the point B, midway between A and C, on one side of a square of any size, and D, midway between C and E, on an adjoining side, the direction of

the cuts is too obvious to need further explanation. Every design in this article is built up from the seven pieces of blackened cardboard. It will at once be understood that the possible combinations are infinite.

The late Mr. Sam Loyd, of New York, who published a small book of very ingenious designs, possessed the manuscripts of the late Mr. Challenor, who made a long and close study of Tangrams. This gentleman, it is said, records that there were originally seven books of Tangrams, compiled in China two thousand years before the Christian era. These books are so rare that, after forty years' residence in the country, he only succeeded in seeing perfect copies of the first and seventh volumes with fragments of the second. Portions of one of the books, printed in gold leaf upon parchment, were found in Peking by an English soldier and sold for three hundred pounds.

A few years ago a little book came into my possession, from the library of the late Lewis Carroll, entitled *The Fashionable Chinese Puzzle.* It contains three hundred and twenty-three Tangram designs, mostly nondescript geometrical figures, to be constructed from the seven pieces. It was " Published by J. and E. Wallis, 42 Skinner Street, and J. Wallis, Jun., Marine Library, Sidmouth " (South Devon). There is no date, but the following note fixes the time of publication pretty closely : " This ingenious contrivance has for some time past been the favourite amusement of the ex-Emperor Napoleon, who, being now in a debilitated state and living very retired, passes many hours a day in thus exercising his patience and ingenuity." The reader will find, as did the great exile, that much amusement, not wholly uninstructive, may be derived from forming the designs of others. He will find many of the illustrations to this article quite easy to build up, and some rather difficult. Every picture may thus be regarded as a puzzle.

But it is another pastime altogether to create new and original designs of a pictorial character, and it is surprising what extraordinary scope the Tangrams afford for producing pictures of real life—angular and often grotesque, it is true, but full of character. I give an example of a recumbent figure (2) that is particularly graceful, and only needs some slight reduction of its angularities to produce an entirely satisfactory outline.

As I have referred to the author of *Alice in Wonderland*, I give also my designs of the March

Hare (3) and the Hatter (4). I also give an attempt at Napoleon (5), and a very excellent Red Indian with his Squaw by Mr. Loyd (6 and 7). A large number of other designs will be found in an article by me in *The Strand Magazine* for November, 1908.

On the appearance of this magazine article, the late Sir James Murray, the eminent philologist, tried, with that amazing industry that characterized all his work, to trace the word "tangram" to its source. At length he wrote as follows :—" One of my sons is a professor in the Anglo-Chinese college at Tientsin. Through him, his colleagues, and his students, I was able to make inquiries as to the alleged Tan among Chinese scholars. Our Chinese professor here (Oxford) also took an interest in the matter and obtained information from the secretary of the Chinese Legation in Lon-

don, who is a very eminent representative of the Chinese literati.

"The result has been to show that the man Tan, the god Tan, and the 'Book of Tan' are entirely unknown to Chinese literature, history, or tradition. By most of the learned men the name, or allegation of the existence, of these had never been heard of. The puzzle is, of course, well known. It is called in Chinese *ch'i ch'iao t'u*; literally, 'seven-ingenious-plan'

or 'ingenious-puzzle figure of seven pieces.' No name approaching 'tangram,' or even 'tan,' occurs in Chinese, and the only suggestions for the latter were the Chinese *t'an*, 'to extend'; or *t'ang*, Cantonese dialect for 'Chinese.' It was suggested that probably some American

or Englishman who knew a little Chinese or Cantonese, wanting a name for the puzzle, might concoct one out of one of these words and the European ending 'gram.' I should say the name 'tangram' was probably invented by an American some little time before 1864 and after 1847, but I cannot find it in print before the 1864 edition of Webster. I have therefore had to deal very shortly with the word in the dictionary, telling what it is applied to and what conjectures or guesses have been made at the name, and giving a few quotations, one from your own article, which has enabled me to make more of the subject than I could otherwise have done."

Several correspondents have informed me that they possess, or had possessed, specimens of the old Chinese books. An American gentleman writes to me as follows :—" I have in my possession a book made of tissue paper, printed in black (with a Chinese inscription on the front page), containing over three hundred designs, which belongs to the box of 'tangrams,' which I also own. The blocks are seven in number, made of mother-of-pearl, highly polished and finely engraved on either side. These are contained in a rosewood box 2¼ in. square. My great uncle, ——, was one of the first missionaries to visit China. This box and book, along with quite a collection of other relics, were sent to my grandfather and descended to myself."

My correspondent kindly supplied me with rubbings of the Tangrams, from which it is clear that they are cut in the exact proportions that I have indicated. I reproduce the Chinese inscription (8) for this reason. The owner of the book informs me that he has submitted it to a number of Chinamen in the United States and offered as much as a dollar for a translation. But they all steadfastly refused to read the words, offering the lame excuse that the inscription is Japanese. Natives of Japan, however, insist that it is Chinese. Is there something occult and esoteric about Tangrams, that it is so difficult to lift the veil? Perhaps this page will come under the eye of some reader acquainted with the Chinese language, who will

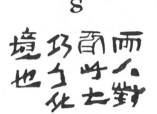

supply the required translation, which may, or may not, throw a little light on this curious question.

By using several sets of Tangrams at the

same time we may construct more ambitious pictures. I was advised by a friend not to send my picture, "A Game of Billiards" (9), to the Academy. He assured me that it would not be accepted because the "judges are so hide-bound by convention." Perhaps he was right, and it will be more appreciated by Post-impressionists and Cubists. The players are considering a very delicate stroke at the top of the

My second picture is named "The Orchestra" (10), and it was designed for the decoration of a large hall of music. Here we have the conductor, the pianist, the fat little cornet-player, the left-handed player of the double-bass, whose attitude is life-like, though he does stand at an unusual distance from his instrument, and the drummer-boy, with his imposing music-stand. The dog at the back of the pianoforte is not howling: he is an appreciative listener.

One remarkable thing about these Tangram pictures is that they suggest to the imagination such a lot that is not really there. Who, for example, can look for a few minutes at Lady Belinda (11) and the Dutch girl (12) without

soon feeling the haughty expression in the one case and the arch look in the other? Then look again at the stork (13), and see how it is suggested to the mind that the leg is actually much more slender than any one of the pieces employed. It is really an optical illusion. Again, notice in the case of the yacht (14) how, by leaving that little angular point at the top, a complete mast is suggested. If you place your Tangrams together on white paper so that they do not quite touch one another, in some cases the effect is improved by the white lines; in other cases it is almost destroyed.

Finally, I give an example from the many

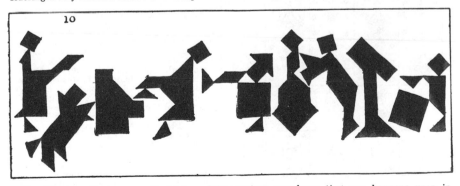

table. Of course, the two men, the table, and the clock are formed from four sets of Tangrams.

curious paradoxes that one happens upon in manipulating Tangrams. I show designs of

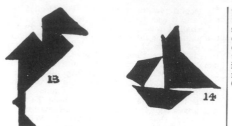

two dignified individuals (15 and 16) who appear to be exactly alike, except for the fact that one **has a** foot and the other has not. Now, both

of these figures are made from the same seven Tangrams. Where does the second man get his foot from ?

PATCHWORK PUZZLES.

" Of shreds and patches."—*Hamlet*, iii. 4.

170.—THE CUSHION COVERS.

THE above represents a square of brocade. A lady wishes to cut it in four pieces so that two pieces will form one perfectly square cushion top, and the remaining two pieces another square cushion top. How is she to do it ? Of course, she can only cut along the lines that divide the twenty-five squares, and the pattern must " match " properly without any irregularity whatever in the design of the material. There is only one way of doing it. Can you find it ?

171.—THE BANNER PUZZLE.

A LADY had a square piece of bunting with two lions on it, of which the illustration is an exactly reproduced reduction. She wished to cut the stuff into pieces that would fit together and form two square banners with a lion on each banner. She discovered that this could be done in as few as four pieces. How did she manage it ? Of course, to cut the British Lion would be an unpardonable offence, so you must be careful that no cut passes through any portion of either of them. Ladies are informed that no allowance whatever has to be made for " turnings," and no part of the material may be wasted. It is quite a simple little dissection puzzle if rightly attacked. Remember that the banners have to be perfect squares, though they need not be both of the same size.

172.—MRS. SMILEY'S CHRISTMAS PRESENT.

MRS. SMILEY'S expression of pleasure was sincere when her six granddaughters sent to her, as a Christmas present, a very pretty patchwork quilt, which they had made with their own hands. It was constructed of square pieces of silk material, all of one size, and as they made a large quilt with fourteen of these little squares on each side, it is obvious that just 196 pieces had been stitched into it. Now, the six granddaughters each contributed a part of the work in the form of a perfect square (all six portions being different in size), but in order to join them up to form the square quilt it was necessary that the work of one girl should be unpicked into three separate parts. Can you show how the joins might have been made ? Of course, no portion can be turned over.

173.—MRS. PERKINS'S QUILT.

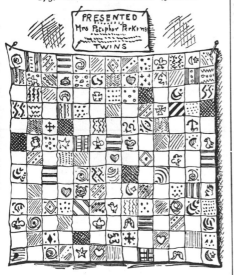

It will be seen that in this case the square patchwork quilt is built up of 169 pieces. The

puzzle is to find the smallest possible number of square portions of which the quilt could be composed and show how they might be joined together. Or, to put it the reverse way, divide the quilt into as few square portions as possible by merely cutting the stitches.

174.—THE SQUARES OF BROCADE.

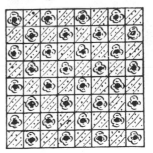

I HAPPENED to be paying a call at the house of a lady, when I took up from a table two lovely squares of brocade. They were beautiful specimens of Eastern workmanship—both of the same design, a delicate chequered pattern. "Are they not exquisite?" said my friend. "They were brought to me by a cousin who has just returned from India. Now, I want you

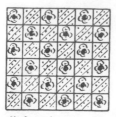

to give me a little assistance. You see, I have decided to join them together so as to make one large square cushion-cover. How should I do this so as to mutilate the material as little as possible? Of course I propose to make my cuts only along the lines that divide the little chequers."

I cut the two squares in the manner desired into four pieces that would fit together and form another larger square, taking care that the pattern should match properly, and when I had finished I noticed that two of the pieces were of exactly the same area; that is, each of the two contained the same number of chequers. Can you show how the cuts were made in accordance with these conditions?

175.—ANOTHER PATCHWORK PUZZLE.

A LADY was presented, by two of her girl friends, with the pretty pieces of silk patchwork shown in our illustration. It will be seen that both

pieces are made up of squares all of the same size—one 12 × 12 and the other 5 × 5. She proposes to join them together and make one square patchwork quilt, 13 × 13, but, of course, she will not cut any of the material—merely cut the stitches where necessary and join together again. What perplexes her is this. A friend assures her that there need be no more than four pieces in all to join up for the new quilt. Could you show her how this little needlework puzzle is to be solved in so few pieces?

176.—LINOLEUM CUTTING.

THE diagram herewith represents two separate pieces of linoleum. The chequered pattern is

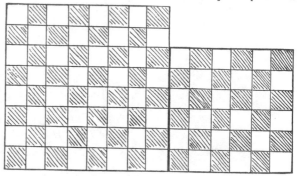

not repeated at the back, so that the pieces cannot be turned over. The puzzle is to cut the two squares into four pieces so that they shall fit together and form one perfect square 10 × 10, so that the pattern shall properly

match, and so that the larger piece shall have as small a portion as possible cut from it.

177.—ANOTHER LINOLEUM PUZZLE.

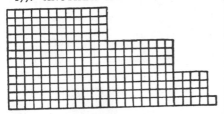

CAN you cut this piece of linoleum into four pieces that will fit together and form a perfect square? Of course the cuts may only be made along the lines.

VARIOUS GEOMETRICAL PUZZLES.

"So various are the tastes of men."
MARK AKENSIDE.

178.—THE CARDBOARD BOX.

THIS puzzle is not difficult, but it will be found entertaining to discover the simple rule for its solution. I have a rectangular cardboard box. The top has an area of 120 square inches, the side 96 square inches, and the end 80 square inches. What are the exact dimensions of the box?

179.—STEALING THE BELL-ROPES.

Two men broke into a church tower one night to steal the bell-ropes. The two ropes passed through holes in the wooden ceiling high above them, and they lost no time in climbing to the top. Then one man drew his knife and cut the rope above his head, in consequence of which he fell to the floor and was badly injured. His fellow-thief called out that it served him right for being such a fool. He said that he should have done as he was doing, upon which he cut the rope below the place at which he held on. Then, to his dismay, he found that he was in no better plight, for, after hanging on as long as his strength lasted, he was compelled to let go and fall beside his comrade. Here they were both found the next morning with their limbs broken. How far did they fall? One of the ropes when they found it was just touching the floor, and when you pulled the end to the wall, keeping the rope taut, it touched a point just three inches above the floor, and the wall was four feet from the rope when it hung at rest. How long was the rope from floor to ceiling?

180.—THE FOUR SONS.

READERS will recognize the diagram as a familiar friend of their youth. A man possessed a square-shaped estate. He bequeathed to his widow the quarter of it that is shaded off. The remainder was to be divided equitably amongst his four sons, so that each should receive land of exactly the same area and exactly similar in shape. We are shown how this was done. But the remainder of the story is not so generally known. In the centre of the estate was a well, indicated by the dark spot, and Benjamin, Charles, and David complained that the division was not "equitable," since Alfred had access to this well, while they could not reach it without trespassing on somebody else's land. The puzzle is to show how the estate is to be appor-

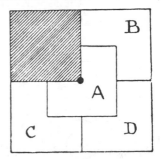

tioned so that each son shall have land of the same shape and area, and each have access to the well without going off his own land.

181.—THE THREE RAILWAY STATIONS.

As I sat in a railway carriage I noticed at the other end of the compartment a worthy squire, whom I knew by sight, engaged in conversation with another passenger, who was evidently a friend of his.

"How far have you to drive to your place from the railway station?" asked the stranger.

"Well," replied the squire, "if I get out at Appleford, it is just the same distance as if I go to Bridgefield, another fifteen miles farther on; and if I changed at Appleford and went thirteen miles from there to Carterton, it would still be the same distance. You see, I am equidistant from the three stations, so I get a good choice of trains."

Now I happened to know that Bridgefield is just fourteen miles from Carterton, so I amused myself in working out the exact distance that the squire had to drive home whichever station he got out at. What was the distance?

182.—THE GARDEN PUZZLE.

PROFESSOR RACKBRANE tells me that he was recently smoking a friendly pipe under a tree in the garden of a country acquaintance. The garden was enclosed by four straight walls, and his friend informed him that he had measured these and found the lengths to be 80, 45, 100, and 63 yards respectively. "Then," said the professor, "we can calculate the exact area of the garden." "Impossible," his host replied,

" because you can get an infinite number of different shapes with those four sides." "But you forget," Rackbrane said, with a twinkle in his eye, "that you told me once you had planted this tree equidistant from all the four corners of the garden." Can you work out the garden's area ?

183.—DRAWING A SPIRAL.

IF you hold the page horizontally and give it a quick rotary motion while looking at the centre of the spiral, it will appear to revolve. Perhaps a good many readers are acquainted with this little optical illusion. But the puzzle is to show how I was able to draw this spiral with so much exactitude without using anything but a pair

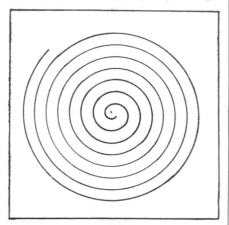

of compasses and the sheet of paper on which the diagram was made. How would you proceed in such circumstances ?

184.—HOW TO DRAW AN OVAL.

CAN you draw a perfect oval on a sheet of paper with one sweep of the compasses ? It is one of the easiest things in the world when you know how.

185.—ST. GEORGE'S BANNER.

AT a celebration of the national festival of St. George's Day I was contemplating the familiar banner of the patron saint of our country. We all know the red cross on a white ground, shown in our illustration. This is the banner of St. George. The banner of St. Andrew (Scotland) is a white "St. Andrew's Cross " on a blue ground. That of St. Patrick (Ireland) is a similar cross in red on a white ground. These three are united in one to form our Union Jack.

Now on looking at St. George's banner it occurred to me that the following question would make a simple but pretty little puzzle. Supposing the flag measures four feet by three feet, how wide must the arm of the cross be if it is

required that there shall be used just the same quantity of red and of white bunting ?

186.—THE CLOTHES LINE PUZZLE.

A BOY tied a clothes line from the top of each of two poles to the base of the other. He then proposed to his father the following question. As one pole was exactly seven feet above the ground and the other exactly five feet, what was the height from the ground where the two cords crossed one another ?

187.—THE MILKMAID PUZZLE.

HERE is a little pastoral puzzle that the reader may, at first sight, be led into supposing is very profound, involving deep calculations. He may even say that it is quite impossible to give any answer unless we are told something definite as to the distances. And yet it is really quite " childlike and bland."

In the corner of a field is seen a milkmaid milking a cow, and on the other side of the field is the dairy where the extract has to be deposited. But it has been noticed that the young woman always goes down to the river with her pail before returning to the dairy. Here the suspicious reader will perhaps ask why she pays these visits to the river. I can only reply that it is no business of ours. The alleged milk is entirely for local consumption.

"Where are you going to, my pretty maid?"
"Down to the river, sir," she said.
"I'll *not* choose your dairy, my pretty maid."
"Nobody axed you, sir," she said.

If one had any curiosity in the matter, such an independent spirit would entirely disarm one. So we will pass from the point of commercial morality to the subject of the puzzle.

Draw a line from the milking-stool down to the river and thence to the door of the dairy, which shall indicate the shortest possible route for the milkmaid. That is all. It is quite easy to indicate the exact spot on the bank of the river to which she should direct her steps if she wants as short a walk as possible. Can you find that spot?

188.—THE BALL PROBLEM.

A STONEMASON was engaged the other day in cutting out a round ball for the purpose of some architectural decoration, when a smart schoolboy came upon the scene.

"Look here," said the mason, "you seem to be a sharp youngster, can you tell me this? If I placed this ball on the level ground, how many balls of the same size could I lay around it (also on the ground) so that every ball should touch this one?"

The boy at once gave the correct answer, and then put this little question to the mason:—

"If the surface of that ball contained just as many square feet as its volume contained cubic feet, what would be the length of its diameter?"

The stonemason could not give an answer. Could you have replied correctly to the mason's and the boy's questions?

189.—THE YORKSHIRE ESTATES.

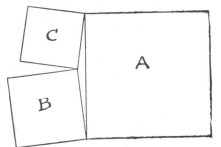

I WAS on a visit to one of the large towns of Yorkshire. While walking to the railway station on the day of my departure a man thrust a handbill upon me, and I took this into the railway carriage and read it at my leisure. It informed me that three Yorkshire neighbouring estates were to be offered for sale. Each estate was square in shape, and they joined one another at their corners, just as shown in the diagram. Estate A contains exactly 370 acres, B contains 116 acres, and C 74 acres.

Now, the little triangular bit of land enclosed by the three square estates was not offered for sale, and, for no reason in particular, I became curious as to the area of that piece. How many acres did it contain?

190.—FARMER WURZEL'S ESTATE.

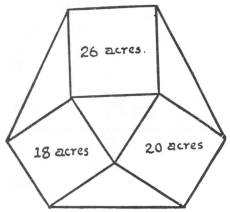

26 acres.

18 acres   20 acres

I WILL now present another land problem. The demonstration of the answer that I shall give will, I think, be found both interesting and easy of comprehension.

Farmer Wurzel owned the three square fields shown in the annexed plan, containing respectively 18, 20, and 26 acres. In order to get a ring-fence round his property he bought the four intervening triangular fields. The puzzle is to discover what was then the whole area of his estate.

191.—THE CRESCENT PUZZLE.

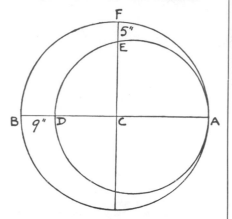

HERE is an easy geometrical puzzle. The crescent is formed by two circles, and C is the centre of the larger circle. The width of the crescent between B and D is 9 inches, and between E and F 5 inches. What are the diameters of the two circles?

192.—THE PUZZLE WALL.

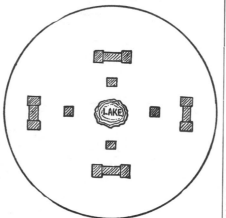

THERE was a small lake, around which four poor men built their cottages. Four rich men afterwards built their mansions, as shown in the illustration, and they wished to have the lake to themselves, so they instructed a builder to put up the shortest possible wall that would exclude the cottagers, but give themselves free access to the lake. How was the wall to be built?

193.—THE SHEEPFOLD.

IT is a curious fact that the answers always given to some of the best-known puzzles that appear in every little book of fireside recreations that has been published for the last fifty or a hundred years are either quite unsatisfactory or clearly wrong. Yet nobody ever seems to detect their faults. Here is an example :—A farmer had a pen made of fifty hurdles, capable of holding a hundred sheep only. Supposing he wanted to make it sufficiently large to hold double that number, how many additional hurdles must he have?

194.—THE GARDEN WALLS.

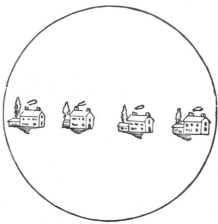

A SPECULATIVE country builder has a circular field, on which he has erected four cottages, as shown in the illustration. The field is surrounded by a brick wall, and the owner undertook to put up three other brick walls, so that the neighbours should not be overlooked by each other, but the four tenants insist that there shall be no favouritism, and that each shall have exactly the same length of wall space for his wall fruit trees. The puzzle is to show how the three walls may be built so that each tenant shall have the same area of ground, and precisely the same length of wall.

Of course, each garden must be entirely enclosed by its walls, and it must be possible to prove that each garden has exactly the same length of wall. If the puzzle is properly solved no figures are necessary.

195.—LADY BELINDA'S GARDEN.

LADY BELINDA is an enthusiastic gardener. In the illustration she is depicted in the act of worrying out a pleasant little problem which I will relate. One of her gardens is oblong in shape, enclosed by a high holly hedge, and she is turning it into a rosary for the cultivation of some of her choicest roses. She wants to devote exactly half of the area of the garden to the flowers, in one large bed, and the other half to be a path going all round it of equal breadth throughout. Such a garden is shown in the diagram at the foot of the picture. How is she

to mark out the garden under these simple conditions? She has only a tape, the length of the garden, to do it with, and, as the holly hedge is so thick and dense, she must make all her measurements inside. Lady Belinda did not know the exact dimensions of the garden, and, as it was not necessary for her to know, I also give no dimensions. It is quite a simple task no matter what the size or proportions of the garden may be. Yet how many lady gardeners would know just how to proceed? The tape may be quite plain—that is, it need not be a graduated measure.

196.—THE TETHERED GOAT.

HERE is a little problem that everybody should know how to solve. The goat is placed in a half-acre meadow, that is in shape an equilateral triangle. It is tethered to a post at one corner of the field. What should be the length of the tether (to the nearest inch) in order that the goat shall be able to eat just half the grass in the field? It is assumed that the goat can feed to the end of the tether.

197.—THE COMPASSES PUZZLE.

IT is curious how an added condition or restriction will sometimes convert an absurdly easy puzzle into an interesting and perhaps difficult one. I remember buying in the street many years ago a little mechanical puzzle that had a tremendous sale at the time. It consisted of a medal with holes in it, and the puzzle was to work a ring with a gap in it from hole to hole until it was finally detached. As I was walking along the street I very soon acquired the trick of taking off the ring with one hand while holding the puzzle in my pocket. A friend to whom I showed the little feat set about accomplishing it himself, and when I met him some days afterwards he exhibited his proficiency in the art. But he was a little taken aback when I then took the puzzle from him and, while simply holding the medal between the finger and thumb of one hand, by a series of little shakes and jerks caused the ring, without my even touching it, to fall off upon the floor. The following little poser will probably prove a rather tough nut for a great many readers, simply on account of the restricted conditions:—

Show how to find exactly the middle of any straight line by means of the compasses only. You are not allowed to use any ruler, pencil, or other article—only the compasses; and no trick or dodge, such as folding the paper, will be permitted. You must simply use the compasses in the ordinary legitimate way.

198.—THE EIGHT STICKS.

I HAVE eight sticks, four of them being exactly half the length of the others. I lay every one of these on the table, so that they enclose three squares, all of the same size. How do I do it? There must be no loose ends hanging over.

199.—PAPA'S PUZZLE.

HERE is a puzzle by Pappus, who lived at Alexandria about the end of the third century. It is the fifth proposition in the eighth book of his *Mathematical Collections*. I give it in the form that I presented it some years ago under the title "Papa's Puzzle," just to see how many readers would discover that it was by Pappus himself. "The little maid's papa has taken two different-sized rectangular pieces of cardboard, and has clipped off a triangular piece from one of them, so that when it is suspended by a thread from the point A it hangs with the long side perfectly horizontal, as shown in the illustration. He has perplexed the child by asking her to find the point A on the other card, so as to produce a similar result when cut and suspended by a thread." Of course, the point must not be

a little calculation that ought to interest my readers. The Professor was paying out the wire to which his kite was attached from a winch on which it had been rolled into a perfectly spherical form. This ball of wire was just two feet in diameter, and the wire had a diameter of one-hundredth of an inch. What was the length of the wire?

Now, a simple little question like this that everybody can perfectly understand will puzzle many people to answer in any way. Let us see whether, without going into any profound mathematical calculations, we can get the answer roughly—say, within a mile of what is correct! We will assume that when the wire is all wound up the ball is perfectly solid throughout, and that no allowance has to be made for the axle that passes through it. With that simplification, I wonder how many readers can state within even a mile of the correct answer the length of that wire.

found by trial clippings. A curious and pretty point is involved in this setting of the puzzle. Can the reader discover it?

201.—HOW TO MAKE CISTERNS.

OUR friend in the illustration has a large sheet of zinc, measuring (before cutting) eight feet by three feet, and he has cut out square pieces (all of the same size) from the four corners and now proposes to fold up the sides, solder the edges,

200.—A KITE-FLYING PUZZLE.

WHILE accompanying my friend Professor Highflite during a scientific kite-flying competition on the South Downs of Sussex I was led into

and make a cistern. But the point that puzzles him is this: Has he cut out those square pieces of the correct size in order that the cistern may hold the greatest possible quantity of water? You see, if you cut them very small you get a very shallow cistern ; if you cut them large you get a tall and slender one. It is all a question of finding a way of cutting out these four square pieces exactly the right size. How are we to avoid making them too small or too large ?

202.—THE CONE PUZZLE.

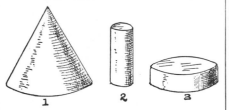

1   2   3

I HAVE a wooden cone, as shown in Fig. 1. How am I to cut out of it the greatest possible cylinder ? It will be seen that I can cut out one that is long and slender, like Fig. 2, or short and thick, like Fig. 3. But neither is the largest possible. A child could tell you where to cut, if he knew the rule. Can you find this simple rule ?

203.—CONCERNING WHEELS.

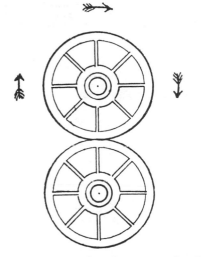

THERE are some curious facts concerning the movements of wheels that are apt to perplex the novice. For example : when a railway train is travelling from London to Crewe certain parts of the train at any given moment are actually moving from Crewe towards London. Can you indicate those parts ? It seems absurd that parts of the same train can at any time travel in opposite directions, but such is the case.

In the accompanying illustration we have two wheels. The lower one is supposed to be fixed and the upper one running round it in the direction of the arrows. Now, how many times does the upper wheel turn on its own axis in making a complete revolution of the other wheel ? Do not be in a hurry with your answer, or you are almost certain to be wrong. Experiment with two pennies on the table and the correct answer will surprise you, when you succeed in seeing it.

204.—A NEW MATCH PUZZLE.

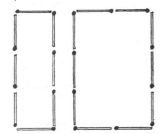

IN the illustration eighteen matches are shown arranged so that they enclose two spaces, one just twice as large as the other. Can you rearrange them (1) so as to enclose two four-sided spaces, one exactly three times as large as the other, and (2) so as to enclose two five-sided spaces, one exactly three times as large as the other ? All the eighteen matches must be fairly used in each case ; the two spaces must be quite detached, and there must be no loose ends or duplicated matches.

205.—THE SIX SHEEP-PENS.

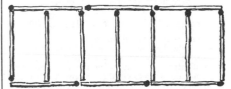

HERE is a new little puzzle with matches. It will be seen in the illustration that thirteen matches, representing a farmer's hurdles, have been so placed that they enclose six sheep-pens all of the same size. Now, one of these hurdles was stolen, and the farmer wanted still to enclose six pens of equal size with the remaining twelve. How was he to do it ? All the twelve matches must be fairly used, and there must be no duplicated matches or loose ends.

POINTS AND LINES PROBLEMS.

" Line upon line, line upon line ; here a little and there a little."—Isa. xxviii. 10.

WHAT are known as " Points and Lines " puzzles are found very interesting by many people. The most familiar example, here given, to plant nine trees so that they shall form ten straight rows with three trees in every row, is attributed to Sir Isaac Newton, but the earliest collection of such puzzles is, I believe, in a rare little book that I possess—published in 1821—*Rational Amusement for Winter Evenings*, by John Jackson. The author gives ten examples of " Trees planted in Rows."

These tree-planting puzzles have always been a matter of great perplexity. They are real " puzzles," in the truest sense of the word, because nobody has yet succeeded in finding a direct and certain way of solving them. They demand the exercise of sagacity, ingenuity, and patience, and what we call " luck " is also sometimes of service. Perhaps some day a genius will discover the key to the whole mystery. Remember that the trees must be regarded as

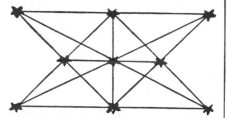

mere points, for if we were allowed to make our trees big enough we might easily " fudge " our diagrams and get in a few extra straight rows that were more apparent than real.

206.—THE KING AND THE CASTLES.

THERE was once, in ancient times, a powerful king, who had eccentric ideas on the subject of military architecture. He held that there was great strength and economy in symmetrical forms, and always cited the example of the bees, who construct their combs in perfect hexagonal cells, to prove that he had nature to support him. He resolved to build ten new castles in his country all to be connected by fortified walls, which should form five lines with four castles in every line. The royal architect presented his preliminary plan in the form I have shown. But the monarch pointed out that every castle could be approached from the outside, and commanded that the plan should be so modified that as many castles as possible should be free from attack from the outside, and could only be reached by crossing the fortified walls. The architect replied that he thought it impossible so to arrange them that even one castle, which the king proposed to use as a royal residence, could be so protected, but his majesty soon enlightened him by pointing out how it might

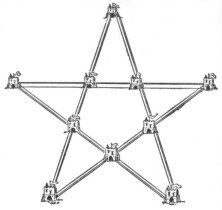

be done. How would you have built the ten castles and fortifications so as best to fulfil the king's requirements ? Remember that they must form five straight lines with four castles in every line.

207.—CHERRIES AND PLUMS.

THE illustration is a plan of a cottage as it stands surrounded by an orchard of fifty-five trees. Ten of these trees are cherries, ten are

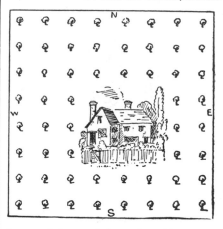

plums, and the remainder apples. The cherries are so planted as to form five straight lines, with four cherry trees in every line. The plum trees are also planted so as to form five straight lines with four plum trees in every line. The puzzle is to show which are the ten cherry trees and which are the ten plums. In order that the cherries and plums should have the most favourable aspect, as few as possible (under the conditions) are planted on the north and east sides of the orchard. Of course in picking out a group of ten trees (cherry or plum, as the case may be) you ignore all intervening trees. That is to say, four trees may be in a straight line irrespective of other trees (or the house) being in between. After the last puzzle this will be quite easy.

208.—A PLANTATION PUZZLE.

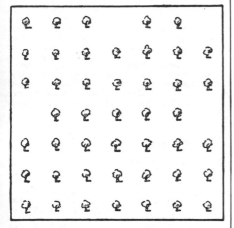

A MAN had a square plantation of forty-nine trees, but, as will be seen by the omissions in the illustration, four trees were blown down and removed. He now wants to cut down all the remainder except ten trees, which are to be so left that they shall form five straight rows with four trees in every row. Which are the ten trees that he must leave ?

209.—THE TWENTY-ONE TREES.

A GENTLEMAN wished to plant twenty-one trees in his park so that they should form twelve straight rows with five trees in every row. Could you have supplied him with a pretty symmetrical arrangement that would satisfy these conditions ?

210.—THE TEN COINS.

PLACE ten pennies on a large sheet of paper or cardboard, as shown in the diagram, five on each edge. Now remove four of the coins, without disturbing the others, and replace them on the paper so that the ten shall form five straight lines with four coins in every line. This in itself is not difficult, but you should try to dis-

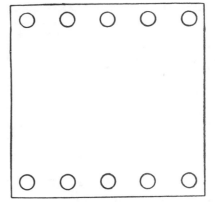

cover in how many different ways the puzzle may be solved, assuming that in every case the two rows at starting are exactly the same.

211.—THE TWELVE MINCE-PIES.

IT will be seen in our illustration how twelve mince-pies may be placed on the table so as to form six straight rows with four pies in every row. The puzzle is to remove only four of them to new positions so that there shall be *seven*

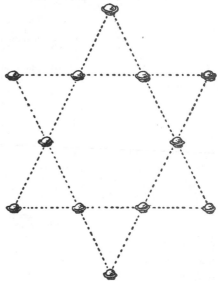

straight rows with four in every row. Which four would you remove, and where would you replace them ?

212.—THE BURMESE PLANTATION.

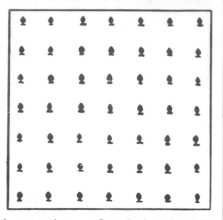

A SHORT time ago I received an interesting communication from the British chaplain at Meiktila, Upper Burma, in which my correspondent informed me that he had found some amusement on board ship on his way out in trying to solve this little poser.

If he has a plantation of forty-nine trees, planted in the form of a square as shown in the accompanying illustration, he wishes to know how he may cut down twenty-seven of the trees so that the twenty-two left standing shall form as many rows as possible with four trees in every row.

Of course there may not be more than four trees in any row.

213.—TURKS AND RUSSIANS.

THIS puzzle is on the lines of the Afridi problem published by me in *Tit-Bits* some years ago.

On an open level tract of country a party of Russian infantry, no two of whom were stationed at the same spot, were suddenly surprised by thirty-two Turks, who opened fire on the Russians from all directions. Each of the Turks simultaneously fired a bullet, and each bullet passed immediately over the heads of three Russian soldiers. As each of these bullets when fired killed a different man, the puzzle is to discover what is the smallest possible number of soldiers of which the Russian party could have consisted and what were the casualties on each side.

MOVING COUNTER PROBLEMS.

> "I cannot do't without counters."
> *Winter's Tale*, iv. 3.

PUZZLES of this class, except so far as they occur in connection with actual games, such as chess, seem to be a comparatively modern introduction. Mathematicians in recent times, notably Vandermonde and Reiss, have devoted some attention to them, but they do not appear to have been considered by the old writers. So far as games with counters are concerned, perhaps the most ancient and widely known in old times is " Nine Men's Morris " (known also, as I shall show, under a great many other names), unless the simpler game, distinctly mentioned in the works of Ovid (No. 110, " Ovid's Game," in *The Canterbury Puzzles*), from which " Noughts and Crosses " seems to be derived, is still more ancient.

In France the game is called Marelle, in Poland Siegen Wulf Myll (She-goat Wolf Mill, or Fight), in Germany and Austria it is called Muhle (the Mill), in Iceland it goes by the name of Mylla, while the Bogas (or native bargees) of South America are said to play it, and on the Amazon it is called Trique, and held to be of Indian origin. In our own country it has different names in different districts, such as Meg Merrylegs, Peg Meryll, Nine Peg o'Merryal, Nine-Pin Miracle, Merry Peg, and Merry Hole. Shakespeare refers to it in " Midsummer Night's Dream " (Act ii., scene 1) :—

> " The nine-men's morris is filled up with mud ;
> And the quaint mazes in the wanton green,
> For lack of tread, are undistinguishable."

It was played by the shepherds with stones in holes cut in the turf. John Clare, the peasant poet of Northamptonshire, in " The Shepherd Boy " (1835) says :—" Oft we track his haunts . . . . By nine-peg-morris nicked upon the green." It is also mentioned by Drayton in his " Polyolbion."

It was found on an old Roman tile discovered during the excavations at Silchester, and cut upon the steps of the Acropolis at Athens. When visiting the Christiania Museum a few years ago I was shown the great Viking ship that was discovered at Gokstad in 1880. On the oak planks forming the deck of the vessel were found holes and lines marking out the game, the holes being made to receive pegs. While inspecting the ancient oak furniture in the Rijks Museum at Amsterdam I became interested in an old catechumen's settle, and was surprised to find the game diagram cut in the centre of the seat—quite conveniently for surreptitious play. It has been discovered cut in the choir stalls of several of our English cathedrals. In the early eighties it was found scratched upon a stone built into a wall (probably about the date 1200), during the restoration of Hargrave church in Northamptonshire. This stone is now in the Northampton Museum. A similar stone has since been found at Sempringham, Lincolnshire. It is to be seen on an ancient tombstone in the Isle of Man, and painted on old Dutch tiles. And in 1901 a stone was dug out of a gravel pit near Oswestry bearing an undoubted diagram of the game.

The game has been played with different

rules at different periods and places. I give a copy of the board. Sometimes the diagonal

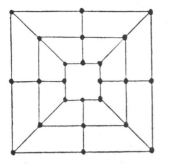

lines are omitted, but this evidently was not intended to affect the play : it simply meant that the angles alone were thought sufficient to indicate the points. This is how Strutt, in *Sports and Pastimes*, describes the game, and it agrees with the way I played it as a boy :— " Two persons, having each of them nine pieces, or men, lay them down alternately, one by one, upon the spots ; and the business of either party is to prevent his antagonist from placing three of his pieces so as to form a row of three, without the intervention of an opponent piece. If a row be formed, he that made it is at liberty to take up one of his competitor's pieces from any part he thinks most to his advantage ; excepting he has made a row, which must not be touched if he have another piece upon the board that is not a component part of that row. When all the pieces are laid down, they are played backwards and forwards, in any direction that the lines run, but only can move from one spot to another (next to it) at one time. He that takes off all his antagonist's pieces is the conqueror."

214.—THE SIX FROGS.

THE SIX educated frogs in the illustration are trained to reverse their order, so that their numbers shall read 6, 5, 4, 3, 2, 1, with the blank square in its present position. They can jump to the next square (if vacant) or leap over one frog to the next square beyond (if vacant), just as we move in the game of draughts, and can go backwards or forwards at pleasure. Can you show how they perform their feat in the fewest possible moves ? It is quite easy, so when you have done it add a seventh frog to the right and try again. Then add more frogs until you are able to give the shortest solution for any number. For it can always be done, with that single vacant square, no matter how many frogs there are.

215.—THE GRASSHOPPER PUZZLE.

IT has been suggested that this puzzle was a great favourite among the young apprentices of the City of London in the sixteenth and seventeenth centuries. Readers will have noticed the curious brass grasshopper on the Royal Exchange. This long-lived creature escaped the fires of 1666 and 1838. The grasshopper, after his kind, was the crest of Sir Thomas Gresham, merchant grocer, who died in 1579, and from this cause it has been used as a sign by grocers in general. Unfortunately for the legend as to its origin, the puzzle was only produced by myself so late as the year 1900. On

twelve of the thirteen black discs are placed numbered counters or grasshoppers. The puzzle is to reverse their order, so that they shall read, 1, 2, 3, 4, etc., in the opposite direction, with the vacant disc left in the same position as at present. Move one at a time in any order, either to the adjoining vacant disc or by jumping over one grasshopper, like the moves in draughts. The moves or leaps may be made in either direction that is at any time possible. What are the fewest possible moves in which it can be done ?

216.—THE EDUCATED FROGS.

OUR six educated frogs have learnt a new and pretty feat. When placed on glass tumblers, as shown in the illustration, they change sides so that the three black ones are to the left and the white frogs to the right, with the unoccupied tumbler at the opposite end—No. 7. They can jump to the next tumbler (if unoccupied), or over one, or two, frogs to an unoccupied

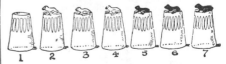

tumbler. The jumps can be made in either direction, and a frog may jump over his own or the opposite colour, or both colours. Four successive specimen jumps will make everything quite plain : 4 to 1, 5 to 4, 3 to 5, 6 to 3. Can you show how they do it in ten jumps ?

217.—THE TWICKENHAM PUZZLE.

IN the illustration we have eleven discs in a circle. On five of the discs we place white counters with black letters—as shown—and on five other discs the black counters with white letters. The bottom disc is left vacant. Starting thus, it is required to get the counters into order so that they spell the word " Twickenham " in a clockwise direction, leaving the vacant disc in the original position. The black counters move in the direction that a clockhand revolves, and the white counters go the opposite way. A counter may jump over one of the opposite colour if the vacant disc is next beyond. Thus, if your first move is with K, then C can jump over K. If then K moves towards E, you may next jump W over C, and so on. The puzzle may be solved in twenty-six moves. Remember a counter cannot jump over one of its own colour.

218.—THE VICTORIA CROSS PUZZLE.

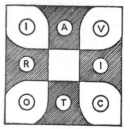

THE puzzle-maker is peculiarly a " snapper-up of unconsidered trifles," and his productions are

often built up with the slenderest materials. Trivialities that might entirely escape the observation of others, or, if they were observed, would be regarded as of no possible moment, often supply the man who is in quest of posers with a pretty theme or an idea that he thinks possesses some " basal value."

When seated opposite to a lady in a railway carriage at the time of Queen Victoria's Diamond Jubilee, my attention was attracted to a brooch that she was wearing. It was in the form of a Maltese or Victoria Cross, and bore the letters of the word VICTORIA. The number and arrangement of the letters immediately gave me the suggestion for the puzzle which I now present.

The diagram, it will be seen, is composed of nine divisions. The puzzle is to place eight counters, bearing the letters of the word VICTORIA, exactly in the manner shown, and then slide one letter at a time from black to white and white to black alternately, until the word reads round in the same direction, only with the initial letter V on one of the black arms of the cross. At no time may two letters be in the same division. It is required to find the shortest method.

Leaping moves are, of course, not permitted. The first move must obviously be made with A, I, T, or R. Supposing you move T to the centre, the next counter played will be O or C, since I or R cannot be moved. There is something a little remarkable in the solution of this puzzle which I will explain.

219.—THE LETTER BLOCK PUZZLE.

HERE is a little reminiscence of our old friend the Fifteen Block Puzzle. Eight wooden blocks are lettered, and are placed in a box, as shown in the illustration. It will be seen that you can only move one block at a time to the place vacant for the time being, as no block may be lifted out of the box. The puzzle is to shift them about until you get them in the order—

A B C
D E F
G H

This you will find by no means difficult if you are allowed as many moves as you like. But the puzzle is to do it in the fewest possible moves. I will not say what this smallest number of moves is, because the reader may like to discover it for himself. In writing down your moves you will find it necessary to record no more than the letters in the order that they are shifted. Thus, your first five moves might be C, H, G, E, F; and this notation can have no possible ambiguity. In practice you only need eight counters and a simple diagram on a sheet of paper.

220.—A LODGING-HOUSE DIFFICULTY.

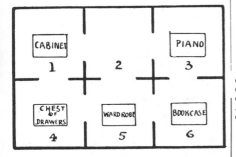

THE Dobsons secured apartments at Slocomb-on-Sea. There were six rooms on the same floor, all communicating, as shown in the diagram. The rooms they took were numbers 4, 5, and 6, all facing the sea. But a little difficulty arose. Mr. Dobson insisted that the piano and the bookcase should change rooms. This was wily, for the Dobsons were not musical, but they wanted to prevent any one else playing the instrument. Now, the rooms were very small and the pieces of furniture indicated were very big, so that no two of these articles could be got into any room at the same time. How was the exchange to be made with the least possible labour? Suppose, for example, you first move the wardrobe into No. 2; then you can move the bookcase to No. 5 and the piano to No. 6, and so on. It is a fascinating puzzle, but the landlady had reasons for not appreciating it. Try to solve her difficulty in the fewest possible removals with counters on a sheet of paper.

221.—THE EIGHT ENGINES.

THE diagram represents the engine-yard of a railway company under eccentric management. The engines are allowed to be stationary only at the nine points indicated, one of which is at present vacant. It is required to move the engines, one at a time, from point to point, in seventeen moves, so that their numbers shall be in numerical order round the circle, with the

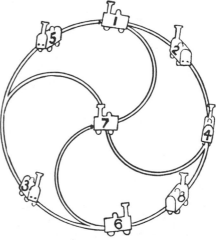

central point left vacant. But one of the engines has had its fire drawn, and therefore cannot move. How is the thing to be done? And which engine remains stationary throughout?

222.—A RAILWAY PUZZLE.

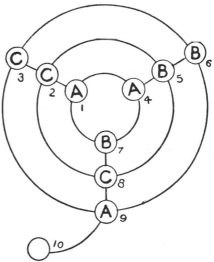

MAKE a diagram, on a large sheet of paper, like the illustration, and have three counters marked A, three marked B, and three marked C. It will be seen that at the intersection of lines there are nine stopping-places, and a tenth stopping-place is attached to the outer circle like the tail of a Q. Place the three counters or engines marked A, the three marked B, and the three marked C at the places indicated. The puzzle is to move the engines, one at a time, along the lines, from stopping-place to stopping-place, until you succeed in getting an A, a B, and a C on each circle, and also A, B, and C on each straight line. You are required to do this in as few moves as possible. How many moves do you need?

223.—A RAILWAY MUDDLE.

THE plan represents a portion of the line of the London, Clodville, and Mudford Railway Company. It is a single line with a loop. There is only room for eight wagons, or seven wagons and an engine, between B and C on either the left line or the right line of the loop. It happened that two goods trains (each consisting of an engine and sixteen wagons) got into the position shown in the illustration. It looked like a hopeless deadlock, and each engine-driver wanted the other to go back to the next station and take off nine wagons. But an ingenious stoker undertook to pass the trains and send them on their respective journeys with their engines properly in front. He also contrived to reverse the engines the fewest times possible. Could you have performed the feat? And how many times would you require to reverse the

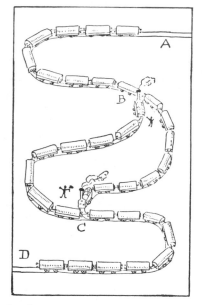

engines? A "reversal" means a change of direction, backward or forward. No rope-shunting, fly-shunting, or other trick is al-

lowed. All the work must be done legitimately by the two engines. It is a simple but interesting puzzle if attempted with counters.

224.—THE MOTOR-GARAGE PUZZLE.

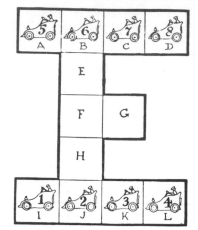

THE difficulties of the proprietor of a motor garage are converted into a little pastime of a kind that has a peculiar fascination. All you need is to make a simple plan or diagram on a sheet of paper or cardboard and number eight counters, 1 to 8. Then a whole family can enter into an amusing competition to find the best possible solution of the difficulty.

The illustration represents the plan of a motor garage, with accommodation for twelve cars. But the premises are so inconveniently restricted that the proprietor is often caused considerable perplexity. Suppose, for example, that the eight cars numbered 1 to 8 are in the positions shown, how are they to be shifted in the quickest possible way so that 1, 2, 3, and 4 shall change places with 5, 6, 7, and 8—that is, with the numbers still running from left to right, as at present, but the top row exchanged with the bottom row? What are the fewest possible moves?

One car moves at a time, and any distance counts as one move. To prevent misunderstanding, the stopping-places are marked in squares, and only one car can be in a square at the same time.

225.—THE TEN PRISONERS.

IF prisons had no other use, they might still be preserved for the special benefit of puzzle-makers. They appear to be an inexhaustible mine of perplexing ideas. Here is a little poser that will perhaps interest the reader for a short period. We have in the illustration a prison of sixteen cells. The locations of the ten prisoners will be seen. The jailer has queer superstitions about odd and even numbers, and he wants to rearrange the ten prisoners so that there shall be as many even rows of men, vertically, horizontally, and diagonally, as possible. At present it will be seen, as indicated by the arrows, that there are only twelve such rows of

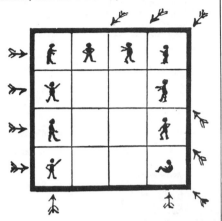

2 and 4. I will state at once that the greatest number of such rows that is possible is sixteen. But the jailer only allows four men to be removed to other cells, and informs me that, as the man who is seated in the bottom right-hand corner is infirm, he must not be moved. Now, how are we to get those sixteen rows of even numbers under such conditions ?

226.—ROUND THE COAST.

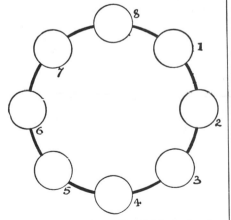

HERE is a puzzle that will, I think, be found as amusing as instructive. We are given a ring of eight circles. Leaving circle 8 blank, we are required to write in the name of a seven-lettered port in the United Kingdom in this manner.

Touch a blank circle with your pencil, then jump over two circles in either direction round the ring, and write down the first letter. Then touch another vacant circle, jump over two circles, and write down your second letter. Proceed similarly with the other letters in their proper order until you have completed the word. Thus, suppose we select " Glasgow," and proceed as follows : 6—1, 7—2, 8—3, 7—4, 8—5, which means that we touch 6, jump over 7 and 8, and write down " G " on 1 ; then touch 7, jump over 8 and 1, and write down " l " on 2 ; and so on. It will be found that after we have written down the first five letters—" Glasg "— as above, we cannot go any further. Either there is something wrong with " Glasgow," or we have not managed our jumps properly. Can you get to the bottom of the mystery ?

227.—CENTRAL SOLITAIRE.

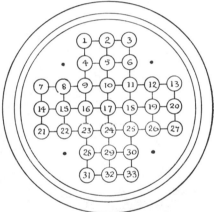

THIS ancient puzzle was a great favourite with our grandmothers, and most of us, I imagine, have on occasions come across a " Solitaire " board—a round polished board with holes cut in it in a geometrical pattern, and a glass marble in every hole. Sometimes I have noticed one on a side table in a suburban front parlour, or found one on a shelf in a country cottage, or had one brought under my notice at a wayside inn. Sometimes they are of the form shown above, but it is equally common for the board to have four more holes, at the points indicated by dots. I select the simpler form.

Though " Solitaire " boards are still sold at the toy shops, it will be sufficient if the reader will make an enlarged copy of the above on a sheet of cardboard or paper, number the " holes," and provide himself with 33 counters, buttons, or beans. Now place a counter in every hole except the central one, No. 17, and the puzzle is to take off all the counters in a series of jumps, except the last counter, which must be left in that central hole. You are allowed to jump one counter over the next one to a vacant hole beyond, just as in the game of draughts, and the counter jumped over is immediately taken off the board. Only remember every move must be a jump; consequently you will take off a counter at each move, and thirty-one single jumps will of course remove all the thirty-one counters. But compound moves are allowed (as in draughts, again), for so long as one counter continues to jump, the jumps all count as one move.

Here is the beginning of an imaginary solution which will serve to make the manner of moving perfectly plain, and show how the solver should write out his attempts: 5–17, 12–10, 26–12, 24–26 (13–11, 11–25), 9–11 (26–24, 24–10, 10–12), etc., etc. The jumps contained within brackets count as one move, because they are made with the same counter. Find the fewest possible moves. Of course, no diagonal jumps are permitted; you can only jump in the direction of the lines.

any diagonal moves — only moves parallel to the sides of the square. It is obvious that as the apples stand no move can be made, but you are permitted to transfer any single apple you like to a vacant plate before starting. Then the moves must be all leaps, taking off the apples leaped over.

229.—THE NINE ALMONDS.

"HERE is a little puzzle," said a Parson, "that I have found peculiarly fascinating. It is so simple, and yet it keeps you interested indefinitely."

The reverend gentleman took a sheet of paper and divided it off into twenty-five squares, like a square portion of a chessboard. Then he placed nine almonds on the central squares, as shown in the illustration, where we have represented numbered counters for convenience in giving the solution.

"Now, the puzzle is," continued the Parson,

228.—THE TEN APPLES.

THE family represented in the illustration are amusing themselves with this little puzzle, which is not very difficult but quite interesting. They have, it will be seen, placed sixteen plates on the table in the form of a square, and put an apple in each of ten plates. They want to find a way of removing all the apples except one by jumping over one at a time to the next vacant square, as in draughts; or, better, as in solitaire, for you are not allowed to make

"to remove eight of the almonds and leave the ninth in the central square. You make the removals by jumping one almond over another to the vacant square beyond and taking off the one jumped over—just as in draughts, only here you can jump in any direction, and not diagonally only. The point is to do the thing in the fewest possible moves."

The following specimen attempt will make everything clear. Jump 4 over 1, 5 over 9, 3 over 6, 5 over 3, 7 over 5 and 2, 4 over 7, 8 over 4. But 8 is not left in the central square, as it

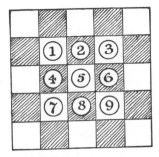

should be. Remember to remove those you jump over. Any number of jumps in succession with the same almond count as one move.

230.—THE TWELVE PENNIES.

HERE is a pretty little puzzle that only requires twelve pennies or counters. Arrange them in a circle, as shown in the illustration. Now take up one penny at a time and, passing it over two pennies, place it on the third penny. Then take up another single penny and do the same thing, and so on, until, in six such moves, you have the

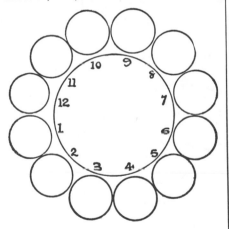

coins in six pairs in the positions 1, 2, 3, 4, 5, 6. You can move in either direction round the circle at every play, and it does not matter whether the two jumped over are separate or a pair. This is quite easy if you use just a little thought.

231.—PLATES AND COINS.

PLACE twelve plates, as shown, on a round table, with a penny or orange in every plate. Start from any plate you like and, always going in one direc-

tion round the table, take up one penny, pass it over two other pennies, and place it in the next plate. Go on again; take up another penny and, having passed it over two pennies, place it in a plate; and so continue your journey. Six coins only are to be removed, and when these have been placed there should be two coins in each of six plates and six plates empty. An important point of the puzzle is to go round the table

as few times as possible. It does not matter whether the two coins passed over are in one or two plates, nor how many empty plates you pass a coin over. But you must always go in one direction round the table and end at the point from which you set out. Your hand, that is to say, goes steadily forward in one direction, without ever moving backwards.

232.—CATCHING THE MICE.

" PLAY fair ! " said the mice. " You know the rules of the game."

" Yes, I know the rules," said the cat. " I've got to go round and round the circle, in the direction that you are looking, and eat every thirteenth mouse, but I must keep the white mouse for a tit-bit at the finish. Thirteen is an unlucky number, but I will do my best to oblige you."

"Hurry up, then !" shouted the mice.

"Give a fellow time to think," said the cat. "I don't know which of you to start at. I must figure it out."

While the cat was working out the puzzle he fell asleep, and, the spell being thus broken, the mice returned home in safety. At which mouse should the cat have started the count in order that the white mouse should be the last eaten ?

When the reader has solved that little puzzle, here is a second one for him. What is the smallest number that the cat can count round and round the circle, if he must start at the white mouse (calling that " one " in the count) and still eat the white mouse last of all ?

And as a third puzzle try to discover what is the smallest number that the cat can count round and round if she must start at the white mouse (calling that " one ") and make the white mouse the third eaten.

He places sixteen cheeses on the floor in a straight row and then makes them into four piles, with four cheeses in every pile, by always passing a cheese over four others. If you use sixteen counters and number them in order from 1 to 16, then you may place 1 on 6, 11 on 1, 7 on 4, and so on, until there are four in every pile. It will be seen that it does not matter whether the four passed over are standing alone or piled; they count just the same, and you can always carry a cheese in either direction. There are a great many different ways of doing it in twelve moves, so it makes a good game of " patience " to try to solve it so that the four piles shall be left in different stipulated places. For example, try to leave the piles at the extreme ends of the row, on Nos. 1, 2, 15 and 16; this is quite easy. Then try to leave three piles together, on Nos. 13, 14, and 15. Then again play so that they shall be left on Nos. 3, 5, 12, and 14.

234.—THE EXCHANGE PUZZLE.

HERE is a rather entertaining little puzzle with

233.—THE ECCENTRIC CHEESEMONGER.

THE cheesemonger depicted in the illustration is an inveterate puzzle lover. One of his favourite puzzles is the piling of cheeses in his warehouse, an amusement that he finds good exercise for the body as well as for the mind.

moving counters. You only need twelve counters—six of one colour, marked A, C, E, G, I, and K, and the other six marked B, D, F, H, J, and L. You first place them on the diagram, as shown in the illustration, and the puzzle is to get them into regular alphabetical order, as follows :—

A	B	C	D
E	F	G	H
I	J	K	L

The moves are made by exchanges of opposite colours standing on the same line. Thus, G and J may exchange places, or F and A, but you cannot exchange G and C, or F and D, because

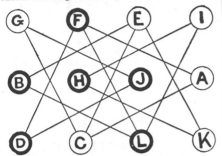

in one case they are both white and in the other case both black. Can you bring about the required arrangement in seventeen exchanges?

vessels and sank the fourth? In the diagram we have arranged the fleet in square formation, where it will be seen that as many as seven ships may be sunk (those in the top row and first column) by firing the torpedoes indicated by arrows. Anchoring the fleet as we like, to what extent can we increase this number? Remember that each successive ship is sunk before another torpedo is launched, and that every torpedo proceeds in a different direction; otherwise, by placing the ships in a straight line, we might sink as many as thirteen! It is an interesting little study in naval warfare, and eminently practical—provided the enemy will allow you to arrange his fleet for your convenience and promise to lie still and do nothing!

236.—THE HAT PUZZLE.

TEN hats were hung on pegs as shown in the illustration—five silk hats and five felt "bowlers," alternately silk and felt. The two pegs at the end of the row were empty.

The puzzle is to remove two contiguous hats to the vacant pegs, then two other adjoining hats to the pegs now unoccupied, and so on until five pairs have been moved and the hats again hang in an unbroken row, but with all

It cannot be done in fewer moves. The puzzle is really much easier than it looks, if properly attacked.

235.—TORPEDO PRACTICE.

If a fleet of sixteen men-of-war were lying at anchor and surrounded by the enemy, how many ships might be sunk if every torpedo, projected in a straight line, passed under three

the silk ones together and all the felt hats together.

Remember, the two hats removed must always be contiguous ones, and you must take one in each hand and place them on their new pegs without reversing their relative position. You are not allowed to cross your hands, nor to hang up one at a time.

Can you solve this old puzzle, which I give as introductory to the next? Try it with counters of two colours or with coins, and remember that the two empty pegs must be left at one end of the row.

237.—BOYS AND GIRLS.

If you mark off ten divisions on a sheet of paper to represent the chairs, and use eight numbered counters for the children, you will have a fascinating pastime. Let the odd numbers represent boys and even numbers girls, or you can use counters of two colours, or coins.

The puzzle is to remove two children who are occupying adjoining chairs and place them in two empty chairs, *making them first change sides*; then remove a second pair of children from adjoining chairs and place them in the two now vacant, making them change sides; and so on, until all the boys are together and all the girls together, with the two vacant chairs at one end as at present. To solve the puzzle you must do this in five moves. The two children must always be taken from chairs that are next to one another; and remember the important point of making the two children change sides.

as this latter is the distinctive feature of the puzzle. By "change sides" I simply mean that if, for example, you first move 1 and 2 to the vacant chairs, then the first (the outside) chair will be occupied by 2 and the second one by 1.

238.—ARRANGING THE JAMPOTS.

I HAPPENED to see a little girl sorting out some jam in a cupboard for her mother. She was putting each different kind of preserve apart on the shelves. I noticed that she took a pot of damson in one hand and a pot of gooseberry in the other and made them change places; then she changed a strawberry with a raspberry, and so on. It was interesting to observe what a lot of unnecessary trouble she gave herself by making more interchanges than there was any need for, and I thought it would work into a good puzzle.

It will be seen in the illustration that little Dorothy has to manipulate twenty-four large jampots in as many pigeon-holes. She wants to get them in correct numerical order—that is, 1, 2, 3, 4, 5, 6 on the top shelf, 7, 8, 9, 10, 11, 12 on the next shelf, and so on. Now, if she always takes one pot in the right hand and another in the left and makes them change places, how **many of these interchanges will be necessary**

to get all the jampots in proper order? She would naturally first change the 1 and the 3, then the 2 and the 3, when she would have the first three pots in their places. How would you advise her to go on then? Place some num-

bered counters on a sheet of paper divided into squares for the pigeon-holes, and you will find it an amusing puzzle.

UNICURSAL AND ROUTE PROBLEMS.

"I see them on their winding way."
REGINALD HEBER.

IT is reasonable to suppose that from the earliest ages one man has asked another such questions as these: "Which is the nearest way home?" "Which is the easiest or pleasantest way?" "How can we find a way that will enable us to dodge the mastodon and the plesiosaurus?" "How can we get there without ever crossing the track of the enemy?" All these are elementary route problems, and they can be turned into good puzzles by the introduction of some conditions that complicate matters. A variety of such complications will be found in the following examples. I have also included some enumerations of more or less difficulty. These afford excellent practice for the reasoning faculties, and enable one to generalize in the case of symmetrical forms in a manner that is most instructive.

239.—A JUVENILE PUZZLE.

FOR years I have been perpetually consulted by my juvenile friends about this little puzzle. Most children seem to know it, and yet, curiously enough, they are invariably unacquainted with the answer. The question they always ask is, "Do, please, tell me whether it is really possible." I believe Houdin the conjurer used to be very fond of giving it to his child friends, but I cannot say whether he invented the little puzzle or not. No doubt a large number of my readers will be glad to have the mystery of the solution cleared up, so I make no apology for introducing this old "teaser."

The puzzle is to draw with three strokes of the pencil the diagram that the little girl is exhibiting in the illustration. Of course, you must not remove your pencil from the paper during a stroke or go over the same line a second time. You will find that you can get in a good deal of the figure with one continuous stroke, but it will always appear as if four strokes are necessary.

Another form of the puzzle is to draw the diagram on a slate and then rub it out in three rubs.

240.—THE UNION JACK.

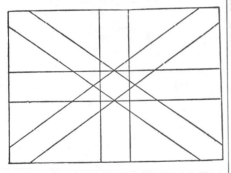

THE illustration is a rough sketch somewhat resembling the British flag, the Union Jack. It is not possible to draw the whole of it without lifting the pencil from the paper or going over the same line twice. The puzzle is to find out just *how much* of the drawing it is possible to make without lifting your pencil or going twice over the same line. Take your pencil and see what is the best you can do.

241.—THE DISSECTED CIRCLE.

How many continuous strokes, without lifting your pencil from the paper, do you require to draw the design shown in our illustration? Directly you change the direction of your pencil it begins a new stroke. You may go over the same line more than once if you like. It re-

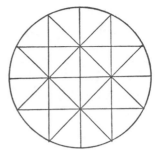

quires just a little care, or you may find yourself beaten by one stroke.

242.—THE TUBE INSPECTOR'S PUZZLE.

THE man in our illustration is in a little dilemma. He has just been appointed inspector of a certain system of tube railways, and it is his duty to inspect regularly, within a stated period, all the company's seventeen lines connecting twelve stations, as shown on the big poster plan that he is contemplating. Now he wants to arrange his route so that it shall take him over all the lines with as little travelling as possible. He may begin where he likes and end where he likes. What is his shortest route?

Could anything be simpler? But the reader will soon find that, however he decides to proceed, the inspector must go over some of the lines more than once. In other words, if we say that the stations are a mile apart, he will

have to travel more than seventeen miles to inspect every line. There is the little difficulty. How far is he compelled to travel, and which route do you recommend?

243.—VISITING THE TOWNS.

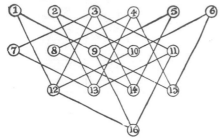

A TRAVELLER, starting from town No. 1, wishes to visit every one of the towns once, and once only, going only by roads indicated by straight lines. How many different routes are there from which he can select? Of course, he must end his journey at No. 1, from which he started, and must take no notice of cross roads, but go straight from town to town. This is an absurdly easy puzzle, if you go the right way to work.

244.—THE FIFTEEN TURNINGS.

HERE is another queer travelling puzzle, the solution of which calls for ingenuity. In this case the traveller starts from the black town and wishes to go as far as possible while making only fifteen turnings and never going along the same road twice. The towns are supposed to be a mile apart. Supposing, for example, that he went straight to A, then straight to B, then

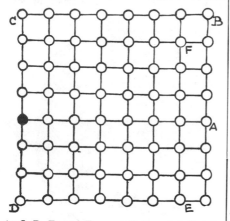

to C, D, E, and F, you will then find that he has travelled thirty-seven miles in five turnings. Now, how far can he go in fifteen turnings?

245.—THE FLY ON THE OCTAHEDRON.

"LOOK here," said the professor to his colleague, "I have been watching that fly on the octa-

hedron, and it confines its walks entirely to the edges. What can be its reason for avoiding the sides?"

"Perhaps it is trying to solve some route problem," suggested the other. "Supposing it to start from the top point, how many different routes are there by which it may walk over all the edges, without ever going twice along the same edge in any route?"

The problem was a harder one than they expected, and after working at it during leisure moments for several days their results did not agree—in fact, they were both wrong. If the

reader is surprised at their failure, let him attempt the little puzzle himself. I will just explain that the octahedron is one of the five regular, or Platonic, bodies, and is contained under eight equal and equilateral triangles. If you cut out the two pieces of cardboard of the shape shown in the margin of the illustration, cut half through along the dotted lines and then bend them and put them together, you will have a perfect octahedron. In any route over all the edges it will be found that the fly must end at the point of departure at the top.

246.—THE ICOSAHEDRON PUZZLE.

THE icosahedron is another of the five regular, or Platonic, bodies having all their sides, angles, and planes similar and equal. It is bounded by twenty similar equilateral triangles. If you cut out a piece of cardboard of the form shown in the smaller diagram, and cut half through along the dotted lines, it will fold up and form a perfect icosahedron.

Now, a Platonic body does not mean a

heavenly body; but it will suit the purpose of our puzzle if we suppose there to be a habitable planet of this shape. We will also suppose that, owing to a superfluity of water, the only dry land is along the edges, and that the inhabitants have no knowledge of navigation. If every one

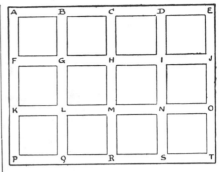

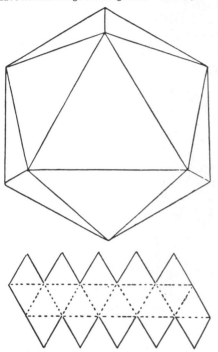

of those edges is 10,000 miles long and a solitary traveller is placed at the North Pole (the highest point shown), how far will he have to travel before he will have visited every habitable part of the planet—that is, have traversed every one of the edges?

247.—INSPECTING A MINE.

THE diagram is supposed to represent the passages or galleries in a mine. We will assume that every passage, A to B, B to C, C to H, H to I, and so on, is one furlong in length. It will be seen that there are thirty-one of these passages. Now, an official has to inspect all of them, and he descends by the shaft to the point A. How far must he travel, and what route do you recommend? The reader may at first say, "As there are thirty-one passages, each a furlong in length, he will have to travel just thirty-one furlongs." But this is assuming that he need never go along a passage more than once, which is not the case. Take your pencil and try to find the shortest route. You will soon discover that there is

room for considerable judgment. In fact, it is a perplexing puzzle.

248.—THE CYCLISTS' TOUR.

Two cyclists were consulting a road map in preparation for a little tour together. The circles represent towns, and all the good roads are represented by lines. They are starting from the town with a star, and must complete their tour at E. But before arriving there they want to visit every other town once, and only once. That is the difficulty. Mr. Spicer said, "I am certain we can find a way of doing it;" but Mr. Maggs replied, "No way, I'm sure." Now, which of them was correct? Take your

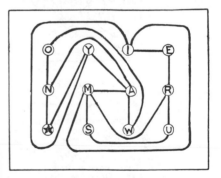

pencil and see if you can find any way of doing it. Of course you must keep to the roads indicated.

249.—THE SAILOR'S PUZZLE.

THE sailor depicted in the illustration stated that he had since his boyhood been engaged in trading with a small vessel among some twenty little islands in the Pacific. He supplied the rough chart of which I have given a copy, and explained that the lines from island to island represented the only routes that he ever adopted. He always started from island A at the beginning of the season, and then visited every island

once, and once only, finishing up his tour at the starting-point A. But he always put off his visit to C as long as possible, for trade reasons that I need not enter into. The puzzle is to discover his exact route, and this can be done with certainty. Take your pencil and, starting at A, try to trace it out. If you write down the islands in the order in which you visit them —thus, for example, A, I, O, L, G, etc.—you can at once see if you have visited an island twice or omitted any. Of course, the crossings of the lines must be ignored—that is, you must continue your route direct, and you are not allowed to switch off at a crossing and proceed in another direction. There is no trick of this kind in the puzzle. The sailor knew the best route. Can you find it ?

250.—THE GRAND TOUR.

ONE of the everyday puzzles of life is the working out of routes. If you are taking a holiday on your bicycle, or a motor tour, there always arises the question of how you are to make the best of your time and other resources. You have determined to get as far as some particular place, to include visits to such-and-such a town, to try to see something of special interest elsewhere, and perhaps to try to look up an old friend at a spot that will not take you much out of your way. Then you have to plan your route so as to avoid bad roads, uninteresting country, and, if possible, the necessity of a return by the same way that you went. With a map before you, the interesting puzzle is attacked and solved. I will present a little poser based on these lines.

I give a rough map of a country—it is not necessary to say what particular country—the circles representing towns and the dotted lines the railways connecting them. Now there lived in the town marked A a man who was born there, and during the whole of his life had never once left his native place. From his youth upwards he had been very industrious, sticking incessantly to his trade, and had no desire whatever to roam abroad. However, on attaining his fiftieth birthday he decided to see something of his country, and especially to pay a visit to a very old friend living at the town marked Z.

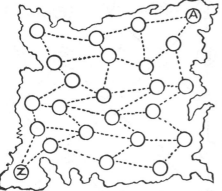

What he proposed was this : that he would start from his home, enter every town once and only once, and finish his journey at Z. As he made up his mind to perform this grand. tour by rail only, he found it rather a puzzle to work

of the three houses, A, B, and C, without any pipe crossing another. Take your pencil and

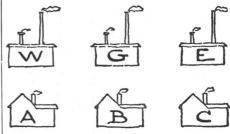

draw lines showing how this should be done. You will soon find yourself landed in difficulties.

252.—A PUZZLE FOR MOTORISTS.

EIGHT motorists drove to church one morning. Their respective houses and churches, together with the only roads available (the dotted lines), are shown. One went from his house A to his

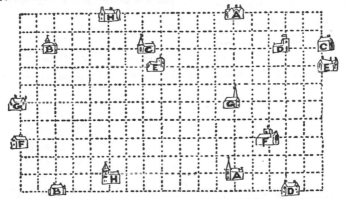

out his route, but he at length succeeded in doing so. How did he manage it ? Do not forget that every town has to be visited once, and not more than once.

251.—WATER, GAS, AND ELECTRICITY.

THERE are some half-dozen puzzles, as old as the hills, that are perpetually cropping up, and there is hardly a month in the year that does not bring inquiries as to their solution. Occasionally one of these, that one had thought was an extinct volcano, bursts into eruption in a surprising manner. I have received an extraordinary number of letters respecting the ancient puzzle that I have called " Water, Gas, and Electricity." It is much older than electric lighting, or even gas, but the new dress brings it up to date. The puzzle is to lay on water, gas, and electricity, from W, G, and E, to each

church A, another from his house B to his church B, another from C to C, and so on, but it was afterwards found that no driver ever crossed the track of another car. Take your pencil and try to trace out their various routes.

253.—A BANK HOLIDAY PUZZLE.

Two friends were spending their bank holiday on a cycling trip. Stopping for a rest at a village inn, they consulted a route map, which is represented in our illustration in an exceedingly simplified form, for the puzzle is interesting enough without all the original complexities. They started from the town in the top left-hand corner marked A. It will be seen that there are one hundred and twenty such towns, all connected by straight roads. Now they discovered that there are exactly 1,365 different routes by which they may reach their destination, always travelling either due south or due east. The puzzle is to discover which town is their destination.

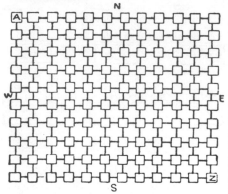

Of course, if you find that there are more than 1,365 different routes to a town it cannot be the right one.

254.—THE MOTOR-CAR TOUR.

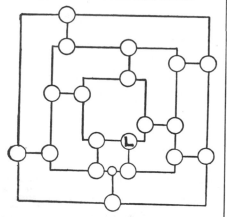

In the above diagram the circles represent towns and the lines good roads. In just how many different ways can a motorist, starting from London (marked with an L), make a tour of all these towns once, visiting every town once, and only once, on a tour, and always coming back to London on the last ride? The exact reverse of any route is not counted as different.

255.—THE LEVEL PUZZLE.

This is a simple counting puzzle. In how many different ways can you spell out the word LEVEL by placing the point of your pencil on an L and then passing along the lines from letter to letter. You may go in any direction, backwards or forwards. Of course you are not allowed to miss letters—that is to say, if you come to a letter you must use it.

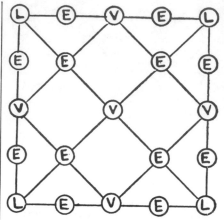

256.—THE DIAMOND PUZZLE.

In how many different ways may the word DIAMOND be read in the arrangement shown? You may start wherever you like at a D and go up or down, backwards or forwards, in and out, in any direction you like, so long as you

always pass from one letter to another that adjoins it. How many ways are there?

257.—THE DEIFIED PUZZLE.

In how many different ways may the word DEIFIED be read in this arrangement under

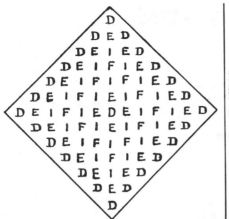

the same conditions as in the last puzzle, with the addition that you can use any letters twice in the same reading?

258.—THE VOTERS' PUZZLE.

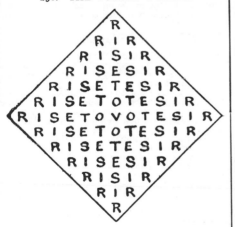

HERE we have, perhaps, the most interesting form of the puzzle. In how many different ways can you read the political injunction, "RISE TO VOTE, SIR," under the same conditions as before? In this case every reading of the palindrome requires the use of the central V as the middle letter.

259.—HANNAH'S PUZZLE.

A MAN was in love with a young lady whose Christian name was Hannah. When he asked her to be his wife she wrote down the letters of her name in this manner :—

```
H  H  H  H  H  H
H  A  A  A  A  H
H  A  N  N  A  H
H  A  N  N  A  H
H  A  A  A  A  H
H  H  H  H  H  H
```

and promised that she would be his if he could tell her correctly in how many different ways it was possible to spell out her name, always passing from one letter to another that was adjacent. Diagonal steps are here allowed. Whether she did this merely to tease him or to test his cleverness is not recorded, but it is satisfactory to know that he succeeded. Would you have been equally successful? Take your pencil and try. You may start from any of the H's and go backwards or forwards and in any direction, so long as all the letters in a spelling are adjoining one another. How many ways are there, no two exactly alike?

260.—THE HONEYCOMB PUZZLE.

HERE is a little puzzle with the simplest possible conditions. Place the point of your pencil on a letter in one of the cells of the honeycomb, and trace out a very familiar proverb by passing always from a cell to one that is contiguous to it. If you take the right route you will have visited every cell once, and only once. The puzzle is much easier than it looks.

261.—THE MONK AND THE BRIDGES.

IN this case I give a rough plan of a river with an island and five bridges. On one side of the river is a monastery, and on the other side is seen a monk in the foreground. Now, the monk has decided that he will cross every bridge once, and only once, on his return to the monastery. This is, of course, quite easy to do, but on the way he thought to himself, "I wonder how many different routes there are from which I might have selected." Could you have told him? That is the puzzle. Take your pencil and trace out a route that will take you once

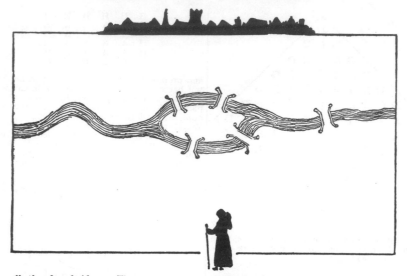

over all the five bridges. Then trace out a second route, then a third, and see if you can count all the variations. You will find that the difficulty is twofold: you have to avoid dropping routes on the one hand and counting the same routes more than once on the other.

COMBINATION AND GROUP PROBLEMS.

"A combination and a form indeed."
Hamlet, iii. 4.

VARIOUS puzzles in this class might be termed problems in the "geometry of situation," but their solution really depends on the theory of combinations which, in its turn, is derived directly from the theory of permutations. It has seemed convenient to include here certain group puzzles and enumerations that might, perhaps, with equal reason have been placed elsewhere; but readers are again asked not to be too critical about the classification, which is very difficult and arbitrary. As I have included my problem of "The Round Table" (No. 273), perhaps a few remarks on another well-known problem of the same class, known by the French as La Problême des Ménages, may be interesting. If n married ladies are seated at a round table in any determined order, in how many different ways may their n husbands be placed so that every man is between two ladies but never next to his own wife?

This difficult problem was first solved by Laisant, and the method shown in the following table is due to Moreau:—

4	0	2
5	3	13
6	13	80
7	83	579
8	592	4738
9	4821	43387
10	43979	439792

The first column shows the number of married couples. The numbers in the second column are obtained in this way: $5 \times 3 + 0 - 2 = 13$; $6 \times 13 + 3 + 2 = 83$; $7 \times 83 + 13 - 2 = 592$; $8 \times 592 + 83 + 2 = 4821$; and so on. Find all the numbers, except 2, in the table, and the method will be evident. It will be noted that the 2 is subtracted when the first number (the number of couples) is odd, and added when that number is even. The numbers in the third column are obtained thus: $13 - 0 = 13$; $83 - 3 = 80$; $592 - 13 = 579$; $4821 - 83 = 4738$; and so on. The numbers in this last column give the required solutions. Thus, four husbands may be seated in two ways, five husbands may be placed in thirteen ways, and six husbands in eighty ways.

The following method, by Lucas, will show the remarkable way in which chessboard analysis may be applied to the solution of a circular problem of this kind. Divide a square into thirty-six cells, six by six, and strike out all the cells in the long diagonal from the bottom left-hand corner to the top right-hand corner, also the five cells in the diagonal next above it and the cell in the bottom right-hand corner. The answer for six couples will be the same as the number of ways in which you can place six rooks (not using the cancelled cells) so that no rook shall ever attack another rook. It will be found that the six rooks may be placed in eighty different ways, which agrees with the above table.

262.—THOSE FIFTEEN SHEEP.

A CERTAIN cyclopædia has the following curious problem, I am told : " Place fifteen sheep in four pens so that there shall be the same number of sheep in each pen." No answer whatever is vouchsafed, so I thought I would investigate the matter. I saw that in dealing with apples or bricks the thing would appear to be quite impossible, since four times any number must be an even number, while fifteen is an odd number. I thought, therefore, that there must be some quality peculiar to the sheep that was not generally known. So I decided to interview some farmers on the subject. The first one pointed out that if we put one pen inside an-

263.—KING ARTHUR'S KNIGHTS.

KING ARTHUR sat at the Round Table on three successive evenings with his knights—Beleobus, Caradoc, Driam, Eric, Floll, and Galahad—but on no occasion did any person have as his neighbour one who had before sat next to him. On the first evening they sat in alphabetical order round the table. But afterwards King Arthur arranged the two next sittings so that he might have Beleobus as near to him as possible and Galahad as far away from him as could be managed. How did he seat the knights to the best advantage, remembering that rule that no knight may have the same neighbour twice ?

other, like the rings of a target, and placed all sheep in the smallest pen, it would be all right. But I objected to this, because you admittedly place all the sheep in one pen, not in four pens. The second man said that if I placed four sheep in each of three pens and three sheep in the last pen (that is fifteen sheep in all), and one of the ewes in the last pen had a lamb during the night, there would be the same number in each pen in the morning. This also failed to satisfy me.

The third farmer said, " I've got four hurdle pens down in one of my fields, and a small flock of wethers, so if you will just step down with me I will show you how it is done." The illustration depicts my friend as he is about to demonstrate the matter to me. His lucid explanation was evidently that which was in the mind of the writer of the article in the cyclopædia. What was it ? Can you place those fifteen sheep ?

264.—THE CITY LUNCHEONS.

TWELVE men connected with a large firm in the City of London sit down to luncheon together every day in the same room. The tables are small ones that only accommodate two persons at the same time. Can you show how these twelve men may lunch together on eleven days in pairs, so that no two of them shall ever sit twice together ? We will represent the men by the first twelve letters of the alphabet, and suppose the first day's pairing to be as follows—

(A B) (C D) (E F) (G H) (I J) (K L).

Then give any pairing you like for the next day, say—

(A C) (B D) (E G) (F H) (I K) (J L),

and so on, until you have completed your eleven lines, with no pair ever occurring twice. There are a good many different arrangements possible. Try to find one of them.

265.—A PUZZLE FOR CARD-PLAYERS.

TWELVE members of a club arranged to play bridge together on eleven evenings, but no player was ever to have the same partner more than once, or the same opponent more than twice. Can you draw up a scheme showing how they may all sit down at three tables every evening ? Call the twelve players by the first twelve letters of the alphabet and try to group them.

266.—A TENNIS TOURNAMENT.

FOUR married couples played a "mixed double" tennis tournament, a man and a lady always playing against a man and a lady. But no person ever played with or against any other person more than once. Can you show how they all could have played together in the two courts on three successive days ? This is a little puzzle of a quite practical kind, and it is just perplexing enough to be interesting.

267.—THE WRONG HATS.

" ONE of the most perplexing things I have come across lately," said Mr. Wilson, " is this. Eight men had been dining not wisely but too well at a certain London restaurant. They were the last to leave, but not one man was in a condition to identify his own hat. Now, considering that they took their hats at random, what are the chances that every man took a hat that did not belong to him ? "

" The first thing," said Mr. Waterson, " is to see in how many different ways the eight hats could be taken."

" That is quite easy," Mr. Stubbs explained. " Multiply together the numbers, 1, 2, 3, 4, 5, 6, 7, and 8. Let me see—half a minute—yes ; there are 40,320 different ways."

" Now all you've got to do is to see in how many of these cases no man has his own hat," said Mr. Waterson.

" Thank you, I'm not taking any," said Mr. Packhurst. " I don't envy the man who attempts the task of writing out all those forty-thousand-odd cases and then picking out the ones he wants."

They all agreed that life is not long enough for that sort of amusement ; and as nobody saw any other way of getting at the answer, the matter was postponed indefinitely. Can you solve the puzzle ?

268.—THE PEAL OF BELLS.

A CORRESPONDENT, who is apparently much interested in campanology, asks me how he is to construct what he calls a " true and correct " peal for four bells. He says that every possible permutation of the four bells must be rung once, and once only. He adds that no bell must move more than one place at a time, that no bell must make more than two successive strokes in either the first or the last place, and that the last change must be able to pass into the first. These fantastic conditions will be found to be observed in the little peal for three bells, as follows :—

1	2	3
2	1	3
2	3	1
3	2	1
3	1	2
1	3	2

How are we to give him a correct solution for his four bells ?

269.—THREE MEN IN A BOAT.

A CERTAIN generous London manufacturer gives his workmen every year a week's holiday at the seaside at his own expense. One year fifteen of his men paid a visit to Herne Bay. On the morning of their departure from London they were addressed by their employer, who expressed the hope that they would have a very pleasant time.

" I have been given to understand," he added, " that some of you fellows are very fond of rowing, so I propose on this occasion to provide you with this recreation, and at the same time give you an amusing little puzzle to solve. During the seven days that you are at Herne Bay every one of you will go out every day at the same time for a row, but there must always be three men in a boat and no more. No two men may ever go out in a boat together more than once, and no man is allowed to go out twice in the same boat. If you can manage to do this, and use as few different boats as possible, you may charge the firm with the expense."

One of the men tells me that the experience he has gained in such matters soon enabled him to work out the answer to the entire satisfaction of themselves and their employer. But the amusing part of the thing is that they never really solved the little mystery. I find their method to have been quite incorrect, and I think it will amuse my readers to discover how the men should have been placed in the boats. As their names happen to have been Andrews, Baker, Carter, Danby, Edwards, Frith, Gay, Hart, Isaacs, Jackson, Kent, Lang, Mason, Napper, and Onslow, we can call them by their initials and write out the five groups for each of the seven days in the following simple way :

| 1 | 2 | 3 | 4 | 5 |

First Day : (ABC) (DEF) (GHI) (JKL) (MNO).

The men within each pair of brackets are here seen to be in the same boat, and therefore A can never go out with B or with C again, and C can never go out again with B. The same applies to the other four boats. The figures show the number on the boat, so that A, B, or C, for example, can never go out in boat No. 1 again.

270.—THE GLASS BALLS.

A NUMBER of clever marksmen were staying at a country house, and the host, to provide a little amusement, suspended strings of glass balls, as shown in the illustration, to be fired at. After they had all put their skill to a sufficient test, somebody asked the following question : " What is the total number of different ways in which these sixteen balls may be broken, if we must always break the lowest ball that remains on any string ? " Thus, one way would be to break all the four balls on each string in succession, taking the strings from left to right. Another would be to break all the fourth balls on the four strings first, then

K, L, M, N, and O, and with them form thirty-five groups of three letters so that the combinations should include the greatest number possible of common English words. No two letters may appear together in a group more than once. Thus, A and L having been together in ALE, must never be found together again ; nor may A appear again in a group with E, nor L with E. These conditions will be found complied with in the above solution, and

break the three remaining on the first string, then take the balls on the three other strings alternately from right to left, and so on. There is such a vast number of different ways (since every little variation of order makes a different way) that one is apt to be at first impressed by the great difficulty of the problem. Yet it is really quite simple when once you have hit on the proper method of attacking it. How many different ways are there ?

271.—FIFTEEN LETTER PUZZLE.

ALE	FOE	HOD	BGN
CAB	HEN	JOG	KFM
HAG	GEM	MOB	BFH
FAN	KIN	JEK	DFL
JAM	HIM	GCL	LJH
AID	JIB	FCJ	NJD
OAK	FIG	HCK	MLN
BED	OIL	MCD	BLK
ICE	CON	DGK	

The above is the solution of a puzzle I gave in *Tit-bits* in the summer of 1896. It was required to take the letters, A, B, C, D, E, F, G, H, I, J,

the number of words formed is twenty-one. Many persons have since tried hard to beat this number, but so far have not succeeded.

More than thirty-five combinations of the fifteen letters cannot be formed within the conditions. Theoretically, there cannot possibly be more than twenty-three words formed, because only this number of combinations is possible with a vowel or vowels in each. And as no English word can be formed from three of the given vowels (A, E, I, and O), we must reduce the number of possible words to twenty-two. This is correct theoretically, but practically that twenty-second word cannot be got in. If JEK, shown above, were a word it would be all right ; but it is not, and no amount of juggling with the other letters has resulted in a better answer than the one shown. I should say that proper nouns and abbreviations, such as Joe, Jim, Alf, Hal, Flo, Ike, etc., are disallowed.

Now, the present puzzle is a variation of the above. It is simply this : Instead of using the fifteen letters given, the reader is allowed to select any fifteen different letters of the alphabet that he may prefer. Then construct thirty-five groups in accordance with the conditions, and show as many good English words as possible.

272.—THE NINE SCHOOLBOYS.

This is a new and interesting companion puzzle to the " Fifteen Schoolgirls " (see solution of No. 269), and even in the simplest possible form in which I present it there are unquestionable difficulties. Nine schoolboys walk out in triplets on the six week days so that no boy ever walks *side by side* with any other boy more than once. How would you arrange them ?

If we represent them by the first nine letters of the alphabet, they might be grouped on the first day as follows :—

A B C
D E F
G H I

Then A can never walk again side by side with B, or B with C, or D with E, and so on. But A can, of course, walk side by side with C. It is here not a question of being together in the same triplet, but of walking side by side in a triplet. Under these conditions they can walk out on six days ; under the " Schoolgirls " conditions they can only walk on four days.

273.—THE ROUND TABLE.

Seat the same *n* persons at a round table on $\frac{(n-1)(n-2)}{2}$ occasions so that no person shall ever have the same two neighbours twice. This is, of course, equivalent to saying that every person must sit once, and once only, between every possible pair.

274.—THE MOUSE-TRAP PUZZLE.

This is a modern version, with a difference, of an old puzzle of the same name. Number twenty-one cards, 1, 2, 3, etc., up to 21, and place them in a circle in the particular order shown in the illustration. These cards represent mice. You start from any card, calling that card " one," and count, " one, two, three," etc., in a clockwise direction, and when your count agrees with the number on the card, you have made a " catch," and you remove the card. Then start at the next card, calling " one," and try again to make another " catch." And so on. Supposing you start at 18, calling that card " one," your first " catch " will be 19. Remove 19 and your next " catch " is 10. Remove 10 and your next " catch " is 1. Remove the 1, and if you count up to 21 (you must never go beyond), you cannot make another " catch." Now, the ideal is to " catch " all the twenty-one mice, but this is not here possible, and if it were it would merely require twenty-one different trials, at the most, to succeed. But the reader may make any two cards change places before he begins. Thus, you can change the 6 with the 2, or the 7 with the 11, or any other pair. This can be done in several ways so as to enable you to " catch " all the twenty-one mice, if you then start at the right place. You may never pass over a " catch " ; you must always remove the card and start afresh.

275.—THE SIXTEEN SHEEP.

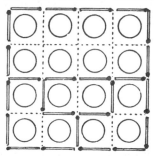

Here is a new puzzle with matches and counters or coins. In the illustration the matches represent hurdles and the counters sheep. The sixteen hurdles on the outside, and the sheep, must be regarded as immovable ; the puzzle has to do entirely with the nine hurdles on the inside. It will be seen that at present these nine hurdles enclose four groups of 8, 3, 3, and 2 sheep. The farmer requires to readjust some of the hurdles so as to enclose 6, 6, and 4 sheep. Can you do it by only replacing two hurdles ? When you have succeeded, then try to do it by replacing three hurdles ; then four, five, six, and seven in succession. Of course, the hurdles must be legitimately laid on the dotted lines, and no such tricks are allowed as leaving unconnected ends of hurdles, or two hurdles placed side by side, or merely making hurdles change places. In fact, the conditions are so simple that any farm labourer will understand it directly.

276.—THE EIGHT VILLAS.

In one of the outlying suburbs of London a man had a square plot of ground on which he

1	6	2		2	6	1		1	6	2
6		6		6		6		4		4
2	6	1		1	6	2		4	2	3
	A				B				C	

decided to build eight villas, as shown in the illustration, with a common recreation ground in the middle. After the houses were completed, and all or some of them let, he discovered that the number of occupants in the three houses forming a side of the square was in every case nine. He did not state how the occupants were distributed, but I have shown by the numbers on the sides of the houses one way in which it might have happened. The puzzle is to dis-

turned round in front of a mirror, four other arrangements. All eight must be counted.

277.—COUNTER CROSSES.

ALL that we need for this puzzle is nine counters, numbered 1, 2, 3, 4, 5, 6, 7, 8, and 9. It will be seen that in the illustration A these are arranged so as to form a Greek cross, while in the case of B they form a Latin cross. In both cases the reader will find that the sum of the numbers in the upright of the cross is the same as the sum of the numbers in the horizontal arm. It is quite easy to hit on such an arrangement by trial, but the problem is to discover in exactly how many different ways it may be done in each case. Remember that reversals and reflections do not count as different. That is to say, if you turn this page round you get four arrangements of the Greek cross, and if you turn it round again in front of a mirror, you will get four more. But these eight are all regarded as one and the same. Now, how many different ways are there in each case ?

278.—A DORMITORY PUZZLE.

IN a certain convent there were eight large dormitories on one floor, approached by a spiral staircase in the centre, as shown in our plan. On an inspection one Monday by the abbess it was found that the south aspect was so much preferred that six times as many nuns slept on the south side as on each of the other three sides. She objected to this overcrowding, and ordered that it should be reduced. On Tues-

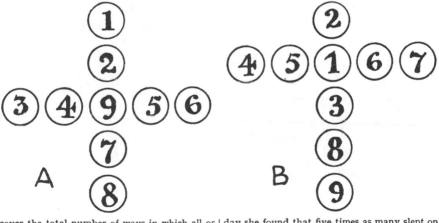

A

B

cover the total number of ways in which all or any of the houses might be occupied, so that there should be nine persons on each side. In order that there may be no misunderstanding, I will explain that although B is what we call a reflection of A, these would count as two different arrangements, while C, if it is turned round, will give four arrangements; and if

day she found that five times as many slept on the south side as on each of the other sides. Again she complained. On Wednesday she found four times as many on the south side, on Thursday three times as many, and on Friday twice as many. Urging the nuns to further efforts, she was pleased to find on Saturday that an equal number slept on each of the four

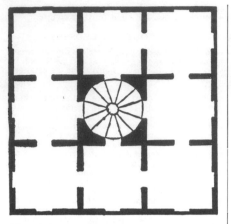

value. So that the best quality was numbered
" 1 " and the worst numbered " 10," and all
the other numbers of graduating values. Now,
the rule of Ahmed Assan, the merchant, was
that he never put a barrel either beneath or
to the right of one of less value. The arrangement shown is, of course, the simplest
way of complying with this condition. But
there are many other ways—such, for example,
as this :—

$$\begin{array}{ccccc} 1 & 2 & 5 & 7 & 8 \\ 3 & 4 & 6 & 9 & 10 \end{array}$$

Here, again, no barrel has a smaller number
than itself on its right or beneath it. The
puzzle is to discover in how many different
ways the merchant of Bagdad might have
arranged his barrels in the two rows without
breaking his rule. Can you count the number
of ways ?

280.—BUILDING THE TETRAHEDRON.

I possess a tetrahedron, or triangular pyramid,
formed of six sticks glued together, as shown in
the illustration. Can you count correctly the
number of different ways in which these six
sticks might have been stuck together so as to
form the pyramid ?

sides of the house. What is the smallest number of nuns there could have been, and how
might they have arranged themselves on each
of the six nights ? No room may ever be
unoccupied.

279.—THE BARRELS OF BALSAM.

A merchant of Bagdad had ten barrels of
precious balsam for sale. They were numbered,
and were arranged in two rows, one on top of
the other, as shown in the picture. The smaller
the number on the barrel, the greater was its

Some friends worked at it together one evening, each person providing himself with six
lucifer matches to aid his thoughts ; but it was
found that no two results were the same. You
see, if we remove one of the sticks and turn it
round the other way, that will be a different
pyramid. If we make two of the sticks change

places the result will again be different. But remember that every pyramid may be made to

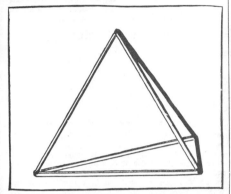

stand on either of its four sides without being a different one. How many ways are there altogether?

281.—PAINTING A PYRAMID.

This puzzle concerns the painting of the four sides of a tetrahedron, or triangular pyramid. If you cut out a piece of cardboard of the triangular shape shown in Fig. 1, and then cut half through along the dotted lines, it will fold up and form a perfect triangular pyramid. And I would first remind my readers that the primary colours of the solar spectrum are seven—violet, indigo, blue, green, yellow, orange, and red. When I was a child I was taught to remember these by the ungainly word formed by the initials of the colours, "Vibgyor."

other side that is out of view is yellow), and then paint another in the order shown in Fig. 3, these are really both the same and count as one way. For if you tilt over No. 2 to the right it will so fall as to represent No. 3. The avoidance of repetitions of this kind is the real puzzle of the thing. If a coloured pyramid cannot be placed so that it exactly resembles in its colours and their relative order another pyramid, then they are different. Remember that one way would be to colour all the four sides red, another to colour two sides green, and the remaining sides yellow and blue; and so on.

282.—THE ANTIQUARY'S CHAIN.

An antiquary possessed a number of curious old links, which he took to a blacksmith, and told him to join together to form one straight piece of chain, with the sole condition that the two circular links were not to be together. The following illustration shows the appearance of the chain and the form of each link. Now, supposing the owner should separate the links

again, and then take them to another smith and repeat his former instructions exactly, what are the chances against the links being put together exactly as they were by the first man?

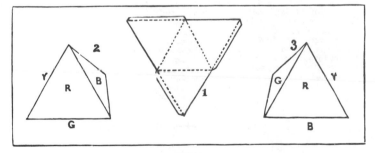

In how many different ways may the triangular pyramid be coloured, using in every case one, two, three, or four colours of the solar spectrum? Of course a side can only receive a single colour, and no side can be left uncoloured. But there is one point that I must make quite clear. The four sides are not to be regarded as individually distinct. That is to say, if you paint your pyramid as shown in Fig. 2 (where the bottom side is green and the

Remember that every successive link can be joined on to another in one of two ways, just as you can put a ring on your finger in two ways, or link your forefingers and thumbs in two ways.

283.—THE FIFTEEN DOMINOES.

In this case we do not use the complete set of twenty-eight dominoes to be found in the ordinary box. We dispense with all those dominoes that have a five or a six on them and limit ourselves to the fifteen that remain, where the double-four is the highest.

In how many different ways may the fifteen dominoes be arranged in a straight line in accordance with the simple rule of the game that a number must always be placed against a similar number—that is, a four against a four, a blank against a blank, and so on? Left to right and right to left of the same arrangement are to be counted as two different ways.

284.—THE CROSS TARGET.

In the illustration we have a somewhat curious target designed by an eccentric sharpshooter. His idea was that in order to score you must hit four circles in as many shots so that those four shots shall form a square. It will be seen by the results recorded on the target that two attempts have been successful. The first man hit the four circles at the top of the cross, and thus formed his square. The second man intended to hit the four in the bottom arm, but his second shot, on the left, went too high. This compelled him to complete his four in a different way than he intended. It will thus be seen that though it is immaterial which circle you hit at the first shot, the second shot may commit you to a definite procedure if you are to get your square. Now, the puzzle is to say in just how many different ways it is possible to form a square on the target with four shots.

285.—THE FOUR POSTAGE STAMPS.

" It is as easy as counting," is an expression one sometimes hears. But mere counting may be

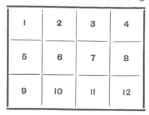

I	2	3	4
5	6	7	8
9	10	II	12

puzzling at times. Take the following simple example. Suppose you have just bought twelve postage stamps, in this form—three by four—and a friend asks you to oblige him with four stamps, all joined together—no stamp hanging on by a mere corner. In how many different ways is it possible for you to tear off those four stamps? You see, you can give him 1, 2, **3**, 4, or 2, 3, 6, 7, or 1, 2, 3, 6, or 1, 2, 3, 7, or 2, 3, 4, 8, and so on. Can you count the number of different ways in which those four stamps might be delivered? There are not many more than fifty ways, so it is not a big count. Can you get the exact number?

286.—PAINTING THE DIE.

In how many different ways may the numbers on a single die be marked, with the only condition that the 1 and 6, the 2 and 5, and the 3 and 4 must be on opposite sides? It is a simple enough question, and yet it will puzzle a good many people.

287.—AN ACROSTIC PUZZLE.

In the making or solving of double acrostics, has it ever occurred to you to consider the variety and limitation of the pair of initial and final letters available for cross words? You may have to find a word beginning with A and ending with B, or A and C, or A and D, and so on. Some combinations are obviously impossible—such, for example, as those with Q at the end. But let us assume that a good English word can be found for every case. Then how many possible pairs of letters are available?

CHESSBOARD PROBLEMS.

" You and I will goe to the chesse."
GREENE'S *Groatsworth of Wit.*

DURING a heavy gale a chimney-pot was hurled through the air, and crashed upon the pavement just in front of a pedestrian. He quite calmly said, " I have no use for it : I do not smoke." Some readers, when they happen to see a puzzle represented on a chessboard with chess pieces, are apt to make the equally inconsequent remark, " I have no use for it : I do not play

chess." This is largely a result of the common, but erroneous, notion that the ordinary chess puzzle with which we are familiar in the press (dignified, for some reason, with the name " problem ") has a vital connection with the game of chess itself. But there is no condition in the game that you shall checkmate your opponent in two moves, in three moves, or in four moves, while the majority of the positions given in these puzzles are such that one player would have so great a superiority in pieces that the other would have resigned before the situations were reached. And the solving of them helps you but little, and that quite indirectly, in playing the game, it being well known that, as a rule, the best " chess problemists " are indifferent players, and *vice versa*. Occasionally a man will be found strong on both subjects, but he is the exception to the rule.

Yet the simple chequered board and the characteristic moves of the pieces lend themselves in a very remarkable manner to the devising of the most entertaining puzzles. There is room for such infinite variety that the true puzzle lover cannot afford to neglect them. It was with a view to securing the interest of readers who are frightened off by the mere presentation of a chessboard that so many puzzles of this class were originally published by me in various fanciful dresses. Some of these posers I still retain in their disguised form ; others I have translated into terms of the chessboard. In the majority of cases the reader will not need any knowledge whatever of chess, but I have thought it best to assume throughout that he is acquainted with the terminology, the moves, and the notation of the game.

I first deal with a few questions affecting the chessboard itself; then with certain statical puzzles relating to the Rook, the Bishop, the Queen, and the Knight in turn ; then dynamical puzzles with the pieces in the same order ; and, finally, with some miscellaneous puzzles on the chessboard. It is hoped that the formulæ and tables given at the end of the statical puzzles will be of interest, as they are, for the most part, published for the first time.

THE CHESSBOARD.

"Good company's a chessboard."
BYRON'S *Don Juan*, xiii. 89.

A CHESSBOARD is essentially a square plane divided into sixty-four smaller squares by straight lines at right angles. Originally it was not chequered (that is, made with its rows and columns alternately black and white, or of any other two colours), and this improvement was introduced merely to help the eye in actual play. The utility of the chequers is unquestionable. For example, it facilitates the operation of the bishops, enabling us to see at the merest glance that our king or pawns on black squares are not open to attack from an opponent's bishop running on the white diagonals. Yet the chequering of the board is not essential to the game of chess. Also, when we are propounding puzzles on the chessboard, it is often well to remember that additional interest may result from " generalizing " for boards containing any number of squares, or from limiting ourselves to some particular chequered arrangement, not necessarily a square. We will give a few puzzles dealing with chequered boards in this general way.

288.—CHEQUERED BOARD DIVISIONS.

I RECENTLY asked myself the question : In how many different ways may a chessboard be divided into two parts of the same size and shape by cuts along the lines dividing the squares ? The problem soon proved to be both fascinating and bristling with difficulties. I present it in a simplified form, taking a board of smaller dimensions.

It is obvious that a board of four squares can only be so divided in one way—by a straight cut down the centre—because we shall not count reversals and reflections as different. In the case of a board of sixteen squares—four by four—there are just six different ways. I have given all these in the diagram, and the reader

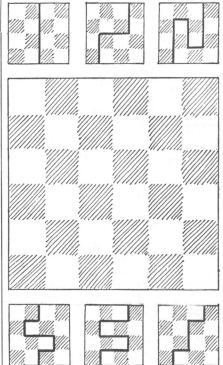

will not find any others. Now, take the larger board of thirty-six squares, and try to discover in how many ways it may be cut into two parts of the same size and shape.

289.—LIONS AND CROWNS.

THE young lady in the illustration is confronted with a little cutting-out difficulty in which the reader may be glad to assist her. She wishes, for some reason that she has not communicated to me, to cut that square piece of valuable material into four parts, all of exactly the same size and shape, but it is important that every piece shall contain a lion and a crown. As she

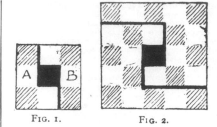

FIG. 1.   FIG. 2.

shape in as many different ways as possible. I have shown in the illustration one way of doing it. How many different ways are there altogether? A piece which when turned over resembles another piece is not considered to be of a different shape.

291.—THE GRAND LAMA'S PROBLEM.

ONCE upon a time there was a Grand Lama who had a chessboard made of pure gold, magnificently engraved, and, of course, of great value. Every year a tournament was held at Lhassa among the priests, and whenever any one beat the Grand Lama it was considered a great honour, and his name was inscribed on the back of the board, and a costly jewel set in the particular square on which the checkmate had been given. After this sovereign pontiff had been defeated on four occasions he died—possibly of chagrin.

Now the new Grand Lama was an inferior chess-player, and preferred other forms of innocent amusement, such as cutting off people's heads. So he discouraged chess as a degrading game, that did not improve either the mind or the morals, and abolished the tournament summarily. Then he sent for the four priests who

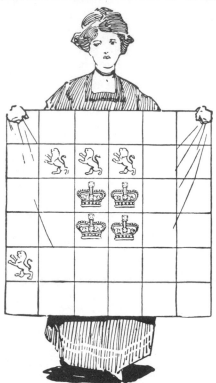

insists that the cuts can only be made along the lines dividing the squares, she is considerably perplexed to find out how it is to be done. Can you show her the way? There is only one possible method of cutting the stuff.

290.—BOARDS WITH AN ODD NUMBER OF SQUARES.

WE will here consider the question of those boards that contain an odd number of squares. We will suppose that the central square is first cut out, so as to leave an even number of squares for division. Now, it is obvious that a square three by three can only be divided in one way, as shown in Fig. 1. It will be seen that the pieces A and B are of the same size and shape, and that any other way of cutting would only produce the same shaped pieces, so remember that these variations are not counted as different ways. The puzzle I propose is to cut the board five by five (Fig. 2) into two pieces of the same size and

had had the effrontery to play better than a Grand Lama, and addressed them as follows:

"Miserable and heathenish men, calling yourselves priests ! Know ye not that to lay claim to a capacity to do anything better than my predecessor is a capital offence ? Take that chessboard and, before day dawns upon the torture chamber, cut it into four equal parts of the same shape, each containing sixteen perfect squares, with one of the gems in each part ! If in this you fail, then shall other sports be devised for your special delectation. Go !" The four priests succeeded in their apparently hopeless task. Can you show how the board may be divided into four equal parts, each of exactly the same shape, by cuts along the lines dividing the squares, each part to contain one of the gems ?

292.—THE ABBOT'S WINDOW.

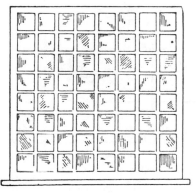

Once upon a time the Lord Abbot of St. Edmondsbury, in consequence of "devotions too strong for his head," fell sick and was unable to leave his bed. As he lay awake, tossing his head restlessly from side to side, the attentive monks noticed that something was disturbing his mind; but nobody dared ask what it might be, for the abbot was of a stern disposition, and never would brook inquisitiveness. Suddenly he called for Father John, and that venerable monk was soon at the bedside.

"Father John," said the Abbot, "dost thou know that I came into this wicked world on a Christmas Even ? "

The monk nodded assent.

"And have I not often told thee that, having been born on Christmas Even, I have

no love for the things that are odd ? Look there ! "

The Abbot pointed to the large dormitory window, of which I give a sketch. The monk looked, and was perplexed.

"Dost thou not see that the sixty-four lights add up an even number vertically and horizontally, but that all the *diagonal* lines, except fourteen, are of a number that is odd ? Why is this ? "

"Of a truth, my Lord Abbot, it is of the very nature of things, and cannot be changed."

"Nay, but it *shall* be changed. I command thee that certain of the lights be closed this day, so that every line shall have an even number of lights. See thou that this be done without delay, lest the cellars be locked up for a month and other grievous troubles befall thee."

Father John was at his wits' end, but after consultation with one who was learned in strange mysteries, a way was found to satisfy the whim of the Lord Abbot. Which lights were blocked up, so that those which remained added up an even number in every line horizontally, vertically, and diagonally, while the least possible obstruction of light was caused ?

293.—THE CHINESE CHESSBOARD.

Into how large a number of different pieces may the chessboard be cut (by cuts along the lines only), no two pieces being exactly alike ? Remember that the arrangement of black and white constitutes a difference. Thus, a single black square will be different from a single white square, a row of three containing two white squares will differ from a row of three containing two black, and so on. If two pieces cannot be placed on the table so as to be exactly alike, they count as different. And as the back of the board is plain, the pieces cannot be turned over.

294.—THE CHESSBOARD SENTENCE.

I once set myself the amusing task of so dissecting an ordinary chessboard into letters of the alphabet that they would form a complete sentence. It will be seen from the illustration that the pieces assembled give the sentence, "CUT THY LIFE," with the stops between." The ideal sentence would, of course, have only

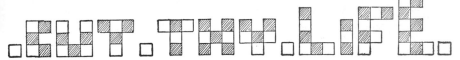

one full stop, but that I did not succeed in obtaining.

The sentence is an appeal to the transgressor to cut himself adrift from the evil life he is living. Can you fit these pieces together to form a perfect chessboard ?

STATICAL CHESS PUZZLES.

"They also serve who only stand and wait."
MILTON.

295.—THE EIGHT ROOKS.

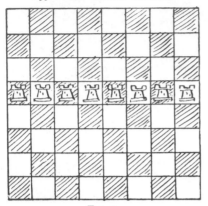

FIG. 1.

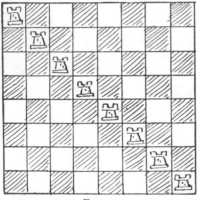

FIG. 2.

It will be seen in the first diagram that every square on the board is either occupied or attacked by a rook, and that every rook is "guarded" (if they were alternately black and white rooks we should say "attacked") by another rook. Placing the eight rooks on any row or file obviously will have the same effect. In diagram 2 every square is again either occupied or attacked, but in this case every rook is unguarded. Now, in how many different ways can you so place the eight rooks on the board that every square shall be occupied or attacked and no rook ever guarded by another? I do not wish to go into the question of reversals and reflections on this occasion, so that placing the rooks on the other diagonal will count as different, and similarly with other repetitions obtained by turning the board round.

296.—THE FOUR LIONS.

The puzzle is to find in how many different ways the four lions may be placed so that there shall never be more than one lion in any row or column. Mere reversals and reflections will not count as different. Thus, regarding the example given, if we place the lions in the other

diagonal, it will be considered the same arrangement. For if you hold the second arrangement in front of a mirror or give it a quarter turn, you merely get the first arrangement. It is a simple little puzzle, but requires a certain amount of careful consideration.

297.—BISHOPS—UNGUARDED.

Place as few bishops as possible on an ordinary chessboard so that every square of the board shall be either occupied or attacked. It will be seen that the rook has more scope than the bishop: for wherever you place the former, it will always attack fourteen other squares; whereas the latter will attack seven, nine, eleven, or thirteen squares, according to the position of the diagonal on which it is placed. And it is well here to state that when we speak of "diagonals" in connection with the chessboard, we do not limit ourselves to the two long diagonals from corner to corner, but include all the shorter lines that are parallel to these. To prevent misunderstanding on future occasions, it will be well for the reader to note carefully this fact.

298.—BISHOPS—GUARDED.

Now, how many bishops are necessary in order that every square shall be either occupied or attacked, and every bishop guarded by another bishop? And how may they be placed?

299.—BISHOPS IN CONVOCATION.

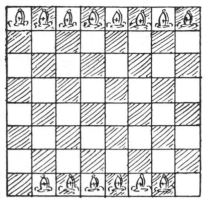

THE greatest number of bishops that can be placed at the same time on the chessboard, without any bishop attacking another, is fourteen. I show, in diagram, the simplest way of doing this. In fact, on a square chequered board of any number of squares the greatest number of bishops that can be placed without attack is always two less than twice the number of squares on the side. It is an interesting puzzle to discover in just how many different ways the fourteen bishops may be so placed without mutual attack. I shall give an exceedingly simple rule for determining the number of ways for a square chequered board of any number of squares.

300.—THE EIGHT QUEENS.

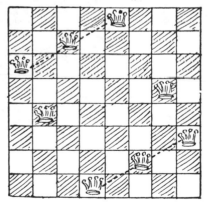

THE queen is by far the strongest piece on the chessboard. If you place her on one of the four squares in the centre of the board, she attacks no fewer than twenty-seven other squares; and if you try to hide her in a corner,

she still attacks twenty-one squares. Eight queens may be placed on the board so that no queen attacks another, and it is an old puzzle (first proposed by Nauck in 1850, and it has quite a little literature of its own) to discover in just how many different ways this may be done. I show one way in the diagram, and there are in all twelve of these fundamentally different ways. These twelve produce ninety-two ways if we regard reversals and reflections as different. The diagram is in a way a symmetrical arrangement. If you turn the page upside down, it will reproduce itself exactly; but if you look at it with one of the other sides at the bottom, you get another way that is not identical. Then if you reflect these two ways in a mirror you get two more ways. Now, all the other eleven solutions are non-symmetrical, and therefore each of them may be presented in eight ways by these reversals and reflections. It will thus be seen why the twelve fundamentally different solutions produce only ninety-two arrangements, as I have said, and not ninety-six, as would happen if all twelve were non-symmetrical. It is well to have a clear understanding on the matter of reversals and reflections when dealing with puzzles on the chessboard.

Can the reader place the eight queens on the board so that no queen shall attack another and so that no three queens shall be in a straight line in any oblique direction? Another glance at the diagram will show that this arrangement will not answer the conditions, for in the two directions indicated by the dotted lines there are three queens in a straight line. There is only one of the twelve fundamental ways that will solve the puzzle. Can you find it?

301.—THE EIGHT STARS.

THE puzzle in this case is to place eight stars in the diagram so that no star shall be in line with

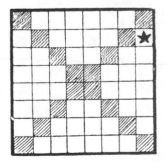

another star horizontally, vertically, or diagonally. One star is already placed, and that must not be moved, so there are only seven for the reader now to place. But you must not place a star on any one of the shaded squares. There is only one way of solving this little puzzle.

302.—A PROBLEM IN MOSAICS.

THE art of producing pictures or designs by means of joining together pieces of hard substances, either naturally or artificially coloured, is of very great antiquity. It was certainly known in the time of the Pharaohs, and we find a reference in the Book of Esther to "a pavement of red, and blue, and white, and black marble." Some of this ancient work that has come down to us, would seem to show clearly, even where design is not at first evident, that much thought was bestowed upon apparently disorderly arrangements. Where, for example, the work has been produced with a very limited number of colours, there are evidences of great ingenuity in preventing the same tints coming in close proximity. Lady readers who are familiar with the construction of patchwork quilts will know how desirable it is sometimes, when they are limited in the choice of material, to prevent pieces of the same stuff coming too near together. Now, this puzzle will apply equally to patchwork quilts or tesselated pavements.

It will be seen from the diagram how a square piece of flooring may be paved with sixty-two square tiles of the eight colours violet, red, yellow, green, orange, purple, white, and blue (indicated by the initial letters), so that no tile is in line with a similarly coloured tile, vertically, horizontally, or diagonally. Sixty-four such tiles could not possibly be placed under these conditions, but the two shaded squares happen to be occupied by iron ventilators. The puzzle is this. These two ventilators

have to be removed to the positions indicated by the darkly bordered tiles, and two tiles placed in those bottom corner squares. Can you readjust the thirty-two tiles so that no two of the same colour shall still be in line?

303.—UNDER THE VEIL.

IF the reader will examine the above diagram, he will see that I have so placed eight V's, eight E's, eight I's, and eight L's in the diagram that no letter is in line with a similar one horizontally, vertically, or diagonally. Thus, no V is in line with another V, no E with another E, and so

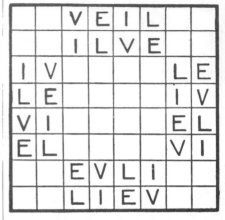

on. There are a great many different ways of arranging the letters under this condition. The puzzle is to find an arrangement that produces the greatest possible number of four-letter words, reading upwards and downwards, backwards and forwards, or diagonally. All repetitions count as different words, and the five variations that may be used are : VEIL, VILE, LEVI, LIVE, and EVIL.

This will be made perfectly clear when I say that the above arrangement scores eight, because the top and bottom row both give VEIL; the second and seventh columns both give VEIL; and the two diagonals, starting from the L in the 5th row and E in the 8th row, both give LIVE and EVIL. There are therefore eight different readings of the words in all.

This difficult word puzzle is given as an example of the use of chessboard analysis in solving such things. Only a person who is familiar with the "Eight Queens" problem could hope to solve it.

304.—BACHET'S SQUARE.

ONE of the oldest card puzzles is by Clauae Gaspar Bachet de Méziriac, first published, I believe, in the 1624 edition of his work. Rearrange the sixteen court cards (including the aces) in a square so that in no row of four cards, horizontal, vertical, or diagonal, shall be found two cards of the same suit or the same value. This in itself is easy enough, but a point of the puzzle is to find in how many different ways this may be done. The eminent French mathematician A. Labosne, in his modern edition of Bachet, gives the answer incorrectly. And yet the puzzle is really quite easy. Any arrangement produces seven others by turning the square round and reflecting it in a mirror. These are counted as different by Bachet.

Note " row of four cards," so that the only diagonals we have here to consider are the two long ones.

305.—THE THIRTY-SIX LETTER-BLOCKS.

A	B	C	D	E	F
A	B	C	D	E	F
A	B	C	D	E	F
A	B	C	D	E	F
A	B	C	D	E	F
A	B	C	D	E	F

THE illustration represents a box containing thirty-six letter-blocks. The puzzle is to rearrange these blocks so that no A shall be in a line vertically, horizontally, or diagonally with another A, no B with another B, no C with another C, and so on. You will find it impossible to get all the letters into the box under these conditions, but the point is to place as many as possible. Of course no letters other than those shown may be used.

306.—THE CROWDED CHESSBOARD.

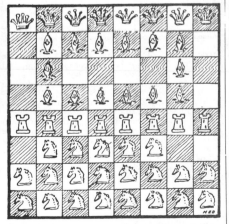

THE puzzle is to rearrange the fifty-one pieces on the chessboard so that no queen shall attack another queen, no rook attack another rook, no bishop attack another bishop, and no knight attack another knight. No notice is to be taken of the intervention of pieces of another type from that under consideration—that is, two queens will be considered to attack one another although there may be, say, a rook, a bishop, and a knight between them. And so with the rooks and bishops. It is not difficult to dispose of each type of piece separately; the difficulty comes in when you have to find room for all the arrangements on the board simultaneously.

307.—THE COLOURED COUNTERS.

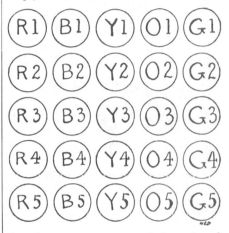

THE diagram represents twenty-five coloured counters, Red, Blue, Yellow, Orange, and Green (indicated by their initials), and there are five of each colour, numbered 1, 2, 3, 4, and 5. The problem is so to place them in a square that neither colour nor number shall be found repeated in any one of the five rows, five columns, and two diagonals. Can you so rearrange them?

308.—THE GENTLE ART OF STAMP-LICKING.

THE Insurance Act is a most prolific source of entertaining puzzles, particularly entertaining if you happen to be among the exempt. One's initiation into the gentle art of stamp-licking suggests the following little poser: If you have a card divided into sixteen spaces (4 × 4), and are provided with plenty of stamps of the values 1d., 2d., 3d., 4d., and 5d., what is the greatest value that you can stick on the card if the Chancellor of the Exchequer forbids you to place any stamp in a straight line (that is, horizontally, vertically, or diagonally) with another stamp of similar value? Of course, only one stamp can be affixed in a space. The reader will probably find, when he sees the solution, that, like the stamps themselves, he is licked

He will most likely be twopence short of the maximum. A friend asked the Post Office how it was to be done; but they sent him to the Customs and Excise officer, who sent him to the Insurance Commissioners, who sent him to an approved society, who profanely sent him —but no matter.

309.—THE FORTY-NINE COUNTERS.

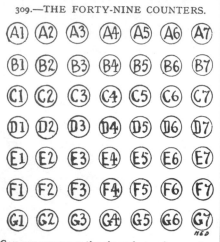

CAN you rearrange the above forty-nine counters in a square so that no letter, and also no number, shall be in line with a similar one, vertically, horizontally, or diagonally? Here I, of course, mean in the lines parallel with the diagonals, in the chessboard sense.

310.—THE THREE SHEEP.

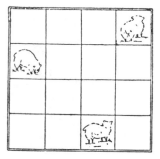

A FARMER had three sheep and an arrangement of sixteen pens, divided off by hurdles in the manner indicated in the illustration. In how many different ways could he place those sheep, each in a separate pen, so that every pen should be either occupied or in line (horizontally, vertically, or diagonally) with at least one sheep? I have given one arrangement that fulfils the conditions. How many others can you find?

Mere reversals and reflections must not be counted as different. The reader may regard the sheep as queens. The problem is then to place the three queens so that every square shall be either occupied or attacked by at least one queen—in the maximum number of different ways.

311.—THE FIVE DOGS PUZZLE.

IN 1863, C. F. de Jaenisch first discussed the "Five Queens Puzzle"—to place five queens on the chessboard so that every square shall be attacked or occupied—which was propounded by his friend, a "Mr. de R." Jaenisch showed that if no queen may attack another there are ninety-one different ways of placing the five queens, reversals and reflections not counting as different. If the queens may attack one another, I have recorded hundreds of ways, but it is not practicable to enumerate them exactly. The illustration is supposed to represent an

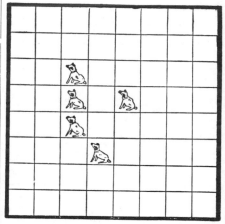

arrangement of sixty-four kennels. It will be seen that five kennels each contain a dog, and on further examination it will be seen that every one of the sixty-four kennels is in a straight line with at least one dog—either horizontally, vertically, or diagonally. Take any kennel you like, and you will find that you can draw a straight line to a dog in one or other of the three ways mentioned. The puzzle is to replace the five dogs and discover in just how many different ways they may be placed in five kennels *in a straight row*, so that every kennel shall always be in line with at least one dog. Reversals and reflections are here counted as different.

312.—THE FIVE CRESCENTS OF BYZANTIUM.

WHEN Philip of Macedon, the father of Alexander the Great, found himself confronted with great difficulties in the siege of Byzantium, he set his men to undermine the walls. His desires, however, miscarried, for no sooner had the operations been begun than a crescent moon suddenly appeared in the heavens and discovered his plans to his adversaries. The Byzantines were naturally elated, and in order to show their gratitude they erected a statue to Diana, and the crescent became thenceforward a symbol of the state. In the temple that contained the statue was a square pavement composed of sixty-four large and costly tiles. These were all plain, with the exception of five, which bore the symbol of the crescent. These five were for occult reasons so placed that every tile should be watched over by (that is, in a straight line, vertically, horizontally, or diagonally with) at least one of the crescents. The arrangement adopted by the Byzantine architect was as follows :—

314.—THE SOUTHERN CROSS.

In the above illustration we have five Planets and eighty-one Fixed Stars, five of the latter being hidden by the Planets. It will be found that every Star, with the exception of the ten that have a black spot in their centres, is in a straight line, vertically, horizontally, or diagonally, with at least one of the Planets. The puzzle is so to rearrange the Planets that all the Stars shall be in line with one or more of them.

In rearranging the Planets, each of the five may be moved once in a straight line, in either of the three directions mentioned. They will, of course, obscure five other Stars in place of those at present covered.

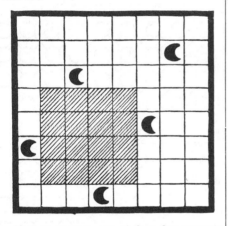

Now, to cover up one of these five crescents was a capital offence, the death being something very painful and lingering. But on a certain occasion it was necessary to lay down on this pavement a square carpet of the largest dimensions possible, and I have shown in the illustration by dark shading the largest dimensions that would be available.

The puzzle is to show how the architect, if he had foreseen this question of the carpet, might have so arranged his five crescent tiles in accordance with the required conditions, and yet have allowed for the largest possible square carpet to be laid down without any one of the five crescent tiles being covered, or any portion of them.

313.—QUEENS AND BISHOP PUZZLE.

It will be seen that every square of the board is either occupied or attacked. The puzzle is to substitute a bishop for the rook on the same square, and then place the four queens on other squares so that every square shall again be either occupied or attacked.

315.—THE HAT-PEG PUZZLE.

Here is a five-queen puzzle that I gave in a fanciful dress in 1897. As the queens were there represented as hats on sixty-four pegs, I will keep to the title, " The Hat-Peg Puzzle." It will be seen that every square is occupied or attacked. The puzzle is to remove one queen

to a different square so that still every square is occupied or attacked, then move a second queen under a similar condition, then a third queen, and finally a fourth queen. After the fourth move every square must be attacked or occupied, but no queen must then attack another. Of course, the moves need not be " queen moves ; " you can move a queen to any part of the board.

316.—THE AMAZONS.

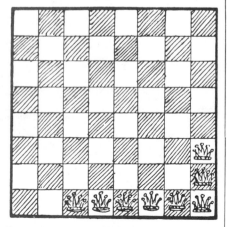

THIS puzzle is based on one by Captain Turton. Remove three of the queens to other squares so that there shall be eleven squares on the

board that are not attacked. The removal of the three queens need not be by "queen moves." You may take them up and place them anywhere. There is only one solution.

317.—A PUZZLE WITH PAWNS.

PLACE two pawns in the middle of the chessboard, one at Q 4 and the other at K 5. Now, place the remaining fourteen pawns (sixteen in all) so that no three shall be in a straight line in any possible direction.

Note that I purposely do not say queens, because by the words " any possible direction " I go beyond attacks on diagonals. The pawns must be regarded as mere points in space—at the centres of the squares. See dotted lines in the case of No. 300, " The Eight Queens."

318.—LION-HUNTING.

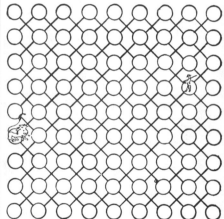

My friend Captain Potham Hall, the renowned hunter of big game, says there is nothing more exhilarating than a brush with a herd—a pack—a team—a flock—a swarm (it has taken me a full quarter of an hour to recall the right word, but I have it at last)—a *pride* of lions. Why a number of lions are called a " pride," a number of whales a " school," and a number of foxes a " skulk " are mysteries of philology into which I will not enter.

Well, the captain says that if a spirited lion crosses your path in the desert it becomes lively, for the lion has generally been looking for the man just as much as the man has sought the king of the forest. And yet when they meet they always quarrel and fight it out. A little contemplation of this unfortunate and long-standing feud between two estimable families has led me to figure out a few calculations as to the probability of the man and the lion crossing one another's path in the jungle. In all these cases one has to start on certain more or less arbitrary assumptions. That is why in the above illustration I have thought it necessary to represent the paths in the desert with such rigid regularity. Though the captain assures me that the tracks of the lions usually run much in this way, I have doubts.

The puzzle is simply to find out in how many different ways the man and the lion may be placed on two different spots that are not on the same path. By "paths" it must be understood that I only refer to the ruled lines. Thus, with the exception of the four corner spots, each combatant is always on two paths and no more. It will be seen that there is a lot of scope for evading one another in the desert, which is just what one has always understood.

319.—THE KNIGHT-GUARDS.

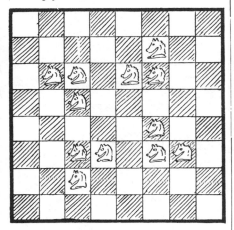

THE knight is the irresponsible low comedian of the chessboard. "He is a very uncertain, sneaking, and demoralizing rascal," says an American writer. "He can only move two squares, but makes up in the quality of his locomotion for its quantity, for he can spring one square sideways and one forward simultaneously, like a cat; can stand on one leg in the middle of the board and jump to any one of eight squares he chooses; can get on one side of a fence and blackguard three or four men on the other; has an objectionable way of inserting himself in safe places where he can scare the king and compel him to move, and then gobble a queen. For pure cussedness the knight has no equal, and when you chase him out of one hole he skips into another." Attempts have been made over and over again to obtain a short, simple, and exact definition of the move of the knight—without success. It really consists in moving one square like a rook, and then another square like a bishop—the two operations being done in one leap, so that it does not matter whether the first square passed over is occupied by another piece or not. It is, in

fact, the only leaping move in chess. But difficult as it is to define, a child can learn it by inspection in a few minutes.

I have shown in the diagram how twelve knights (the fewest possible that will perform the feat) may be placed on the chessboard so that every square is either occupied or attacked by a knight. Examine every square in turn, and you will find that this is so. Now, the puzzle in this case is to discover what is the smallest possible number of knights that is required in order that every square shall be either occupied or attacked, and every knight protected by another knight. And how would you arrange them? It will be found that of the twelve shown in the diagram only four are thus protected by being a knight's move from another knight.

THE GUARDED CHESSBOARD.

ON an ordinary chessboard, 8 by 8, every square can be guarded—that is, either occupied or attacked—by 5 queens, the fewest possible. There are exactly 91 fundamentally different arrangements in which no queen attacks another queen. If every queen must attack (or be protected by) another queen, there are at fewest 41 arrangements, and I have recorded some 150 ways in which some of the queens are attacked and some not, but this last case is very difficult to enumerate exactly.

On an ordinary chessboard every square can be guarded by 8 rooks (the fewest possible) in 40,320 ways, if no rook may attack another rook, but it is not known how many of these are fundamentally different. (See solution to No. 295, "The Eight Rooks.") I have not enumerated the ways in which every rook shall be protected by another rook.

On an ordinary chessboard every square can be guarded by 8 bishops (the fewest possible), if no bishop may attack another bishop. Ten bishops are necessary if every bishop is to be protected. (See Nos. 297 and 298, "Bishops unguarded" and "Bishops guarded.")

On an ordinary chessboard every square can be guarded by 12 knights if all but 4 are unprotected. But if every knight must be protected, 14 are necessary. (See No. 319, "The Knight-Guards.")

Dealing with the queen on n^2 boards generally, where n is less than 8, the following results will be of interest :—

1 queen guards 2^2 board in 1 fundamental way.

1 queen guards 3^2 board in 1 fundamental way.

2 queens guard 4^2 board in 3 fundamental ways (protected).

3 queens guard 4^2 board in 2 fundamental ways (not protected).

3 queens guard 5^2 board in 37 fundamental ways (protected).

3 queens guard 5^2 board in 2 fundamental ways (not protected).

3 queens guard 6^2 board in 1 fundamental way (protected).

4 queens guard 6^2 board in 17 fundamental ways (not protected).

4 queens guard 7^2 board in 5 fundamental ways (protected).

4 queens guard 7^2 board in 1 fundamental way (not protected).

NON-ATTACKING CHESSBOARD ARRANGEMENTS.

WE know that n queens may always be placed on a square board of n^2 squares (if n be greater than 3) without any queen attacking another queen. But no general formula for enumerating the number of different ways in which it may be done has yet been discovered; probably it is undiscoverable. The known results are as follows :—

Where $n = 4$ there is 1 fundamental solution and 2 in all.

Where $n = 5$ there are 2 fundamental solutions and 10 in all.

Where $n = 6$ there is 1 fundamental solution and 4 in all.

Where $n = 7$ there are 6 fundamental solutions and 40 in all.

Where $n = 8$ there are 12 fundamental solutions and 92 in all.

Where $n = 9$ there are 46 fundamental solutions.

Where $n = 10$ there are 92 fundamental solutions.

Where $n = 11$ there are 341 fundamental solutions.

Obviously n rooks may be placed without attack on an n^2 board in $\lfloor n$ ways, but how many of these are fundamentally different I have only worked out in the four cases where n equals 2, 3, 4, and 5. The answers here are respectively 1, 2, 7, and 23. (See No. 296, " The Four Lions.")

We can place $2n-2$ bishops on an n^2 board in 2^n ways. (See No. 299, " Bishops in Convocation.") For boards containing 2, 3, 4, 5, 6, 7, 8 squares on a side there are respectively 1, 2, 3, 6, 10, 20, 36 fundamentally different arrangements. Where n is odd there are $2^{\frac{1}{2}(n-1)}$ such arrangements, each giving 4 by reversals and reflections, and $2^{n-3} - 2^{\frac{1}{2}(n-3)}$ giving 8. Where n is even there are $2^{\frac{1}{2}(n-2)}$, each giving 4 by reversals and reflections, and $2^{n-3} - 2^{\frac{1}{2}(n-4)}$, each giving 8.

We can place $\frac{1}{2}(n^2 + 1)$ knights on an n^2 board without attack, when n is odd, in 1 fundamental way ; and $\frac{1}{2}n^2$ knights on an n^2 board, when n is even, in 1 fundamental way. In the first case we place all the knights on the same colour as the central square ; in the second case we place them all on black, or all on white, squares.

THE TWO PIECES PROBLEM.

ON a board of n^2 squares, two queens, two rooks, two bishops, or two knights can always be placed, irrespective of attack or not, in

$\dfrac{n^4 - n^2}{2}$ ways. The following formulæ will show in how many of these ways the two pieces may be placed with attack and without :—

	With Attack.	Without Attack.
2 Queens	$\dfrac{5n^3 - 6n^2 + n}{3}$	$\dfrac{3n^4 - 10n^3 + 9n^2 - 2n}{6}$
2 Rooks	$n^3 - n^2$	$\dfrac{n^4 - 2n^3 + n^2}{2}$
2 Bishops	$\dfrac{4n^3 - 6n^2 + 2n}{6}$	$\dfrac{3n^4 - 4n^3 + 3n^2 - 2n}{6}$
2 Knights	$4n^2 - 12n + 8$	$\dfrac{n^4 - 9n^2 + 24n}{2}$

(See No. 318, " Lion Hunting.")

DYNAMICAL CHESS PUZZLES.

" Push on—keep moving."
THOS. MORTON : *Cure for the Heartache.*

320.—THE ROOK'S TOUR.

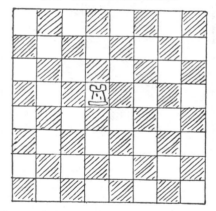

THE puzzle is to move the single rook over the whole board, so that it shall visit every square of the board once, and only once, and end its tour on the square from which it starts. You have to do this in as few moves as possible, and unless you are very careful you will take just one move too many. Of course, a square is regarded equally as " visited " whether you merely pass over it or make it a stopping-place, and we will not quibble over the point whether the original square is actually visited twice. We will assume that it is not.

321.—THE ROOK'S JOURNEY.

THIS puzzle I call " The Rook's Journey," because the word " tour " (derived from a turner's wheel) implies that we return to the point from which we set out, and we do not do this in the present case. We should not be satisfied with a personally conducted holiday tour that ended by leaving us, say, in the middle of the Sahara. The rook here makes twenty-one moves, in the

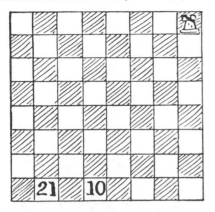

course of which journey it visits every square of the board once and only once, stopping at the square marked 10 at the end of its tenth move, and ending at the square marked 21. Two consecutive moves cannot be made in the same direction—that is to say, you must make a turn after every move.

322.—THE LANGUISHING MAIDEN.

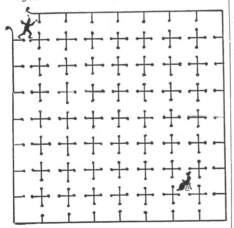

A WICKED baron in the good old days imprisoned an innocent maiden in one of the deepest dungeons beneath the castle moat. It will be seen from our illustration that there were sixty-three cells in the dungeon, all connected by open doors, and the maiden was chained in the cell in which she is shown. Now, a valiant knight, who loved the damsel, succeeded in

rescuing her from the enemy. Having gained an entrance to the dungeon at the point where he is seen, he succeeded in reaching the maiden after entering every cell once and only once. Take your pencil and try to trace out such a route. When you have succeeded, then try to discover a route in twenty-two straight paths through the cells. It can be done in this number without entering any cell a second time.

323.—A DUNGEON PUZZLE.

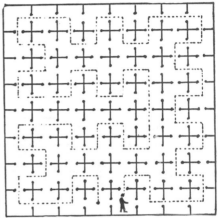

A FRENCH prisoner, for his sins (or other people's), was confined in an underground dungeon containing sixty-four cells, all communicating with open doorways, as shown in our illustration. In order to reduce the tedium of his restricted life, he set himself various puzzles, and this is one of them. Starting from the cell in which he is shown, how could he visit every cell once, and only once, and make as many turnings as possible? His first attempt is shown by the dotted track. It will be found that there are as many as fifty-five straight lines in his path, but after many attempts he improved upon this. Can you get more than fifty-five? You may end your path in any cell you like. Try the puzzle with a pencil on chessboard diagrams, or you may regard them as rooks' moves on a board.

324.—THE LION AND THE MAN.

IN a public place in Rome there once stood a prison divided into sixty-four cells, all open to the sky and all communicating with one another, as shown in the illustration. The sports that here took place were watched from a high tower. The favourite game was to place a Christian in one corner cell and a lion in the diagonally opposite corner and then leave them with all the inner doors open. The consequent effect was sometimes most laughable. On one occasion the man was given a sword. He was no coward, and was as anxious to find the lion as the lion undoubtedly was to find him.

The man visited every cell once and only once in the fewest possible straight lines until he reached the lion's cell. The lion, curiously enough, also visited every cell once and only once in the fewest possible straight lines until he finally reached the man's cell. They started together and went at the same speed; yet,

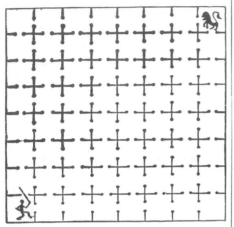

although they occasionally got glimpses of one another, they never once met. The puzzle is to show the route that each happened to take.

325.—AN EPISCOPAL VISITATION.

THE white squares on the chessboard represent the parishes of a diocese. Place the bishop on any square you like, and so contrive that (using the ordinary bishop's move of chess) he shall visit every one of his parishes in the fewest possible moves. Of course, all the parishes passed through on any move are regarded as "visited." You can visit any squares more than once, but you are not allowed to move twice between the same two adjoining squares. What are the fewest possible moves? The bishop need not end his visitation at the parish from which he first set out.

326.—A NEW COUNTER PUZZLE.

HERE is a new puzzle with moving counters, or coins, that at first glance looks as if it must be absurdly simple. But it will be found quite a little perplexity. I give it in this place for a reason that I will explain when we come to the next puzzle. Copy the simple diagram, enlarged, on a sheet of paper; then place two white counters on the points 1 and 2, and two red counters on 9 and 10. The puzzle is to make the red and white change places. You may move the counters one at a time in

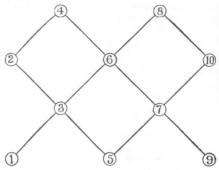

any order you like, along the lines from point to point, with the only restriction that a red and a white counter may never stand at once on the same straight line. Thus the first move can only be from 1 or 2 to 3, or from 9 or 10 to 7.

327.—A NEW BISHOP'S PUZZLE.

THIS is quite a fascinating little puzzle. Place eight bishops (four black and four white) on the reduced chessboard, as shown in the illustration. The problem is to make the black bishops change places with the white ones, no bishop ever attacking another of the opposite colour. They must move alternately—first a white, then a black, then a white, and so on. When you have succeeded in doing it at all, try to find the fewest possible moves.

If you leave out the bishops standing on black squares, and only play on the white squares, you will discover my last puzzle turned on its side.

328.—THE QUEEN'S TOUR.

THE puzzle of making a complete tour of the chessboard with the queen in the fewest possible moves (in which squares may be visited more than once) was first given by the late Sam Loyd in his *Chess Strategy*. But the solution shown below is the one he gave in *American Chess-Nuts* in 1868. I have recorded at least six

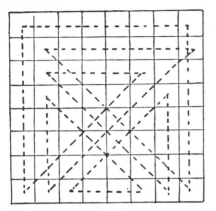

different solutions in the minimum number of moves—fourteen—but this one is the best of all, for reasons I will explain.

If you will look at the lettered square you will understand that there are only ten really differently placed squares on a chessboard—those enclosed by a dark line—all the others are mere reversals or reflections. For example, every A is a corner square, and every J a central square. Consequently, as the solution shown has a turning-point at the enclosed D square, we can obtain a solution starting from and

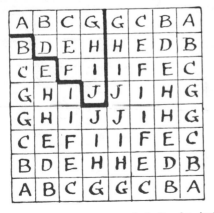

ending at any square marked D — by just turning the board about. Now, this scheme will give you a tour starting from any A, B, C, D, E, F, or H, while no other route that I know can be adapted to more than five different starting-points. There is no Queen's Tour in fourteen moves (remember a tour must be re-entrant) that may start from a G, I, or J. But we can have a non-re-entrant path over the whole board in fourteen moves, starting from any given square. Hence the following puzzle:—

Start from the J in the enclosed part of the lettered diagram and visit every square of the board in fourteen moves, ending wherever you like.

329.—THE STAR PUZZLE.

PUT the point of your pencil on one of the white stars and (without ever lifting your pencil from the paper) strike out all the stars in fourteen continuous straight strokes, ending at the second white star. Your straight strokes may be in any direction you like, only every turning must be made on a star. There is no objection to striking out any star more than once.

In this case, where both your starting and ending squares are fixed inconveniently, you cannot obtain a solution by breaking a Queen's Tour, or in any other way by queen moves alone. But you are allowed to use oblique straight lines—such as from the upper white star direct to a corner star.

330.—THE YACHT RACE.

Now then, ye land-lubbers, hoist your baby-jib-topsails, break out your spinnakers, ease off your balloon sheets, and get your head-sails set !

Our race consists in starting from the point at which the yacht is lying in the illustration and touching every one of the sixty-four buoys in fourteen straight courses, returning in the final tack to the buoy from which we start. The seventh course must finish at the buoy from which a flag is flying.

This puzzle will call for a lot of skilful seamanship on account of the sharp angles at which it will occasionally be necessary to tack.

The point of a lead pencil and a good nautical eye are all the outfit that we require.

This is difficult, because of the condition as to the flag-buoy, and because it is a re-entrant tour. But again we are allowed those oblique lines.

331.—THE SCIENTIFIC SKATER.

It will be seen that this skater has marked on the ice sixty-four points or stars, and he proposes to start *from his present position* near the corner and enter every one of the points in fourteen straight lines. How will he do it? Of course there is no objection to his passing over any point more than once, but his last straight stroke must bring him back to the position from which he started.

It is merely a matter of taking your pencil and starting from the spot on which the skater's foot is at present resting, and striking out all the stars in fourteen continuous straight lines, returning to the point from which you set out.

332.—THE FORTY-NINE STARS.

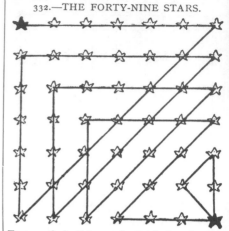

The puzzle in this case is simply to take your pencil and, starting from one black star, strike out all the stars in twelve straight strokes, ending at the other black star. It will be seen that the attempt shown in the illustration requires fifteen strokes. Can you do it in twelve? Every turning must be made on a star, and the lines must be parallel to the sides and diagonals of the square, as shown. In this case we are dealing with a chessboard of reduced dimensions, but only queen moves (without going outside the boundary as in the last case) are required.

333.—THE QUEEN'S JOURNEY.

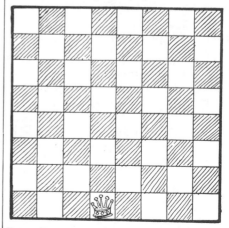

Place the queen on her own square, as shown in the illustration, and then try to discover the greatest distance that she can travel over the board in five queen's moves without passing over any square a second time. Mark the queen's path on the board, and note carefully also that she must never cross her own track. It seems simple enough, but the reader may find that he has tripped.

334.—ST. GEORGE AND THE DRAGON.

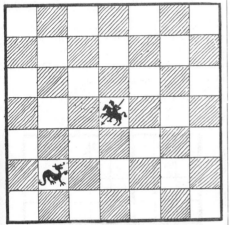

HERE is a little puzzle on a reduced chessboard of forty-nine squares. St. George wishes to kill the dragon. Killing dragons was a well-known pastime of his, and, being a knight, it was only natural that he should desire to perform the feat in a series of knight's moves. Can you show how, starting from that central square, he may visit once, and only once, every square of the board in a chain of chess knight's moves, and end by capturing the dragon on his last move? Of course a variety of different ways are open to him, so try to discover a route that forms some pretty design when you have marked each successive leap by a straight line from square to square.

335.—FARMER LAWRENCE'S CORN-FIELDS.

ONE of the most beautiful districts within easy distance of London for a summer ramble is that part of Buckinghamshire known as the Valley of the Chess—at least, it was a few years ago, before it was discovered by the speculative builder. At the beginning of the present century there lived, not far from Latimers, a worthy but eccentric farmer named Lawrence. One of his queer notions was that every person who lived near the banks of the river Chess ought to be in some way acquainted with the noble game of the same name, and in order to impress this fact on his men and his neighbours he adopted at times strange terminology. For example, when one of his ewes presented him

with a lamb, he would say that it had "queened a pawn"; when he put up a new barn against the highway, he called it "castling on the king's side"; and when he sent a man with a gun to keep his neighbour's birds off his fields, he spoke of it as "attacking his opponent's rooks." Everybody in the neighbourhood used to be amused at Farmer Lawrence's little jokes, and one boy (the wag of the village) who got his ears pulled by the old gentleman for stealing his "chestnuts" went so far as to call him "a silly old chess-protector!"

One year he had a large square field divided into forty-nine square plots, as shown in the illustration. The white squares were sown with wheat and the black squares with barley. When the harvest time came round he gave orders that his men were first to cut the corn in the patch marked 1, and that each successive cutting should be exactly a knight's move from the last one, the thirteenth cutting being in

the patch marked 13, the twenty-fifth in the patch marked 25, the thirty-seventh in the one marked 37, and the last, or forty-ninth cutting, in the patch marked 49. This was too much for poor Hodge, and each day Farmer Lawrence had to go down to the field and show which piece had to be operated upon. But the problem will perhaps present no difficulty to my readers.

336.—THE GREYHOUND PUZZLE.

IN this puzzle the twenty kennels do not communicate with one another by doors, but are divided off by a low wall. The solitary occupant is the greyhound which lives in the kennel in the top left-hand corner. When he is allowed his liberty he has to obtain it by visiting every kennel once and only once in a series of knight's moves, ending at the bottom right-hand corner, which is open to the world. The lines in the

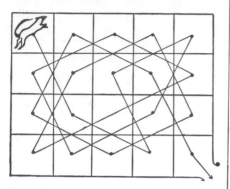

above diagram show one solution. The puzzle is to discover in how many different ways the greyhound may thus make his exit from his corner kennel.

337.—THE FOUR KANGAROOS.

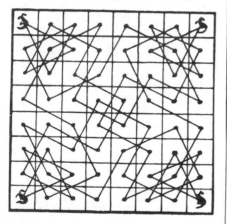

IN introducing a little Commonwealth problem, I must first explain that the diagram represents the sixty-four fields, all properly fenced off from one another, of an Australian settlement, though I need hardly say that our kith and kin "down under" always *do* set out their land in this methodical and exact manner. It will be seen that in every one of the four corners is a kangaroo. Why kangaroos have a marked preference for corner plots has never been satisfactorily explained, and it would be out of place to discuss the point here. I should also add that kangaroos, as is well known, always leap in what we call "knight's moves." In fact,

chess players would probably have adopted the better term "kangaroo's move" had not chess been invented before kangaroos.

The puzzle is simply this. One morning each kangaroo went for his morning hop, and in sixteen consecutive knight's leaps visited just fifteen different fields and jumped back to his corner. No field was visited by more than one of the kangaroos. The diagram shows how they arranged matters. What you are asked to do is to show how they might have performed the feat without any kangaroo ever crossing the horizontal line in the middle of the square that divides the board into two equal parts.

338.—THE BOARD IN COMPARTMENTS.

WE cannot divide the ordinary chessboard into four equal square compartments, and describe a complete tour, or even path, in each compartment. But we may divide it into four compartments, as in the illustration, two containing

each twenty squares, and the other two each twelve squares, and so obtain an interesting puzzle. You are asked to describe a complete re-entrant tour on this board, starting where you like, but visiting every square in each successive compartment before passing into another one, and making the final leap back to the square from which the knight set out. It is not difficult, but will be found very entertaining and not uninstructive.

Whether a re-entrant "tour" or a complete knight's "path" is possible or not on a rectangular board of given dimensions depends not only on its dimensions, but also on its shape. A tour is obviously not possible on a board containing an odd number of cells, such as 5 by 5 or 7 by 7, for this reason: Every successive leap of the knight must be from a white square to a black and a black to a white alternately. But if there be an odd number of cells or squares there must be one more square of one colour than of the other, therefore the path must begin from a square of the colour that is in excess, and end on a similar colour, and as a knight's move from one colour to a similar colour is impossible the path cannot be re-entrant. But a perfect tour may be made on a rectangular board of any dimensions provided the number of squares be even, and that the number of squares on one side be not less than 6 and on the other not less than 5. In other words, the smallest rectangular board on which a re-entrant tour is possible is one that is 6 by 5.

A complete knight's path (not re-entrant) over all the squares of a board is never possible if there be only two squares on one side; nor is it possible on a square board of smaller dimensions than 5 by 5. So that on a board 4 by 4 we can neither describe a knight's tour nor a complete knight's path; we must leave one square unvisited. Yet on a board 4 by 3 (containing four squares fewer) a complete path may be described in sixteen different ways. It may interest the reader to discover all these. Every path that starts from and ends at different squares is here counted as a different solution, and even reverse routes are called different.

339.—THE FOUR KNIGHTS' TOURS.

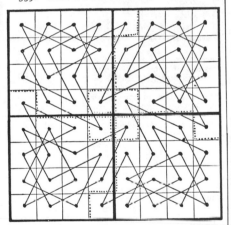

I WILL repeat that if a chessboard be cut into four equal parts, as indicated by the dark lines in the illustration, it is not possible to perform a knight's tour, either re-entrant or not, on one of the parts. The best re-entrant attempt is shown, in which each knight has to trespass twice on other parts. The puzzle is to cut the board differently into four parts, each of the same size and shape, so that a re-entrant knight's tour may be made on each part. Cuts along the dotted lines will not do, as the four central squares of the board would be either detached or hanging on by a mere thread.

340.—THE CUBIC KNIGHT'S TOUR.

SOME few years ago I happened to read somewhere that Abnit Vandermonde, a clever mathe-

matician, who was born in 1736 and died in 1793, had devoted a good deal of study to the question of knight's tours. Beyond what may be gathered from a few fragmentary references, I am not aware of the exact nature or results of his investigations, but one thing attracted my attention, and that was the statement that he had proposed the question of a tour of the knight over the six surfaces of a cube, each surface being a chessboard. Whether he obtained a solution or not I do not know, but I have never seen one published. So I at once set to work to master this interesting problem. Perhaps the reader may like to attempt it.

341.—THE FOUR FROGS.

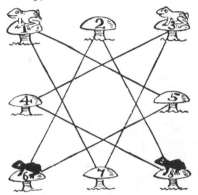

IN the illustration we have eight toadstools, with white frogs on 1 and 3 and black frogs on 6 and 8. The puzzle is to move one frog at a time, in any order, along one of the straight lines from toadstool to toadstool, until they have exchanged places, the white frogs being left on 6 and 8 and the black ones on 1 and 3. If you use four counters on a simple diagram, you will find this quite easy, but it is a little more puzzling to do it in only seven plays, any number of successive moves by one frog counting as one play. Of course, more than one frog cannot be on a toadstool at the same time.

342.—THE MANDARIN'S PUZZLE.

THE following puzzle has an added interest from the circumstance that a correct solution of it secured for a certain young Chinaman the hand of his charming bride. The wealthiest mandarin within a radius of a hundred miles of Peking was Hi-Chum-Chop, and his beautiful daughter, Peeky-Bo, had innumerable admirers. One of her most ardent lovers was Winky-Hi, and when he asked the old mandarin for his consent to their marriage, Hi-Chum-Chop presented him with the following puzzle and promised him his consent if the youth brought him the correct answer within a week. Winky-Hi, following a habit which obtains among certain solvers to this day, gave it to all his friends, and when he had compared their solutions he handed in the best one as his own. Luckily it was quite right. The mandarin thereupon fulfilled his promise. The fatted pup was killed for the wedding feast, and when Hi-Chum-Chop passed Winky-Hi the liver wing all present knew that it was a token of eternal goodwill, in accordance with Chinese custom from time immemorial.

The mandarin had a table divided into twenty-five squares, as shown in the diagram. On each of twenty-four of these squares was placed a numbered counter, just as I have indicated. The puzzle is to get the counters in numerical order by moving them one at a time in what we call "knight's moves." Counter 1 should be where 16 is, 2 where 11 is, 4 where 13 now is, and so on. It will be seen that all the counters on shaded squares are in their proper positions. Of course, two counters may never be on a square at the same time. Can you perform the feat in the fewest possible moves?

In order to make the manner of moving perfectly clear I will point out that the first knight's move can only be made by 1 or by 2 or by 10. Supposing 1 moves, then the next move must

be by 23, 4, 8, or 21. As there is never more than one square vacant, the order in which the counters move may be written out as follows: 1—21—14—18—22, etc. A rough diagram should be made on a larger scale for practice, and numbered counters or pieces of cardboard used.

343.—EXERCISE FOR PRISONERS.

The following is the plan of the north wing of a certain gaol, showing the sixteen cells all communicating by open doorways. Fifteen prisoners were numbered and arranged in the cells as shown. They were allowed to change their cells as much as they liked, but if two prisoners were ever in the same cell together there was a severe punishment promised them.

Now, in order to reduce their growing obesity, and to combine physical exercise with mental

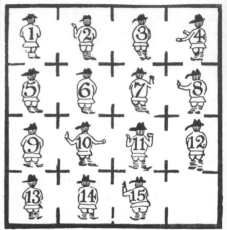

recreation, the prisoners decided, on the suggestion of one of their number who was interested in knight's tours, to try to form themselves into a perfect knight's path without breaking the prison regulations, and leaving the bottom right-hand corner cell vacant, as originally. The joke of the matter is that the arrangement at which they arrived was as follows:—

8	3	12	1
11	14	9	6
4	7	2	13
15	10	5	

The warders failed to detect the important fact that the men could not possibly get into this position without two of them having been at some time in the same cell together. Make the attempt with counters on a ruled diagram, and you will find that this is so. Otherwise the solution is correct enough, each member being, as required, a knight's move from the preceding number, and the original corner cell vacant.

The puzzle is to start with the men placed as in the illustration and show how it might have been done in the fewest moves, while giving a complete rest to as many prisoners as possible.

As there is never more than one vacant cell for a man to enter, it is only necessary to write down the numbers of the men in the order in which they move. It is clear that very few men can be left throughout in their cells undisturbed, but I will leave the solver to discover just how many, as this is a very essential part of the puzzle.

344.—THE KENNEL PUZZLE.

A MAN has twenty-five dog kennels all communicating with each other by doorways, as shown in the illustration. He wishes to arrange his twenty dogs so that they shall form a knight's string from dog No. 1 to dog No. 20, the bottom row of five kennels to be left empty, as at present. This is to be done by moving one dog at a time into a vacant kennel. The dogs are well trained to obedience, and may be trusted to remain in the kennels in which they are placed, except that if two are placed in the same kennel together they will fight it out to the death. How is the puzzle to be solved in the fewest possible moves without two dogs ever being together?

345.—THE TWO PAWNS.

BLACK

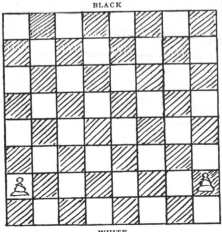

WHITE

HERE is a neat little puzzle in counting. In how many different ways may the two pawns advance to the eighth square? You may move them in any order you like to form a different sequence. For example, you may move the Q R P (one or two squares) first, or the K R P first, or one pawn as far as you like before touching the other. Any sequence is permissible, only in this puzzle as soon as a pawn reaches the eighth square it is dead, and remains there unconverted. Can you count the number of different sequences? At first it will strike you as being very difficult, but I will show that it is really quite simple when properly attacked.

VARIOUS CHESS PUZZLES.

"Chesse-play is a good and wittie exercise of the minde for some kinde of men."
BURTON'S *Anatomy of Melancholy.*

346.—SETTING THE BOARD.

I HAVE a single chessboard and a single set of chessmen. In how many different ways may the men be correctly set up for the beginning of a game? I find that most people slip at a particular point in making the calculation.

347.—COUNTING THE RECTANGLES.

CAN you say correctly just how many squares and other rectangles the chessboard contains? In other words, in how great a number of different ways is it possible to indicate a square or other rectangle enclosed by lines that separate the squares of the board?

348.—THE ROOKERY.

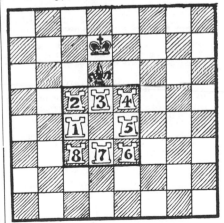

THE White rooks cannot move outside the little square in which they are enclosed except on the final move, in giving checkmate. The puzzle is how to checkmate Black in the fewest possible moves with No. 8 rook, the other rooks being left in numerical order round the sides of their square with the break between 1 and 7.

349.—STALEMATE.

SOME years ago the puzzle was proposed to construct an imaginary game of chess, in which White shall be stalemated in the fewest possible moves with all the thirty-two pieces on the board. Can you build up such a position in fewer than twenty moves?

350.—THE FORSAKEN KING.

SET up the position shown in the diagram. Then the condition of the puzzle is—White to play and checkmate in six moves. Notwithstanding the complexities, I will show how the

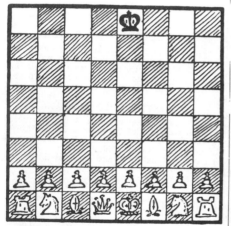

manner of play may be condensed into quite a few lines, merely stating here that the first two moves of White cannot be varied.

351.—THE CRUSADER.

THE following is a prize puzzle propounded by me some years ago. Produce a game of chess which, after sixteen moves, shall leave White with all his sixteen men on their original squares and Black in possession of his king alone (not necessarily on his own square). White is then to *force* mate in three moves.

352.—IMMOVABLE PAWNS.

STARTING from the ordinary arrangement of the pieces as for a game, what is the smallest possible number of moves necessary in order to arrive at the following position? The moves for both sides must, of course, be played strictly in accordance with the rules of the game, though the result will necessarily be a very weird kind of chess.

BLACK

WHITE

353.—THIRTY-SIX MATES.

PLACE the remaining eight White pieces in such a position that White shall have the choice of thirty-six different mates on the move.

Every move that checkmates and leaves a different position is a different mate. The pieces already placed must not be moved.

354.—AN AMAZING DILEMMA.

IN a game of chess between Mr. Black and Mr. White, Black was in difficulties, and as usual was obliged to catch a train. So he proposed that White should complete the game in his absence on condition that no moves whatever should be made for Black, but only with the White pieces. Mr. White accepted, but to his dismay found it utterly impossible to win the game under such conditions. Try as he would, he could not checkmate his opponent. On

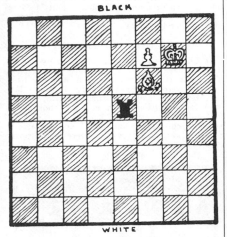

BLACK

WHITE

which square did Mr. Black leave his king? The other pieces are in their proper positions in the diagram. White may leave Black in check as often as he likes, for it makes no difference, as he can never arrive at a checkmate position.

355.—CHECKMATE!

BLACK

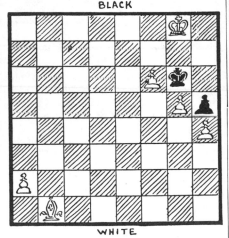

WHITE

STROLLING into one of the rooms of a London club, I noticed a position left by two players

who had gone. This position is shown in the diagram. It is evident that White has checkmated Black. But how did he do it? That is the puzzle.

356.—QUEER CHESS.

CAN you place two White rooks and a White knight on the board so that the Black king (who must be on one of the four squares in the middle of the board) shall be in check with no possible move open to him? "In other words," the reader will say, "the king is to be shown checkmated." Well, you can use the term if you wish, though I intentionally do not employ it myself. The mere fact that there is no White king on the board would be a sufficient reason for my not doing so.

357.—ANCIENT CHINESE PUZZLE.

BLACK

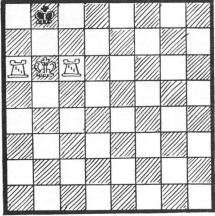

WHITE

My next puzzle is supposed to be Chinese, many hundreds of years old, and never fails to interest. White to play and mate, moving each of the three pieces once, and once only.

358.—THE SIX PAWNS.

IN how many different ways may I place six pawns on the chessboard so that there shall be an even number of unoccupied squares in every row and every column? We are not here considering the diagonals at all, and every different six squares occupied makes a different solution, so we have not to exclude reversals or reflections.

359.—COUNTER SOLITAIRE.

HERE is a little game of solitaire that is quite easy, but not so easy as to be uninteresting. You can either rule out the squares on a sheet of cardboard or paper, or you can use a portion of your chessboard. I have shown numbered counters in the illustration so as to make the

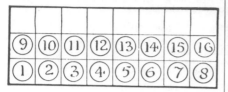

solution easy and intelligible to all, but chess pawns or draughts will serve just as well in practice.

The puzzle is to remove all the counters except one, and this one that is left must be No. 1. You remove a counter by jumping over another counter to the next space beyond, if that square is vacant, but you cannot make a leap in a diagonal direction. The following moves will make the play quite clear : 1—9, 2—10, 1—2, and so on. Here 1 jumps over 9, and you remove 9 from the board ; then 2 jumps over 10, and you remove 10 ; then 1 jumps over 2, and you remove 2. Every move is thus a capture, until the last capture of all is made by No. 1.

360.—CHESSBOARD SOLITAIRE.

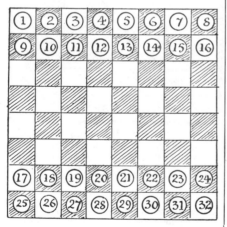

HERE is an extension of the last game of solitaire. All you need is a chessboard and the thirty-two pieces, or the same number of draughts or counters. In the illustration numbered counters are used. The puzzle is to remove all the counters except two, and these two must have originally been on the same side of the board ; that is, the two left must either belong to the group 1 to 16 or to the other group, 17 to 32. You remove a counter by jumping over it with another counter to the

next square beyond, if that square is vacant, but you cannot make a leap in a diagonal direction. The following moves will make the play quite clear : 3—11, 4—12, 3—4, 13—3. Here 3 jumps over 11, and you remove 11 ; 4 jumps over 12, and you remove 12 ; and so on. It will be found a fascinating little game of patience, and the solution requires the exercise of some ingenuity.

361.—THE MONSTROSITY.

ONE Christmas Eve I was travelling by rail to a little place in one of the southern counties. The compartment was very full, and the passengers were wedged in very tightly. My neighbour in one of the corner seats was closely studying a position set up on one of those little folding chessboards that can be carried conveniently in the pocket, and I could scarcely avoid looking at it myself. Here is the position :—

BLACK

WHITE

My fellow-passenger suddenly turned his head and caught the look of bewilderment on my face.

" Do you play chess ? " he asked.

" Yes, a little. What is that ? A problem ? "

" Problem ? No ; a game."

" Impossible ! " I exclaimed rather rudely. " The position is a perfect monstrosity ! "

He took from his pocket a postcard and handed it to me. It bore an address at one side and on the other the words " 43. K to Kt 8."

" It is a correspondence game." he exclaimed. " That is my friend's last move, and I am considering my reply."

" But you really must excuse me ; the position seems utterly impossible. How on earth, for example——"

" Ah ! " he broke in smilingly. " I see ; you are a beginner ; you play to win."

"Of course you wouldn't play to lose or draw!"

He laughed aloud.

"You have much to learn. My friend and myself do not play for results of that antiquated kind. We seek in chess the wonderful, the whimsical, the weird. Did you ever see a position like that?"

I inwardly congratulated myself that I never had.

"That position, sir, materializes the sinuous evolvements and syncretic, synthetic, and synchronous concatenations of two cerebral individualities. It is the product of an amphoteric and intercalatory interchange of——"

"Have you seen the evening paper, sir?" interrupted the man opposite, holding out a newspaper. I noticed on the margin beside his thumb some pencilled writing. Thanking him, I took the paper and read—"Insane, but quite harmless. He is in my charge."

After that I let the poor fellow run on in his wild way until both got out at the next station.

But that queer position became fixed indelibly in my mind, with Black's last move 43. K to Kt 8; and a short time afterwards I found it actually possible to arrive at such a position in forty-three moves. Can the reader construct such a sequence? How did White get his rooks and king's bishop into their present positions, considering Black can never have moved his king's bishop? No odds were given, and every move was perfectly legitimate.

MEASURING, WEIGHING, AND PACKING PUZZLES.

"Measure still for measure."
Measure for Measure, v. 1.

APPARENTLY the first printed puzzle involving the measuring of a given quantity of liquid by pouring from one vessel to others of known capacity was that propounded by Niccola Fontana, better known as "Tartaglia" (the stammerer), 1500-1559. It consists in dividing 24 oz. of valuable balsam into three equal parts, the only measures available being vessels holding 5, 11, and 13 ounces respectively. There are many different solutions to this puzzle in six manipulations, or pourings from one vessel to another. Bachet de Méziriac reprinted this and other of Tartaglia's puzzles in his *Problèmes plaisans et délectables* (1612). It is the general opinion that puzzles of this class can only be solved by trial, but I think formulæ can be constructed for the solution generally of certain related cases. It is a practically unexplored field for investigation.

The classic weighing problem is, of course, that proposed by Bachet. It entails the determination of the least number of weights that would serve to weigh any integral number of pounds from 1 lb. to 40 lbs. inclusive, when we are allowed to put a weight in either of the two pans. The answer is 1, 3, 9, and 27 lbs. Tartaglia had previously propounded the same puzzle with the condition that the weights may only be placed in one pan. The answer in that case is 1, 2, 4, 8, 16, 32 lbs. Major MacMahon has solved the problem quite generally. A full account will be found in Ball's *Mathematical Recreations* (5th edition).

Packing puzzles, in which we are required to pack a maximum number of articles of given dimensions into a box of known dimensions, are, I believe, of quite recent introduction. At least I cannot recall any example in the books of the old writers. One would rather expect to find in the toy shops the idea presented as a mechanical puzzle, but I do not think I have ever seen such a thing. The nearest approach to it would appear to be the puzzles of the jig-saw character, where there is only one depth of the pieces to be adjusted.

362.—THE WASSAIL BOWL.

ONE Christmas Eve three Weary Willies came into possession of what was to them a veritable wassail bowl, in the form of a small barrel, containing exactly six quarts of fine ale. One of the men possessed a five-pint jug and another a three-pint jug, and the problem for them was to divide the liquor equally amongst them without waste. Of course, they are not to use any other vessels or measures. If you can show how it was to be done at all, then try to find the way that requires the fewest possible manipulations, every separate pouring from one vessel to another, or down a man's throat, counting as a manipulation.

363.—THE DOCTOR'S QUERY.

"A CURIOUS little point occurred to me in my dispensary this morning," said a doctor. "I had a bottle containing ten ounces of spirits of wine, and another bottle containing ten ounces of water. I poured a quarter of an ounce of spirits into the water and shook them up together. The mixture was then clearly forty to one. Then I poured back a quarter-ounce of the mixture, so that the two bottles should again each contain the same quantity of fluid. What proportion of spirits to water did the spirits of wine bottle then contain?"

364.—THE BARREL PUZZLE.

THE men in the illustration are disputing over the liquid contents of a barrel. What the particular liquid is it is impossible to say, for we are unable to look into the barrel; so we will call it water. One man says that the barrel is more than half full, while the other insists that it is not half full. What is their easiest way of settling the point? It is not necessary to use stick, string, or implement of any kind for measuring. I

give this merely as one of the simplest possible examples of the value of ordinary sagacity in the solving of puzzles. What are apparently very difficult problems may frequently be solved in a similarly easy manner if we only use a little common sense.

365.—NEW MEASURING PUZZLE.

HERE is a new poser in measuring liquids that will be found interesting. A man has two ten-quart vessels full of wine, and a five-quart and a four-quart measure. He wants to put exactly three quarts into each of the two measures. How is he to do it? And how many manipulations (pourings from one vessel to another) do you require? Of course, waste of wine, tilting, and other tricks are not allowed.

366.—THE HONEST DAIRYMAN.

AN honest dairyman in preparing his milk for public consumption employed a can marked B, containing milk, and a can marked A, containing water. From can A he poured enough to double the contents of can B. Then he poured from can B into can A enough to double its contents. Then he finally poured from can A into can B until their contents were exactly equal. After these operations he would send the can A to London, and the puzzle is to discover what are the relative proportions of milk and water that he provides for the Londoners' breakfast-tables. Do they get equal proportions of milk and water—or two parts of milk and one of water—or what? It is an interesting question, though, curiously enough, we are not told how much milk or water he puts into the cans at the start of his operations.

367.—WINE AND WATER.

MR. GOODFELLOW has adopted a capital idea of late. When he gives a little dinner party and the time arrives to smoke, after the departure of the ladies, he sometimes finds that the conversation is apt to become too political, too personal, too slow, or too scandalous. Then he always manages to introduce to the company some new poser that he has secreted up his sleeve for the occasion. This invariably results in no end of interesting discussion and debate, and puts everybody in a good humour.

Here is a little puzzle that he propounded the other night, and it is extraordinary how the company differed in their answers. He filled a wine-glass half full of wine, and another glass twice the size one-third full of wine. Then he filled up each glass with water and emptied the contents of both into a tumbler. "Now," he said, "what part of the mixture is wine and what part water?" Can you give the correct answer?

368.—THE KEG OF WINE.

HERE is a curious little problem. A man had a ten-gallon keg full of wine and a jug. One day he drew off a jugful of wine and filled up the keg with water. Later on, when the wine and water had got thoroughly mixed, he drew off another jugful and again filled up the keg with water. It was then found that the keg contained equal proportions of wine and water. Can you find from these facts the capacity of the jug ?

369.—MIXING THE TEA.

" MRS. SPOONER called this morning," said the honest grocer to his assistant. " She wants twenty pounds of tea at 2s. 4½d. per lb. Of course we have a good 2s. 6d. tea, a slightly inferior at 2s. 3d., and a cheap Indian at 1s. 9d., but she is very particular always about her prices."

" What do you propose to do ? " asked the innocent assistant.

" Do ? " exclaimed the grocer. " Why, just mix up the three teas in different proportions so that the twenty pounds will work out fairly at the lady's price. Only don't put in more of the best tea than you can help, as we make less profit on that, and of course you will use only our complete pound packets. Don't do any weighing."

How was the poor fellow to mix the three teas ? Could you have shown him how to do it ?

370.—A PACKING PUZZLE.

As we all know by experience, considerable ingenuity is often required in packing articles into a box if space is not to be unduly wasted. A man once told me that he had a large number of iron balls, all exactly two inches in diameter, and he wished to pack as many of these as possible into a rectangular box 24$\frac{9}{10}$ inches long, 22$\frac{4}{5}$ inches wide, and 14 inches deep. Now, what is the greatest number of the balls that he could pack into that box ?

371.—GOLD PACKING IN RUSSIA.

THE editor of the *Times* newspaper was invited by a high Russian official to inspect the gold stored in reserve at St. Petersburg, in order that he might satisfy himself that it was not another " Humbert safe." He replied that it would be of no use whatever, for although the gold might appear to be there, he would be quite unable from a mere inspection to declare that what he saw was really gold. A correspondent of the *Daily Mail* thereupon took up the challenge, but, although he was greatly impressed by what he saw, he was compelled to confess his incompetence (without emptying and counting the contents of every box and sack, and assaying every piece of gold) to give any assurance on the subject. In presenting the following little puzzle, I wish it to be also understood that I do not guarantee the real existence of the gold, and the point is not at all material to our purpose. Moreover, if the reader says that gold is not usually " put up " in slabs of the dimensions that I give, I can only claim problematic licence.

Russian officials were engaged in packing 800 gold slabs, each measuring 12½ inches long, 11 inches wide, and 1 inch deep. What are the interior dimensions of a box of equal length and width, and necessary depth, that will exactly contain them without any space being left over ? Not more than twelve slabs may be laid on edge, according to the rules of the government. It is an interesting little problem in packing, and not at all difficult.

372.—THE BARRELS OF HONEY.

FULL ½ FULL EMPTY

ONCE upon a time there was an aged merchant of Bagdad who was much respected by all who knew him. He had three sons, and it was a rule of his life to treat them all exactly alike. Whenever one received a present, the other two were each given one of equal value. One day this worthy man fell sick and died, bequeathing all his possessions to his three sons in equal shares.

The only difficulty that arose was over the stock of honey. There were exactly twenty-one barrels. The old man had left instructions that not only should every son receive an equal quantity of honey, but should receive exactly the same number of barrels, and that no honey should be transferred from barrel to barrel on account of the waste involved. Now, as seven of these barrels were full of honey, seven were half-full, and seven were empty, this was found to be quite a puzzle, especially as each brother objected to taking more than four barrels of the same description—full, half-full, or empty. Can you show how they succeeded in making a correct division of the property ?

CROSSING RIVER PROBLEMS

> "My boat is on the shore."
>   BYRON.

THIS is another mediæval class of puzzles. Probably the earliest example was by Abbot Alcuin, who was born in Yorkshire in 735 and died at Tours in 804. And everybody knows the story of the man with the wolf, goat, and basket of cabbages whose boat would only take one of the three at a time with the man himself. His difficulties arose from his being unable to leave the wolf alone with the goat, or the goat alone with the cabbages. These puzzles were considered by Tartaglia and Bachet, and have been later investigated by Lucas, De Fonteney, Delannoy, Tarry, and others. In

sight, because she had to come back again with the boat, so nothing was gained by that operation. How did they all succeed in getting across? The reader will find it much easier than the Softleigh family did, for their greatest enemy could not have truthfully called them a brilliant quartette—while the dog was a perfect fool.

374.—CROSSING THE RIVER AXE.

MANY years ago, in the days of the smuggler known as "Rob Roy of the West," a piratical band buried on the coast of South Devon a quantity of treasure which was, of course, abandoned by them in the usual inexplicable way. Some time afterwards its whereabouts

the puzzles I give there will be found one or two new conditions which add to the complexity somewhat. I also include a pulley problem that practically involves the same principles.

373.—CROSSING THE STREAM.

DURING a country ramble Mr. and Mrs. Softleigh found themselves in a pretty little dilemma. They had to cross a stream in a small boat which was capable of carrying only 150 lbs. weight. But Mr. Softleigh and his wife each weighed exactly 150 lbs., and each of their sons weighed 75 lbs. And then there was the dog, who could not be induced on any terms to swim. On the principle of "ladies first," they at once sent Mrs. Softleigh over; but this was a stupid over-

was discovered by three countrymen, who visited the spot one night and divided the spoil between them, Giles taking treasure of the value of £800, Jasper £500 worth, and Timothy £300 worth. In returning they had to cross the river Axe at a point where they had left a small boat in readiness. Here, however, was a difficulty they had not anticipated. The boat would only carry two men, or one man and a sack, and they had so little confidence in one another that no person could be left alone on the land or in the boat with more than his share of the spoil, though two persons (being a check on each other) might be left with more than their shares. The puzzle is to show how they got over the river in the fewest possible crossings, taking their treasure with them. No tricks, such as ropes, "flying bridges," currents, swimming, or similar dodges, may be employed.

375.—FIVE JEALOUS HUSBANDS.

DURING certain local floods five married couples found themselves surrounded by water, and had to escape from their unpleasant position in a boat that would only hold three persons at a time. Every husband was so jealous that he would not allow his wife to be in the boat or on either bank with another man (or with other men) unless he was himself present. Show the quickest way of getting these five men and their wives across into safety.

Call the men A, B, C, D, E, and their respective wives a, b, c, d, e. To go over and return counts as two crossings. No tricks such as ropes, swimming, currents, etc., are permitted.

376.—THE FOUR ELOPEMENTS.

COLONEL B—— was a widower of a very taciturn disposition. His treatment of his four daughters was unusually severe, almost cruel, and they not unnaturally felt disposed to resent it. Being charming girls with every virtue and many accomplishments, it is not surprising that each had a fond admirer. But the father forbade the young men to call at his house, intercepted all letters, and placed his daughters under stricter supervision than ever. But love, which scorns locks and keys and garden walls, was equal to the occasion, and the four youths conspired together and planned a general elopement.

At the foot of the tennis lawn at the bottom of the garden ran the silver Thames, and one night, after the four girls had been safely conducted from a dormitory window to *terra firma*, they all crept softly down to the bank of the river, where a small boat belonging to the Colonel was moored. With this they proposed to cross to the opposite side and make their way to a lane where conveyances were waiting to carry them in their flight. Alas! here at the water's brink their difficulties already began. The young men were so extremely jealous that not one of them would allow his prospective bride to remain at any time in the company of another man, or men, unless he himself were present also. Now, the boat would only hold two persons, though it could, of course, be rowed by one, and it seemed impossible that the four couples would ever get across. But midway in the stream was a small island, and this seemed to present a way out of the difficulty, because a person or persons could be left there while the boat was rowed back or to the opposite shore. If they had been prepared for their difficulty they could have easily worked out a solution to the little poser at any other time. But they were now so hurried and excited in their flight that the confusion they soon got into was exceedingly amusing—or would have been to any one except themselves.

As a consequence they took twice as long and crossed the river twice as often as was really necessary. Meanwhile, the Colonel, who was a very light sleeper, thought he heard a splash of oars. He quickly raised the alarm among his household, and the young ladies were found to be missing. Somebody was sent to the police-station, and a number of officers soon aided in the pursuit of the fugitives, who, in consequence of that delay in crossing the river, were quickly overtaken. The four girls returned sadly to their homes, and afterwards broke off their engagements in disgust.

For a considerable time it was a mystery how the party of eight managed to cross the river in that little boat without any girl being ever left with a man, unless her betrothed was also present. The favourite method is to take eight counters or pieces of cardboard and mark them A, B, C, D, a, b, c, d, to represent the four men and their prospective brides, and carry them from one side of a table to the other in a matchbox (to represent the boat), a penny being placed in the middle of the table as the island.

Readers are now asked to find the quickest method of getting the party across the river. How many passages are necessary from land to land? By "land" is understood either shore or island. Though the boat would not necessarily call at the island every time of crossing, the possibility of its doing so must be provided for. For example, it would not do for a man to be alone in the boat (though it were understood that he intended merely to cross from one bank to the opposite one) if there happened to be a girl alone on the island other than the one to whom he was engaged.

377.—STEALING THE CASTLE TREASURE.

THE ingenious manner in which a box of treasure, consisting principally of jewels and precious stones, was stolen from Gloomhurst Castle has been handed down as a tradition in the De Gourney family. The thieves consisted of a man, a youth, and a small boy, whose only mode of escape with the box of treasure was by means of a high window. Outside the window was fixed a pulley, over which ran a rope with a basket at each end. When one basket was on the ground the other was at the window. The rope was so disposed that the persons in the basket could neither help themselves by means of it nor receive help from others. In short, the only way the baskets could be used was by placing a heavier weight in one than in the other.

Now, the man weighed 195 lbs., the youth 105 lbs., the boy 90 lbs., and the box of treasure 75 lbs. The weight in the descending basket could not exceed that in the other by more than 15 lbs. without causing a descent so rapid as to be most dangerous to a human being, though it would not injure the stolen property. Only two persons, or one person and the treasure, could be placed in the same basket at one time. How did they all manage to escape and take the box of treasure with them?

The puzzle is to find the shortest way of performing the feat, which in itself is not difficult. Remember, a person cannot help himself by hanging on to the rope, the only way being to go down "with a bump," with the weight in the other basket as a counterpoise.

PROBLEMS CONCERNING GAMES.

"The little pleasure of the game."
                MATTHEW PRIOR.

EVERY game lends itself to the propounding of a variety of puzzles. They can be made, as we have seen, out of the chessboard and the peculiar moves of the chess pieces. I will now give just a few examples of puzzles with playing cards and dominoes, and also go out of doors and consider one or two little posers in the cricket field, at the football match, and the horse race and motor-car race.

378.—DOMINOES IN PROGRESSION.

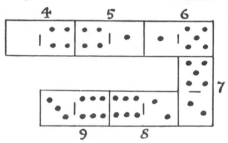

It will be seen that I have played six dominoes, in the illustration, in accordance with the ordinary rules of the game, 4 against 4, 1 against 1, and so on, and yet the sum of the spots on the successive dominoes, 4, 5, 6, 7, 8, 9, are in arithmetical progression; that is, the numbers taken in order have a common difference of 1. In how many different ways may we play six dominoes, from an ordinary box of twenty-eight, so that the numbers on them may lie in arithmetical progression? We must always play from left to right, and numbers in decreasing arithmetical progression (such as 9, 8, 7, 6, 5, 4) are not admissible.

379.—THE FIVE DOMINOES.

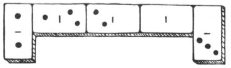

HERE is a new little puzzle that is not difficult, but will probably be found entertaining by my readers. It will be seen that the five dominoes are so arranged in proper sequence

(that is, with 1 against 1, 2 against 2, and so on), that the total number of pips on the two end dominoes is five, and the sum of the pips on the three dominoes in the middle is also five. There are just three other arrangements giving five for the additions. They are :—

(1—0)	(0—0)	(0—2)	(2—1)	(1—3)
(4—0)	(0—0)	(0—2)	(2—1)	(1—0)
(2—0)	(0—0)	(0—1)	(1—3)	(3—0)

Now, how many similar arrangements are there of five dominoes that shall give six instead of five in the two additions?

380.—THE DOMINO FRAME PUZZLE.

It will be seen in the illustration that the full set of twenty-eight dominoes is arranged in the form of a square frame, with 6 against 6, 2 against 2, blank against blank, and so on, as in the game. It will be found that the pips in the top row and left-hand column both add up 44. The pips in the other two sides sum to 59 and 32 respectively. The puzzle is to rearrange the dominoes in the same form so that all of the four sides shall sum to 44. Remember that the dominoes must be correctly placed one against another as in the game.

381.—THE CARD FRAME PUZZLE.

IN the illustration we have a frame constructed from the ten playing cards, ace to ten of diamonds. The children who made it wanted the pips on all four sides to add up alike, but they failed in their attempt and gave it up as impossible. It will be seen that the pips in the top row, the bottom row, and the left-hand side all add up 14, but the right-hand side sums to 23. Now, what they were trying to do is quite

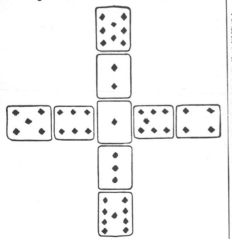

possible. Can you rearrange the ten cards in the same formation so that all four sides shall add up alike? Of course they need not add up 14, but any number you choose to select.

382.—THE CROSS OF CARDS.

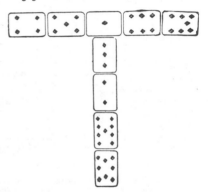

In this case we use only nine cards—the ace to nine of diamonds. The puzzle is to arrange them in the form of a cross, exactly in the way

shown in the illustration, so that the pips in the vertical bar and in the horizontal bar add up alike. In the example given it will be found that both directions add up 23. What I want to know is, how many different ways are there of rearranging the cards in order to bring about this result? It will be seen that, without affecting the solution, we may exchange the 5 with the 6, the 5 with the 7, the 8 with the 3, and so on. Also we may make the horizontal and the vertical bars change places. But such obvious manipulations as these are not to be regarded as different solutions. They are all mere variations of one fundamental solution. Now, how many of these fundamentally different solutions are there? The pips need not, of course, always add up 23.

383.—THE "T" CARD PUZZLE.

An entertaining little puzzle with cards is to take the nine cards of a suit, from ace to nine inclusive, and arrange them in the form of the letter " T," as shown in the illustration, so that the pips in the horizontal line shall count the same as those in the column. In the example given they add up twenty-three both ways. Now, it is quite easy to get a single correct arrangement. The puzzle is to discover in just how many different ways it may be done. Though the number is high, the solution is not really difficult if we attack the puzzle in the right manner. The reverse way obtained by reflecting the illustration in a mirror will not count as different, but all other changes in the relative positions of the cards will here count. How many different ways are there?

384.—CARD TRIANGLES.

Here you pick out the nine cards, ace to nine of diamonds, and arrange them in the form of a triangle, exactly as shown in the illustration, so that the pips add up the same on the three sides. In the example given it will be seen that they sum to 20 on each side, but the particular number is of no importance so long as it is the same on all three sides. The puzzle is to find out in just how many different ways this can be done.

If you simply turn the cards round so that one of the other two sides is nearest to you this will not count as different, for the order will be the same. Also, if you make the 4, 9, 5 change places with the 7, 3, 8, and at the same time exchange the 1 and the 6, it will not be different. But if you only change the 1 and the 6 it

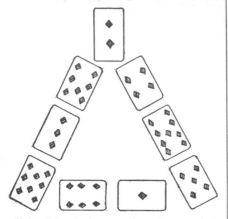

will be different, because the order round the triangle is not the same. This explanation will prevent any doubt arising as to the conditions.

385.—"STRAND" PATIENCE.

THE idea for this came to me when considering the game of Patience that I gave in the *Strand Magazine* for December, 1910, which has been reprinted in Ernest Bergholt's *Second Book of Patience Games*, under the new name of "King Albert."

Make two piles of cards as follows : 9 D, 8 S, 7 D, 6 S, 5 D, 4 S, 3 D, 2 S, 1 D, and 9 H, 8 C, 7 H, 6 C, 5 H, 4 C, 3 H, 2 C, 1 H, with the 9 of diamonds at the bottom of one pile and the 9 of hearts at the bottom of the other. The point is to exchange the spades with the clubs, so that the diamonds and clubs are still in numerical order in one pile and the hearts and spades in the other. There are four vacant spaces in addition to the two spaces occupied by the piles, and any card may be laid on a space, but a card can only be laid on another of the next higher value—an ace on a two, a two on a three, and so on. Patience is required to discover the shortest way of doing this. When there are four vacant spaces you can pile four cards in seven moves, with only three spaces you can pile them in nine moves, and with two spaces you cannot pile more than two cards. When you have a grasp of these and similar facts you will be able to remove a number of cards bodily and write down 7, 9, or whatever the number of moves may be. The gradual shortening of play is fascinating, and first attempts are surprisingly lengthy.

386.—A TRICK WITH DICE.

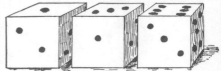

HERE is a neat little trick with three dice. I ask you to throw the dice without my seeing them. Then I tell you to multiply the points of the first die by 2 and add 5 ; then multiply the result by 5 and add the points of the second die ; then multiply the result by 10 and add the points of the third die. You then give me the total, and I can at once tell you the points thrown with the three dice. How do I do it ? As an example, if you threw 1, 3, and 6, as in the illustration, the result you would give me would be 386, from which I could at once say what you had thrown.

387.—THE VILLAGE CRICKET MATCH.

IN a cricket match, Dingley Dell *v.* All Muggleton, the latter had the first innings. Mr. Dumkins and Mr. Podder were at the wickets, when the wary Dumkins made a splendid late cut, and Mr. Podder called on him to run. Four runs were apparently completed, but the vigilant umpires at each end called, "three short," making six short runs in all. What number did Mr. Dumkins score ? When Dingley Dell took their turn at the wickets their champions were Mr. Luffey and Mr. Struggles. The latter made a magnificent off-drive, and invited his colleague to "come along," with the result that the observant spectators applauded them for what was supposed to have been three sharp runs. But the umpires declared that there had been two short runs at each end—four in all. To what extent, if any, did this manœuvre increase Mr. Struggles's total ?

388.—SLOW CRICKET.

IN the recent county match between Wessex and Nincomshire the former team were at the wickets all day, the last man being put out a few minutes before the time for drawing stumps. The play was so slow that most of the spectators were fast asleep, and, on being awakened by one of the officials clearing the ground, we learnt that two men had been put leg-before-wicket for a combined score of 19 runs; four men were caught for a combined score of 17 runs ; one man was run out for a duck's egg ; and the others were all bowled for 3 runs each. There were no extras. We were not told which of the men was the captain, but he made exactly 15 more than the average of his team. What was the captain's score ?

389.—THE FOOTBALL PLAYERS.

"IT is a glorious game !" an enthusiast was heard to exclaim. "At the close of last season, of the footballers of my acquaintance four had broken their left arm, five had broken their right arm, two had the right arm sound, and three had sound left arms." Can you discover from that statement what is the smallest number of players that the speaker could be acquainted with ?

It does not at all follow that there were as many as fourteen men, because, for example, two of the men who had broken the left arm might also be the two who had sound right arms.

390.—THE HORSE-RACE PUZZLE.

THERE are no morals in puzzles. When we are solving the old puzzle of the captain who, having to throw half his crew overboard in a storm, arranged to draw lots, but so placed the men that only the Turks were sacrificed, and all the Christians left on board, we do not stop to discuss the questionable morality of the proceeding. And when we are dealing with a measuring problem, in which certain thirsty pilgrims are to make an equitable division of a barrel of beer, we do not object that, as total abstainers, it is against our conscience to have anything to do with intoxicating liquor. Therefore I make no apology for introducing a puzzle that deals with betting.

Three horses—Acorn, Bluebottle, and Capsule—start in a race. The odds are 4 to 1, Acorn ; 3 to 1, Bluebottle ; 2 to 1, Capsule. Now, how much must I invest on each horse in order to win £13, no matter which horse comes in first ? Supposing, as an example, that I betted £5 on each horse. Then, if Acorn won, I should receive £20 (four times £5), and have to pay £5 each for the other two horses ; thereby winning £10. But it will be found that if Bluebottle was first I should only win £5, and if Capsule won I should gain nothing and lose nothing. This will make the question perfectly clear to the novice, who, like myself, is not interested in the calling of the fraternity who profess to be engaged in the noble task of " improving the breed of horses."

391.—THE MOTOR-CAR RACE.

SOMETIMES a quite simple statement of fact, if worded in an unfamiliar manner, will cause considerable perplexity. Here is an example, and it will doubtless puzzle some of my more youthful readers just a little. I happened to be at a motor-car race at Brooklands, when one spectator said to another, while a number of cars were whirling round and round the circular track :—

" There's Gogglesmith—that man in the white car ! "

" Yes, I see," was the reply ; " but how many cars are running in this race ? "

Then came this curious rejoinder :—

" One-third of the cars in front of Gogglesmith added to three-quarters of those behind him will give you the answer."

Now, can you tell how many cars were running in the race ?

PUZZLE GAMES.

" He that is beaten may be said
To lie in honour's truckle bed."
HUDIBRAS.

IT may be said generally that a game is a contest of skill for two or more persons, into which we enter either for amusement or to win a prize. A puzzle is something to be done or solved by the individual. For example, if it were possible for us so to master the complexities of the game of chess that we could be assured of always winning with the first or second move, as the case might be, or of always drawing, then it would cease to be a game and would become a puzzle. Of course among the young and uninformed, when the correct winning play is not understood, a puzzle may well make a very good game. Thus there is no doubt children will continue to play " Noughts and Crosses," though I have shown (No. 109, " Canterbury Puzzles ") that between two players who both thoroughly understand the play, every game should be drawn. Neither player could ever win except through the blundering of his opponent. But I am writing from the point of view of the student of these things.

The examples that I give in this class are apparently games, but, since I show in every case how one player may win if he only play correctly, they are in reality puzzles. Their interest, therefore, lies in attempting to discover the leading method of play.

392.—THE PEBBLE GAME.

HERE is an interesting little puzzle game that I used to play with an acquaintance on the beach at Slocomb-on-Sea. Two players place an odd number of pebbles, we will say fifteen, between them. Then each takes in turn one, two, or three pebbles (as he chooses), and the winner is the one who gets the odd number. Thus, if you get seven and your opponent eight, you win. If you get six and he gets nine, he wins. Ought the first or second player to win, and how ? When you have settled the question with fifteen pebbles try again with, say, thirteen.

393.—THE TWO ROOKS.

THIS is a puzzle game for two players. Each player has a single rook. The first player places his rook on any square of the board that he may choose to select, and then the second player does the same. They now play in turn, the point of each play being to capture the opponent's rook. But in this game you cannot play through a line of attack without being captured. That is to say, if in the diagram it is Black's turn to play, he cannot move his rook to his king's knight's square, or to his king's rook's square, because he would enter the "line of fire" when passing his king's bishop's square. For the

BLACK.

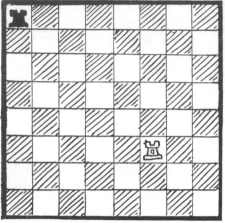

WHITE.

same reason he cannot move to his queen's rook's seventh or eighth squares. Now, the game can never end in a draw. Sooner or later one of the rooks must fall, unless, of course, both players commit the absurdity of not trying to win. The trick of winning is ridiculously simple when you know it. Can you solve the puzzle?

394.—PUSS IN THE CORNER.

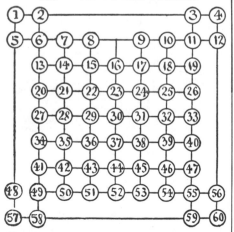

THIS variation of the last puzzle is also played by two persons. One puts a counter on No. 6,

and the other puts one on No. 55, and they play alternately by removing the counter to any other number in a line. If your opponent moves at any time on to one of the lines you occupy, or even crosses one of your lines, you immediately capture him and win. We will take an illustrative game.

A moves from 55 to 52; B moves from 6 to 13; A advances to 23; B goes to 15; A retreats to 26; B retreats to 13; A advances to 21; B retreats to 2; A advances to 7; B goes to 3; A moves to 6; B must now go to 4; A establishes himself at 11, and B must be captured next move because he is compelled to cross a line on which A stands. Play this over and you will understand the game directly. Now, the puzzle part of the game is this: Which player should win, and how many moves are necessary?

395.—A WAR PUZZLE GAME.

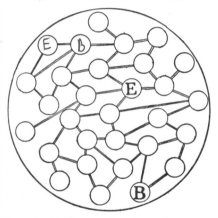

HERE is another puzzle game. One player, representing the British general, places a counter at B, and the other player, representing the enemy, places his counter at E. The Britisher makes the first advance along one of the roads to the next town, then the enemy moves to one of his nearest towns, and so on in turns, until the British general gets into the same town as the enemy and captures him. Although each must always move along a road to the next town only, and the second player may do his utmost to avoid capture, the British general (as we should suppose, from the analogy of real life) must infallibly win. But how? That is the question.

396.—A MATCH MYSTERY.

HERE is a little game that is childishly simple in its conditions. But it is well worth investigation.

Mr. Stubbs pulled a small table between himself and his friend, Mr. Wilson, and took a box of matches, from which he counted out thirty.

"Here are thirty matches," he said. "I

divide them into three unequal heaps. Let me see. We have 14, 11, and 5, as it happens. Now, the two players draw alternately any number from any one heap, and he who draws the last match loses the game. That's all ! I will play with you, Wilson. I have formed the heaps, so you have the first draw."

" As I can draw any number," Mr. Wilson said, " suppose I exhibit my usual moderation and take all the 14 heap."

" That is the worst you could do, for it loses right away. I take 6 from the 11, leaving two equal heaps of 5, and to leave two equal heaps is a certain win (with the single exception of 1, 1), because whatever you do in one heap I can repeat in the other. If you leave 4 in one heap, I leave 4 in the other. If you then leave 2 in one heap, I leave 2 in the other. If you leave only 1 in one heap, then I take all the other heap. If you take all one heap, I take all but one in the other. No, you must never leave two heaps, unless they are equal heaps and more than 1, 1. Let's begin again."

" Very well, then," said Mr. Wilson. " I will take 6 from the 14, and leave you 8, 11, 5."

Mr. Stubbs then left 8, 11, 3 ; Mr. Wilson, 8, 5, 3 ; Mr. Stubbs, 6, 5, 3 ; Mr. Wilson, 4, 5, 3 ; Mr. Stubbs, 4, 5, 1 ; Mr. Wilson, 4, 3, 1 ; Mr. Stubbs, 2, 3, 1 ; Mr. Wilson, 2, 1, 1 ; which Mr. Stubbs reduced to 1, 1, 1.

" It is now quite clear that I must win," said Mr. Stubbs, because you must take 1, and then I take 1, leaving you the last match. You never had a chance. There are just thirteen different ways in which the matches may be grouped at the start for a certain win. In fact, the groups selected, 14, 11, 5, are a certain win, because for whatever your opponent may play there is another winning group you can secure, and so on and on down to the last match."

397.—THE MONTENEGRIN DICE GAME.

It is said that the inhabitants of Montenegro

have a little dice game that is both ingenious and well worth investigation. The two players first select two different pairs of odd numbers (always higher than 3) and then alternately toss three dice. Whichever first throws the dice so that they add up to one of his selected numbers wins. If they are both successful in two successive throws it is a draw and they try again. For example, one player may select 7 and 15 and the other 5 and 13. Then if the first player throws so that the three dice add up 7 or 15 he wins, unless the second man gets either 5 or 13 on his throw.

The puzzle is to discover which two pairs of numbers should be selected in order to give both players an exactly even chance.

398.—THE CIGAR PUZZLE.

I once propounded the following puzzle in a London club, and for a considerable period it absorbed the attention of the members. They could make nothing of it, and considered it quite impossible of solution. And yet, as I shall show, the answer is remarkably simple.

Two men are seated at a square-topped table. One places an ordinary cigar (flat at one end, pointed at the other) on the table, then the other does the same, and so on alternately, a condition being that no cigar shall touch another. Which player should succeed in placing the last cigar, assuming that they each will play in the best possible manner ? The size of the table top and the size of the cigar are not given, but in order to exclude the ridiculous answer that the table might be so diminutive as only to take one cigar, we will say that the table must not be less than 2 feet square and the cigar not more than 4½ inches long. With those restrictions you may take any dimensions you like. Of course we assume that all the cigars are exactly alike in every respect. Should the first player, or the second player, win ?

MAGIC SQUARE PROBLEMS.

" By magic numbers."
CONGREVE, *The Mourning Bride.*

THIS is a very ancient branch of mathematical puzzledom, and it has an immense, though scattered, literature of its own. In their simple form of consecutive whole numbers arranged in a square so that every column, every row, and each of the two long diagonals shall add up alike, these magic squares offer three main lines of investigation : Construction, Enumeration, and Classification. Of recent years many ingenious methods have been devised for the construction of magics, and the law of their formation is so well understood that all the ancient mystery has evaporated and there is no longer any difficulty in making squares of any dimensions. Almost the last word has been said on this subject.
The question of the enumeration of all the

possible squares of a given order stands just where it did over two hundred years ago. Everybody knows that there is only one solution for the third order, three cells by three ; and Frénicle published in 1693 diagrams of all the arrangements of the fourth order—880 in number—and his results have been verified over and over again. I may here refer to the general solution for this order, for numbers not necessarily consecutive, by E. Bergholt in *Nature*, May 26, 1910, as it is of the greatest importance to students of this subject. The enumeration of the examples of any higher order is a completely unsolved problem.

As to classification, it is largely a matter of individual taste—perhaps an æsthetic question, for there is beauty in the law and order of numbers. A man once said that he divided the human race into two great classes : those who take snuff and those who do not. I am not sure that some of our classifications of magic squares are not almost as valueless. However, lovers of these things seem somewhat agreed that Nasik magic squares (so named by Mr. Frost, a student of them, after the town in India where he lived, and also called Diabolique and Pandiagonal) and Associated magic squares are of special interest, so I will just explain what these are for the benefit of the novice.

I published in *The Queen* for January 15, 1910, an article that would enable the reader to write out, if he so desired, all the 880 magics of the fourth order, and the following is the complete classification that I gave. The first example is that of a Simple square that fulfils the simple conditions and no more. The second example is a Semi-Nasik, which has the additional property that the opposite short diagonals of two cells each together sum to 34. Thus, $14+4+11+5=34$ and $12+6+13+3=34$. The third example is not only Semi-Nasik but

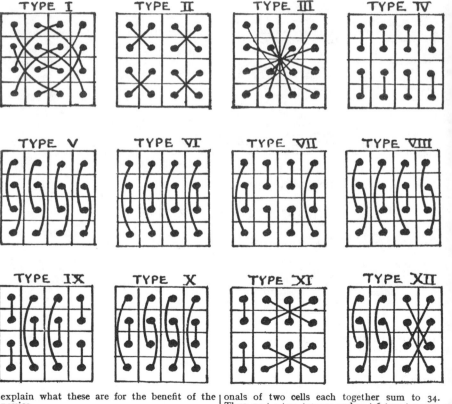

SIMPLE			
1	12	14	7
4	15	9	6
13	2	8	11
16	5	3	10

SEMI-NASIK			
1	14	12	7
4	15	9	6
13	2	8	11
16	3	5	10

ASSOCIATED			
1	14	12	7
8	11	13	2
15	4	6	9
10	5	3	16

NASIK			
1	14	7	12
15	4	9	6
10	5	16	3
8	11	2	13

also Associated, because in it every number, if added to the number that is equidistant, in a straight line, from the centre gives 17. Thus, 1 + 16, 2 + 15, 3 + 14, etc. The fourth example, considered the most "perfect" of all, is a Nasik. Here all the broken diagonals sum to 34. Thus, for example, 15 + 14 + 2 + 3, and 10 + 4 + 7 + 13, and 15 + 5 + 2 + 12. As a consequence, its properties are such that if you repeat the square in all directions you may mark off a square, 4 × 4, wherever you please, and it will be magic.

The following table not only gives a complete enumeration under the four forms described, but also a classification under the twelve graphic types indicated in the diagrams. The dots at the end of each line represent the relative positions of those complementary pairs, 1 + 16, 2 + 15, etc., which sum to 17. For example, it will be seen that the first and second magic squares given are of Type VI., that the third square is of Type III., and that the fourth is of Type I. Édouard Lucas indicated these types, but he dropped exactly half of them and did not attempt the classification.

NASIK (Type I.)	. .	. .	.	48
SEMI-NASIK (Type II., Transpositions of Nasik)	.	48		
„ (Type III., Associated)		48		
„ (Type IV.)	. .	96		
„ (Type V.)	. . .	96	192	
„ (Type VI.)	. . .		96	384
SIMPLE. (Type VI.)	. . .		208	
„ (Type VII.)	. . :	56		
„ (Type VIII.)	. . .	56		
„ (Type IX.)	. . .	56		
„ (Type X.)	. . .	56	224	
„ (Type XI.)	. . .	8		
„ (Type XII.)	. . :	8	16	448
				880

It is hardly necessary to say that every one of these squares will produce seven others by mere reversals and reflections, which we do not count as different. So that there are 7,040 squares of this order, 880 of which are fundamentally different.

An infinite variety of puzzles may be made introducing new conditions into the magic square. In *The Canterbury Puzzles* I have given examples of such squares with coins, with postage stamps, with cutting-out conditions, and other tricks. I will now give a few variants involving further novel conditions.

399.—THE TROUBLESOME EIGHT.

NEARLY everybody knows that a "magic square" is an arrangement of numbers in the form of a square so that every row, every column, and each of the two long diagonals adds up alike. For example, you would find little difficulty in merely placing a different number in each of the nine cells in the illustration so that the rows, columns, and diagonals shall all add up 15. And at your first attempt you will probably find that you have an 8 in

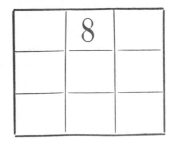

one of the corners. The puzzle is to construct the magic square, under the same conditions, with the 8 in the position shown.

400.—THE MAGIC STRIPS.

I HAPPENED to have lying on my table a number of strips of cardboard, with numbers printed on them from 1 upwards in numerical order. The idea suddenly came to me, as ideas have a way of unexpectedly coming, to make a little puzzle of this. I wonder whether many readers will arrive at the same solution that I did.

Take seven strips of cardboard and lay them together as above. Then write on each of them the numbers 1, 2, 3, 4, 5, 6, 7, as shown, so that the numbers shall form seven rows and seven columns.

Now, the puzzle is to cut these strips into the fewest possible pieces so that they may be placed together and form a magic square, the seven rows, seven columns, and two diagonals adding up the same number. No figures may

be turned upside down or placed on their sides—
that is, all the strips must lie in their original
direction.

Of course you could cut each strip into seven
separate pieces, each piece containing a number,
and the puzzle would then be very easy, but I
need hardly say that forty-nine pieces is a long
way from being the fewest possible.

401.—EIGHT JOLLY GAOL BIRDS.

THE illustration shows the plan of a prison of
nine cells all communicating with one another
by doorways. The eight prisoners have their
numbers on their backs, and any one of them
is allowed to exercise himself in whichever cell
may happen to be vacant, subject to the rule
that at no time shall two prisoners be in the
same cell. The merry monarch in whose do-
minions the prison was situated offered them
special comforts one Christmas Eve if, without
breaking that rule, they could so place themselves that their numbers should form a magic
square.

Now, prisoner No. 7 happened to know a good
deal about magic squares, so he worked out a
scheme and naturally selected the method that
was most expeditious—that is, one involving
the fewest possible moves from cell to cell.
But one man was a surly, obstinate fellow (quite
unfit for the society of his jovial companions),
and he refused to move out of his cell or take
any part in the proceedings. But No. 7 was
quite equal to the emergency, and found that
he could still do what was required in the
fewest possible moves without troubling the
brute to leave his cell. The puzzle is to show
how he did it and, incidentally, to discover
which prisoner was so stupidly obstinate. Can
you find the fellow ?

402.—NINE JOLLY GAOL BIRDS.

SHORTLY after the episode recorded in the last
puzzle occurred, a ninth prisoner was placed in

the vacant cell, and the merry monarch then
offered them all complete liberty on the following strange conditions. They were required so
to rearrange themselves in the cells that their
numbers formed a magic square without their
movements causing any two of them ever to
be in the same cell together, except that at the
start one man was allowed to be placed on the
shoulders of another man, and thus add their
numbers together, and move as one man. For
example, No. 8 might be placed on the shoulders of No. 2, and then they would move about
together as 10. The reader should seek first to
solve the puzzle in the fewest possible moves,
and then see that the man who is burdened has
the least possible amount of work to do.

403.—THE SPANISH DUNGEON.

NOT fifty miles from Cadiz stood in the middle
ages a castle, all traces of which have for centuries disappeared. Among other interesting
features, this castle contained a particularly
unpleasant dungeon divided into sixteen cells,
all communicating with one another, as shown
in the illustration.

Now, the governor was a merry wight, and
very fond of puzzles withal. One day he went
to the dungeon and said to the prisoners, " By
my halidame ! " (or its equivalent in Spanish)
" you shall all be set free if you can solve this
puzzle. You must so arrange yourselves in the
sixteen cells that the numbers on your backs
shall form a magic square in which every
column, every row, and each of the two diagonals shall add up the same. Only remember
this : that in no case may two of you ever be
together in the same cell."

One of the prisoners, after working at the
problem for two or three days, with a piece of
chalk, undertook to obtain the liberty of himself and his fellow-prisoners if they would follow
his directions and move through the doorway

from cell to cell in the order in which he should call out their numbers.

He succeeded in his attempt, and, what is

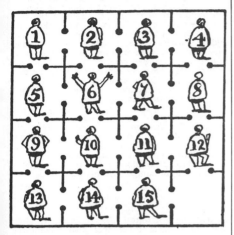

more remarkable, it would seem from the account of his method recorded in the ancient manuscript lying before me, that he did so in the fewest possible moves. The reader is asked to show what these moves were.

404.—THE SIBERIAN DUNGEONS.

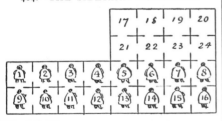

THE above is a trustworthy plan of a certain Russian prison in Siberia. All the cells are numbered, and the prisoners are numbered the same as the cells they occupy. The prison diet is so fattening that these political prisoners are in perpetual fear lest, should their pardon arrive, they might not be able to squeeze themselves through the narrow doorways and get out. And of course it would be an unreasonable thing to ask any government to pull down the walls of a prison just to liberate the prisoners, however innocent they might be. Therefore these men take all the healthy exercise they can in order to retard their increasing obesity, and one of their recreations will serve to furnish us with the following puzzle.

Show, in the fewest possible moves, how the sixteen men may form themselves into a magic square, so that the numbers on their backs shall

add up the same in each of the four columns, four rows, and two diagonals without two prisoners having been at any time in the same cell together. I had better say, for the information of those who have not yet been made acquainted with these places, that it is a peculiarity of prisons that you are not allowed to go outside their walls. Any prisoner may go any distance that is possible in a single move.

405.—CARD MAGIC SQUARES.

TAKE an ordinary pack of cards and throw out the twelve court cards. Now, with nine of the remainder (different suits are of no consequence) form the above magic square. It will be seen that the pips add up fifteen in every row in every column, and in each of the two long diagonals. The puzzle is with the remaining cards (without disturbing this arrangement) to form three more such magic squares, so that each of the four shall add up to a different sum. There will, of course, be four cards in the reduced pack that will not be used. These four may be any that you choose. It is not a difficult puzzle, but requires just a little thought.

406.—THE EIGHTEEN DOMINOES.

THE illustration shows eighteen dominoes arranged in the form of a square so that the pips in every one of the six columns, six rows, and two long diagonals add up 13. This is the smallest summation possible with any selection of dominoes from an ordinary box of twenty-eight. The greatest possible summation is 23, and a solution for this number may be easily obtained by substituting for every number its complement to 6. Thus for every blank substitute a 6, for every 1 a 5, for every 2 a 4, for

3 a 3, for 4 a 2, for 5 a 1, and for 6 a blank. But the puzzle is to make a selection of eighteen dominoes and arrange them (in exactly the form

shown) so that the summations shall be 18 in all the fourteen directions mentioned.

SUBTRACTING, MULTIPLYING, AND DIVIDING MAGICS.

ALTHOUGH the adding magic square is of such great antiquity, curiously enough the multiplying magic does not appear to have been mentioned until the end of the eighteenth century, when it was referred to slightly by one writer and then forgotten until I revived it in *Tit-Bits* in 1897. The dividing magic was apparently first discussed by me in *The Weekly Dispatch* in June 1898. The subtracting magic is here introduced for the first time. It will now be convenient to deal with all four kinds of magic squares together.

second, and the result from the third. You can, of course, perform the operation in either direction; but, in order to avoid negative numbers, it is more convenient simply to deduct the middle number from the sum of the two extreme numbers. This is, in effect, the same thing. It will be seen that the constant of the adding square is n times that of the subtracting square derived from it, where n is the number of cells in the side of square. And the manner of derivation here is simply to reverse the two diagonals. Both squares are "associated"—a term I have explained in the introductory article to this department.

The third square is a multiplying magic. The constant, 216, is obtained by multiplying together the three numbers in any line. It is "associated" by multiplication, instead of by addition. It is here necessary to remark that in an adding square it is not essential that the nine numbers should be consecutive. Write down any nine numbers in this way—

$$
\begin{array}{ccc}
1 & 3 & 5 \\
4 & 6 & 8 \\
7 & 9 & 11
\end{array}
$$

so that the horizontal differences are all alike and the vertical differences also alike (here 2 and 3), and these numbers will form an adding magic square. By making the differences 1 and 3 we, of course, get consecutive numbers—a particular case, and nothing more. Now, in the case of the multiplying square we must take these numbers in geometrical instead of arithmetical progression, thus—

$$
\begin{array}{ccc}
1 & 3 & 9 \\
2 & 6 & 18 \\
4 & 12 & 36
\end{array}
$$

Here each successive number in the rows is multiplied by 3, and in the columns by 2. Had we multiplied by 2 and 8 we should get the regular geometrical progression, 1, 2, 4, 8, 16, 32, 64, 128, and 256, but I wish to avoid high numbers. The numbers are arranged in the square in the same order as in the adding square.

ADDING		
8	1	6
3	5	7
4	9	2

SUBTRACTING		
2	1	4
3	5	7
6	9	8

MULTIPLYING		
12	1	18
9	6	4
2	36	3

DIVIDING		
3	1	2
9	6	4
18	36	12

In these four diagrams we have examples in the third order of adding, subtracting, multiplying, and dividing squares. In the first the constant, 15, is obtained by the addition of the rows, columns, and two diagonals. In the second case you get the constant, 5, by subtracting the first number in a line from the

The fourth diagram is a dividing magic square. The constant 6 is here obtained by dividing the second number in a line by the first (in either direction) and the third number by the quotient. But, again, the process is simplified by dividing the product of the two extreme numbers by the middle number. This square is also " associated " by multiplication. It is derived from the multiplying square by merely reversing the diagonals, and the constant of the multiplying square is the cube of that of the dividing square derived from it.

The next set of diagrams shows the solutions for the fifth order of square. They are all "associated " in the same way as before. The subtracting square is derived from the adding square by reversing the diagonals and exchanging opposite numbers in the centres of the borders, and the constant of one is again n times that of the other. The dividing square is derived from the multiplying square in the same way, and the constant of the latter is the 5th power (that is the nth) of that of the former.

ADDING

17	24	1	8	15
23	5	7	14	16
4	6	13	20	22
10	12	19	21	3
11	18	25	2	9

SUBTRACTING

9	24	25	8	11
23	21	7	12	16
22	6	13	20	4
10	14	19	5	3
15	18	1	2	17

MULTIPLYING

54	648	1	12	144
324	16	6	72	27
8	3	36	432	162
48	18	216	81	4
9	108	1296	2	24

DIVIDING

24	648	1296	12	9
324	81	6	18	27
162	3	36	432	8
48	72	216	16	4
144	108	1	2	54

These squares are thus quite easy for odd orders. But the reader will probably find some difficulty over the even orders, concerning which I will leave him to make his own researches, merely propounding two little problems.

407.—TWO NEW MAGIC SQUARES.

CONSTRUCT a subtracting magic square with the first sixteen whole numbers that shall be " associated " by *subtraction*. The constant is, of course, obtained by subtracting the first number from the second in line, the result from the third, and the result again from the fourth. Also construct a dividing magic square of the same order that shall be " associated " by *division*. The constant is obtained by dividing the second number in a line by the first, the third by the quotient, and the fourth by the next quotient.

408.—MAGIC SQUARES OF TWO DEGREES.

WHILE reading a French mathematical work I happened to come across the following statement : " A very remarkable magic square of 8, in two degrees, has been constructed by M. Pfeffermann. In other words, he has managed to dispose the sixty-four first numbers on the squares of a chessboard in such a way that the sum of the numbers in every line, every column, and in each of the two diagonals, shall be the same ; and more, that if one substitutes for all the numbers their squares, the square still remains magic." I at once set to work to solve this problem, and, although it proved a very hard nut, one was rewarded by the discovery of some curious and beautiful laws that govern it. The reader may like to try his hand at the puzzle.

MAGIC SQUARES OF PRIMES.

THE problem of constructing magic squares with prime numbers only was first discussed by myself in *The Weekly Dispatch* for 22nd July and 5th August 1900; but during the last three or four years it has received great attention from American mathematicians. First, they have sought to form these squares with the lowest possible constants. Thus, the first nine prime numbers, 1 to 23 inclusive, sum to 99, which (being divisible by 3) is theoretically a suitable series ; yet it has been demonstrated that the lowest possible constant is 111, and the required series is as follows : 1, 7, 13, 31, 37, 43, 61, 67, and 73. Similarly, in the case of the fourth order, the lowest series of primes that are " theoretically suitable " will not serve. But in every other order, up to the 12th inclusive, magic squares have been constructed with the lowest series of primes theoretically possible. And the 12th is the lowest order in which a straight series of prime numbers, unbroken, from 1 upwards has been made to work. In other words, the first 144 odd prime numbers have actually been arranged in magic form. The following summary is taken from *The Monist* (Chicago) for October 1913 :—

Order of Square.	Totals of Series.	Lowest Constants.	Squares made by—
3rd	333	111	Henry E. Dudeney (1900).
4th	408	102	Ernest Bergholt and C. D. Shuldham.
5th	1065	213	H. A. Sayles.
6th	2448	408	C. D. Shuldham and J. N. Muncey.
7th	4893	699	do.
8th	8912	1114	do.
9th	15129	1681	do.
10th	24160	2416	J. N. Muncey.
11th	36095	3355	do.
12th	54168	4514	do.

For further details the reader should consult the article itself, by W. S. Andrews and H. A. Sayles.

These same investigators have also performed notable feats in constructing associated and bordered prime magics, and Mr. Shuldham has sent me a remarkable paper in which he gives examples of Nasik squares constructed with primes for all orders from the 4th to the 10th, with the exception of the 3rd (which is clearly impossible) and the 9th, which, up to the time of writing, has baffled all attempts.

409.—THE BASKETS OF PLUMS.

THIS is the form in which I first introduced the

question of magic squares with prime numbers. I will here warn the reader that there is a little trap.

A fruit merchant had nine baskets. Every basket contained plums (all sound and ripe), and the number in every basket was different. When placed as shown in the illustration they formed a magic square, so that if he took any three baskets in a line in the eight possible directions there would always be the same number of plums. This part of the puzzle is easy enough to understand. But what follows seems at first sight a little queer.

The merchant told one of his men to distribute the contents of any basket he chose among some children, giving plums to every child so that each should receive an equal number. But the man found it quite impossible, no matter which basket he selected and no matter how many children he included in the treat. Show, by giving contents of the nine baskets, how this could come about.

410.—THE MANDARIN'S "T" PUZZLE.

BEFORE Mr. Beauchamp Cholmondely Marjoribanks set out on his tour in the Far East, he prided himself on his knowledge of magic squares, a subject that he had made his special hobby; but he soon discovered that he had never really touched more than the fringe of the subject, and that the wily Chinee could beat him easily. I present a little problem that one learned mandarin propounded to our traveller, as depicted on the last page.

The Chinaman, after remarking that the construction of the ordinary magic square of twenty-five cells is "too velly muchee easy," asked our countryman so to place the numbers 1 to 25 in the square that every column, every row, and each of the two diagonals should add up 65, with only prime numbers on the shaded "T." Of course the prime numbers available are 1, 2, 3, 5, 7, 11, 13, 17, 19, and 23, so you are at liberty to select any nine of these that will serve your purpose. Can you construct this curious little magic square?

411.—A MAGIC SQUARE OF COMPOSITES.

As we have just discussed the construction of magic squares with prime numbers, the following forms an interesting companion problem. Make a magic square with nine consecutive composite numbers—the smallest possible.

412.—THE MAGIC KNIGHT'S TOUR.

HERE is a problem that has never yet been solved, nor has its impossibility been demonstrated. Play the knight once to every square of the chessboard in a complete tour, numbering the squares in the order visited, so that when completed the square shall be "magic," adding up to 260 in every column, every row, and each of the two long diagonals. I shall give the best answer that I have been able to obtain, in which there is a slight error in the diagonals alone. Can a perfect solution be found? I am convinced that it cannot, but it is only a "pious opinion."

MAZES AND HOW TO THREAD THEM.

" In wandering mazes lost."
Paradise Lost.

THE Old English word "maze," signifying a labyrinth, probably comes from the Scandinavian, but its origin is somewhat uncertain. The late Professor Skeat thought that the substantive was derived from the verb, and as in old times to be mazed or amazed was to be "lost in thought," the transition to a maze in whose tortuous windings we are lost is natural and easy.

The word "labyrinth" is derived from a Greek word signifying the passages of a mine. The ancient mines of Greece and elsewhere inspired fear and awe on account of their darkness and the danger of getting lost in their intricate passages. Legend was afterwards built round these mazes. The most familiar instance is the labyrinth made by Dædalus in Crete for King Minos. In the centre was placed the Minotaur, and no one who entered could find his way out again, but became the prey of the monster. Seven youths and seven maidens were sent regularly by the Athenians, and were duly devoured, until Theseus slew the monster and escaped from the maze by aid of the clue of thread provided by Ariadne; which accounts for our using to-day the expression "threading a maze."

The various forms of construction of mazes include complicated ranges of caverns, architectural labyrinths, or sepulchral buildings, tortuous devices indicated by coloured marbles and tiled pavements, winding paths cut in the turf, and topiary mazes formed by clipped hedges. As a matter of fact, they may be said to have descended to us in precisely this order of variety.

Mazes were used as ornaments on the state robes of Christian emperors before the ninth century, and were soon adopted in the decoration of cathedrals and other churches. The original idea was doubtless to employ them as symbols of the complicated folds of sin by which man is surrounded. They began to abound in the early part of the twelfth century, and I give an illustration of one of this period in the parish church at St. Quentin (Fig. 1). It formed a pave-

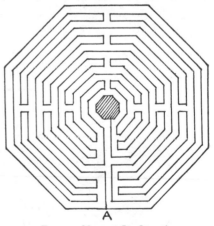

FIG. 1.—Maze at St. Quentin.

ment of the nave, and its diameter is 34½ feet. The path here is the line itself. If you place your pencil at the point A and ignore the enclosing line, the line leads you to the centre by a long route over the entire area; but you never have any option as to direction during your course. As we shall find in similar cases, these early ecclesiastical mazes were generally not of a puzzle nature, but simply long, winding paths that took you over practically all the ground enclosed.

In the abbey church of St. Bertin, at St. Omer, is another of these curious floors, representing the Temple of Jerusalem, with stations for pilgrims. These mazes were actually visited

FIG. 2.—Maze in Chartres Cathedral.

and traversed by them as a compromise for not going to the Holy Land in fulfilment of a vow. They were also used as a means of penance, the penitent frequently being directed to go the whole course of the maze on hands and knees.

The maze in Chartres Cathedral, of which I give an illustration (Fig. 2), is 40 feet across, and was used by penitents following the pro-

FIG. 3.—Maze in Lucca Cathedral.

cession of Calvary. A labyrinth in Amiens Cathedral was octagonal, similar to that at St.

Quentin, measuring 42 feet across. It bore the date 1288, but was destroyed in 1708. In the chapter-house at Bayeux is a labyrinth formed of tiles, red, black, and encaustic, with a pattern of brown and yellow. Dr. Ducarel, in his *Tour through Part of Normandy*" (printed in 1767), mentions the floor of the great guard-chamber in the abbey of St. Stephen, at Caen, "the middle whereof represents a maze or labyrinth about 10 feet diameter, and so artfully contrived that, were we to suppose a man following all the intricate meanders of its volutes, he could not travel less than a mile before he got from one end to the other."

Then these mazes were sometimes reduced in size and represented on a single tile (Fig. 3). I give an example from Lucca Cathedral. It is on one of the porch piers, and is 19½ inches in diameter. A writer in 1858 says that, "from the continual attrition it has received from thousands of tracing fingers, a central group of Theseus and the Minotaur has now been very

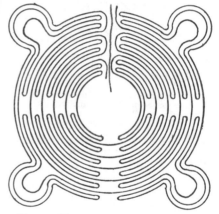

FIG. 4.—Maze at Saffron Walden, Essex.

nearly effaced." Other examples were, and perhaps still are, to be found in the Abbey of Toussarts, at Châlons-sur-Marne, in the very ancient church of St. Michele at Pavia, at Aix in Provence, in the cathedrals of Poitiers, Rheims, and Arras, in the church of Santa Maria in Aquiro in Rome, in San Vitale at Ravenna, in the Roman mosaic pavement found at Salzburg, and elsewhere. These mazes were sometimes called "Chemins de Jerusalem," as being emblematical of the difficulties attending a journey to the earthly Jerusalem and of those encountered by the Christian before he can reach the heavenly Jerusalem—where the centre was frequently called "Ciel."

Common as these mazes were upon the Continent, it is probable that no example is to be found in any English church; at least I am not aware of the existence of any. But almost every county has, or has had, its specimens of mazes cut in the turf. Though these are frequently known as "miz-mazes" or "mize-mazes," it is not uncommon to find them locally called "Troy-towns," "shepherds' races," or "Julian's Bowers"—names that are misleading, as suggesting a false origin. From the facts alone that many of these English turf mazes are clearly copied from those in the Continental churches, and practically all are found close to some ecclesiastical building or near the site of an ancient one, we may regard it as certain that they were of church origin and not invented by the shepherds or other rustics. And curiously enough, these turf mazes are apparently un-

FIG. 5.—Maze at Sneinton, Nottinghamshire.

known on the Continent. They are distinctly mentioned by Shakespeare :—

> "The nine men's morris is filled up with mud,
> And the quaint mazes in the wanton green
> For lack of tread are undistinguishable."
> *A Midsummer Night's Dream*, ii. 1.

> "My old bones ache : here's a maze trod indeed,
> Through forth-rights and meanders ! "
> *The Tempest*, iii. 3.

There was such a maze at Comberton, in Cambridgeshire, and another, locally called the "miz-maze," at Leigh, in Dorset. The latter was on the highest part of a field on the top of a hill, a quarter of a mile from the village, and was slightly hollow in the middle and enclosed by a bank about 3 feet high. It was circular, and was thirty paces in diameter. In 1868 the turf had grown over the little trenches, and it was then impossible to trace the paths of the maze. The Comberton one was at the same date believed to be perfect, but whether either or both have now disappeared I cannot say. Nor have I been able to verify the existence or non-existence of the other examples of which I am able

to give illustrations. I shall therefore write of them all in the past tense, retaining the hope that some are still preserved.

FIG. 6.—Maze at Alkborough, Lincolnshire.

In the next two mazes given—that at Saffron Walden, Essex (110 feet in diameter, Fig. 4), and the one near St. Anne's Well, at Sneinton, Nottinghamshire (Fig. 5), which was ploughed up on February 27th, 1797 (51 feet in diameter, with a path 535 yards long)—the paths must in each case be understood to be on the lines, black or white, as the case may be.

I give in Fig. 6 a maze that was at Alk-

FIG. 7.—Maze at Boughton Green, Nottinghamshire.

borough, Lincolnshire, overlooking the Humber This was 44 feet in diameter, and the resem blance between it and the mazes at Chartres and Lucca (Figs. 2 and 3) will be at once perceived.

FIG. 8.—Maze at Wing, Rutlandshire.

A maze at Boughton Green, in Nottinghamshire, a place celebrated at one time for its fair (Fig. 7), was 37 feet in diameter. I also include the plan (Fig. 8) of one that used to be on the outskirts of the village of Wing, near Uppingham, Rutlandshire. This maze was 40 feet in diameter.

The maze that was on St. Catherine's Hill, Winchester, in the parish of Chilcombe, was a poor specimen (Fig. 9), since, as will be seen,

line itself from end to end. This maze was 86 feet square, cut in the turf, and was locally known as the "Mize-maze." It became very indistinct about 1858, and was then recut by the Warden of Winchester, with the aid of a plan possessed by a lady living in the neighbourhood.

FIG. 10.—Maze on Ripon Common.

A maze formerly existed on Ripon Common, in Yorkshire (Fig. 10). It was ploughed up in 1827, but its plan was fortunately preserved. This example was 20 yards in diameter, and its path is said to have been 407 yards long.

FIG. 9.—Maze on St. Catherine's Hill, Winchester.

FIG. 11.—Maze at Theobalds, Hertfordshire.

there was one short direct route to the centre, unless, as in Fig. 10 again, the path is the

In the case of the maze at Theobalds, Hertfordshire, after you have found the entrance within the four enclosing hedges, the path is

FIG. 12.—Italian Maze of Sixteenth Century.

forced (Fig. 11). As further illustrations of this class of maze, I give one taken from an Italian work on architecture by Serlio, pub-

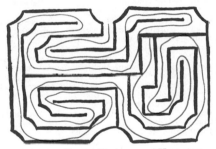

FIG. 13.—By the Designers of Hampton Court Maze.

lished in 1537 (Fig. 12), and one by London and Wise, the designers of the Hampton Court maze, from their book, *The Retir'd Gard'ner*, published

in 1706 (Fig. 13). Also, I add a Dutch maze (Fig. 14).

So far our mazes have been of historical interest, but they have presented no difficulty in threading. After the Reformation period we find mazes converted into mediums for re-creation, and they generally consisted of laby-rinthine paths enclosed by thick and carefully trimmed hedges. These topiary hedges were known to the Romans, with whom the *topiarius* was the ornamental gardener. This type of maze has of late years degenerated into the seaside " Puzzle Gardens. Teas, sixpence, in-cluding admission to the Maze." The Hampton

FIG. 14.—A Dutch Maze.

Court Maze, sometimes called the " Wilderness," at the royal palace, was designed, as I have said, by London and Wise for William III., who had a liking for such things (Fig. 15). I have before me some three or four versions of it, all slightly different from one another ; but the plan I select is taken from an old guide-book to the palace, and therefore ought to be trustworthy. The meaning of the dotted lines, etc., will be ex-plained later on.

The maze at Hatfield House (Fig. 16), the seat of the Marquis of Salisbury, like so many labyrinths, is not difficult on paper ; but both

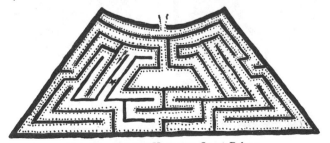

FIG. 15.—Maze at Hampton Court Palace.

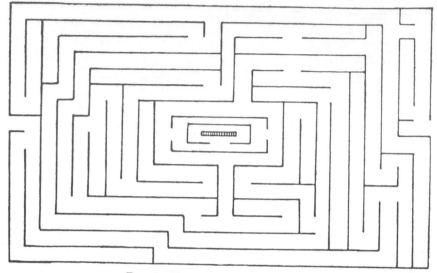

FIG. 16.—Maze at Hatfield House, Herts.

this and the Hampton Court Maze may prove very puzzling to actually thread without know-

FIG. 17.—Maze formerly at South Kensington.

ing the plan. One reason is that one is so apt to go down the same blind alleys over and over again, if one proceeds without method. The maze planned by the desire of the Prince Consort for the Royal Horticultural Society's Gardens at South Kensington was allowed to go to ruin, and was then destroyed—no great loss, for it was a feeble thing. It will be seen that there were three entrances from the outside (Fig. 17), but the way to the centre is very easy to discover. I include a German maze that is curious, but not difficult to thread on paper (Fig. 18). The example of a labyrinth

FIG. 18.—A German Maze.

formerly existing at Pimperne, in Dorset, is in a class by itself (Fig. 19). It was formed of small ridges about a foot high, and covered nearly an acre of ground; but it was, unfortunately, ploughed up in 1730.

FIG. 19.—Maze at Pimperne, Dorset.

We will now pass to the interesting subject of how to thread any maze. While being necessarily brief, I will try to make the matter clear to readers who have no knowledge of mathematics. And first of all we will assume that we are trying to enter a maze (that is, get to the "centre") of which we have no plan and about which we know nothing. The first rule is this: If a maze has no parts of its hedges detached from the rest, then if we always keep in touch with the hedge with the right hand (or always touch it with the left), going down to the stop in every blind alley and coming back on the other side, we shall pass through every part of

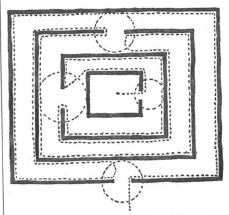

FIG. 20.—M. Trémaux's Method of Solution.

the maze and make our exit where we went in. Therefore we must at one time or another enter the centre, and every alley will be traversed twice.

Now look at the Hampton Court plan. Follow, say to the right, the path indicated by the

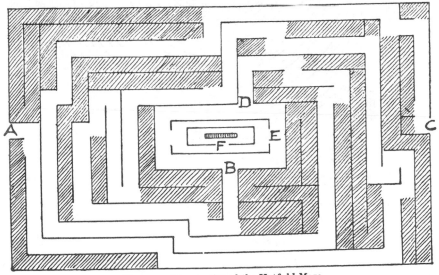

FIG. 21.—How to thread the Hatfield Maze.

dotted line, and what I have said is clearly correct if we obliterate the two detached parts, or "islands," situated on each side of the star. But as these islands are there, you cannot by this method traverse every part of the maze; and if it had been so planned that the "centre" was, like the star, between the two islands, you would never pass through the "centre" at all. A glance at the Hatfield maze will show that there are three of these detached hedges or islands at the centre, so this method will never

This blunder happened to me a few years ago in a little maze on the isle of Caldy, South Wales. I knew the maze was a small one, but after a very long walk I was amazed to find that I did not either reach the "centre" or get out again. So I threw a piece of paper on the ground, and soon came round to it; from which I knew that I had blundered over a supposed blind alley and was going round and round an island. Crossing to the opposite hedge and using more care, I was quickly at the centre and

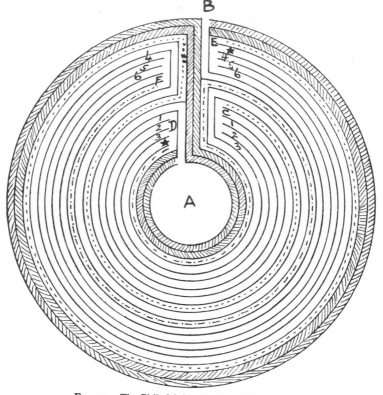

FIG. 22.—The Philadelphia Maze, and its Solution.

take you to the "centre" of that one. But the rule will at least always bring you safely out again unless you blunder in the following way. Suppose, when you were going in the direction of the arrow in the Hampton Court Maze, that you could not distinctly see the turning at the bottom, that you imagined you were in a blind alley and, to save time, crossed at once to the opposite hedge, then you would go round and round that U-shaped island with your right hand still always on the hedge—for ever after!

out again. Now, if I had made a similar mistake at Hampton Court, and discovered the error when at the star, I should merely have passed from one island to another! And if I had again discovered that I was on a detached part, I might with ill luck have recrossed to the first island again! We thus see that this "touching the hedge" method should always bring us safely out of a maze that we have entered; it may happen to take us through the "centre," and if we miss the centre we shall know there must be islands. But it has to be done with a little care, and in no case can we be sure that we have traversed every alley or that there are no detached parts.

FIG. 23.—Simplified Diagram of Fig. 22.

If the maze has many islands, the traversing of the whole of it may be a matter of considerable difficulty. Here is a method for solving character that will serve our purpose just as well as something more complex (Fig. 20). The circles at the regions where we have a choice of turnings we may call nodes. A " new " path or node is one that has not been entered before on the route; an "old" path or node is one that has already been entered. 1. No path may be traversed more than twice. 2. When you come to a new node, take any path you like. 3. When by a new path you come to an old node or to the stop of a blind alley, return by the path you came. 4. When by an old path you come to an old node, take a new path if there is one; if not, an old path. The route indicated by the

FIG. 24.—Can you find the Shortest Way to Centre?

any maze, due to M. Trémaux, but it necessitates carefully marking in some way your entrances and exits where the galleries fork. I give a diagram of an imaginary maze of a very simple dotted line in the diagram is taken in accordance with these simple rules, and it will be seen that it leads us to the centre, although the maze consists of four islands.

FIG. 25.—Rosamund's Bower.

Neither of the methods I have given will disclose to us the shortest way to the centre, nor the number of the different routes. But we can easily settle these points with a plan. Let us take the Hatfield maze (Fig. 21). It will be seen that I have suppressed all the blind alleys by the shading. I begin at the stop and work backwards until the path forks. These shaded parts, therefore, can never be entered without our having to retrace our steps. Then it is very clearly seen that if we enter at A we must come out at B ; if we enter at C we must come out at D. Then we have merely to determine whether A, B, E, or C, D, E, is the shorter route. As a matter of fact, it will be found by rough measurement or calculation that the shortest route to the centre is by way of C, D, E, F.

I will now give three mazes that are simply puzzles on paper, for, so far as I know, they have never been constructed in any other way. The first I will call the Philadelphia maze (Fig. 22). Fourteen years ago a travelling salesman, living in Philadelphia, U.S.A., developed a curiously unrestrained passion for puzzles. He neglected his business, and soon his position was taken from him. His days and nights were now passed with the subject that fascinated him, and this little maze seems to have driven him into insanity. He had been puzzling over it for some time, and finally it sent him mad and caused him to fire a bullet through his brain. Goodness knows what his difficulties could have been! But there can be little doubt that he had a disordered mind, and that if this little puzzle had not caused him to lose his mental balance some other more or less trivial thing would in time have done so. There is no moral in the story, unless it be that of the Irish maxim, which applies to every occupation of life as much as to the solving of puzzles : " Take things aïsy ; if you can't take them aisy, take them as aisy as you can." And it is a bad and empirical way of solving any puzzle—by blowing your brains out.

Now, how many different routes are there from A to B in this maze if we must never in any route go along the same passage twice ? The four open spaces where four passages end are not reckoned as "passages." In the diagram (Fig. 22) it will be seen that I have again suppressed the blind alleys. It will be found that, in any case, we must go from A to C, and also from F to B. But when we

have arrived at C there are three ways, marked 1, 2, 3, of getting to D. Similarly, when we get to E there are three ways, marked 4, 5, 6, of getting to F. We have also the dotted route from C to E, the other dotted route from D to F, and the passage from D to E, indicated by stars. We can, therefore, express the position of affairs by the little diagram annexed (Fig. 23). Here every condition of route exactly corresponds to that in the circular maze, only it is much less confusing to the eye. Now, the number of routes, under the conditions, from A to B on this simplified diagram is 640, and that is the required answer to the maze puzzle.

Finally, I will leave two easy maze puzzles (Figs. 24, 25) for my readers to solve for themselves. The puzzle in each case is to find the shortest possible route to the centre. Everybody knows the story of Fair Rosamund and the Woodstock maze. What the maze was like or whether it ever existed except in imagination is not known, many writers believing that it was simply a badly-constructed house with a large number of confusing rooms and passages. At any rate, my sketch lacks the authority of the other mazes in this article. My " Rosamund's Bower " is simply designed to show that where you have the plan before you it often happens that the easiest way to find a route into a maze is by working backwards and first finding a way out.

THE PARADOX PARTY.

" Is not life itself a paradox ? "
C. L. Dodgson, *Pillow Problems.*

" It is a wonderful age ! " said Mr. Allgood, and everybody at the table turned towards him and assumed an attitude of expectancy.

This was an ordinary Christmas dinner of the Allgood family, with a sprinkling of local friends. Nobody would have supposed that the above remark would lead, as it did, to a succession of curious puzzles and paradoxes, to which every member of the party contributed something of interest. The little symposium was quite unpremeditated, so we must not be too critical respecting a few of the posers that were forthcoming. The varied character of the contributions is just what we would expect on such an occasion, for it was a gathering not of expert mathematicians and logicians, but of quite ordinary folk.

" It is a wonderful age ! " repeated Mr. Allgood. " A man has just designed a square house in such a cunning manner that all the windows on the four sides have a south aspect."

" That would appeal to me," said Mrs. Allgood, " for I cannot endure a room with a north aspect."

" I cannot conceive how it is done," Uncle John confessed. " I suppose he puts bay windows on the east and west sides ; but how on earth can he contrive to look south from the

north side ? Does he use mirrors, or something of that kind ? "

" No," replied Mr. Allgood, " nothing of the sort. All the windows are flush with the walls, and yet you get a southerly prospect from every one of them. You see, there is no real difficulty in designing the house if you select the proper spot for its erection. Now, this house is designed for a gentleman who proposes to build it exactly at the North Pole. If you think a moment you will realize that when you stand at the North Pole it is impossible, no matter which way you may turn, to look elsewhere than due south ! There are no two directions as north, east, or west when you are exactly at the North Pole. Everything is due south ! "

" I am afraid, mother," said her son George, after the laughter had subsided, " that, however much you might like the aspect, the situation would be a little too bracing for you."

" Ah, well ! " she replied. " Your Uncle John fell also into the trap. I am no good at catches and puzzles. I suppose I haven't the right sort of brain. Perhaps some one will explain this to me. Only last week I remarked to my hairdresser that it had been said that there are more persons in the world than any one of them has hairs on his head. He replied, ' Then it follows, madam, that two persons, at least, must have exactly the same number of hairs on their heads.' If this is a fact, I confess I cannot see it."

" How do the bald-headed affect the question ? " asked Uncle John.

" If there are such persons in existence," replied Mrs. Allgood, " who haven't a solitary hair on their heads discoverable under a magnifying-glass, we will leave them out of the question. Still, I don't see how you are to prove that at least two persons have exactly the same number to a hair."

" I think I can make it clear," said Mr. Filkins, who had dropped in for the evening. " Assume the population of the world to be only one million. Any number will do as well as another. Then your statement was to the effect that no person has more than nine hundred and ninety-nine thousand nine hundred and ninety-nine hairs on his head. Is that so ? "

" Let me think," said Mrs. Allgood. " Yes —yes—that is correct."

" Very well, then. As there are only nine hundred and ninety-nine thousand nine hundred and ninety-nine *different* ways of bearing hair, it is clear that the millionth person must repeat one of those ways. Do you see ? "

" Yes ; I see that—at least I think I see it."

" Therefore two persons at least must have the same number of hairs on their heads ; and as the number of people on the earth so greatly exceeds the number of hairs on any one person's head, there must, of course, be an immense number of these repetitions."

" But, Mr. Filkins," said little Willie Allgood, " why could not the millionth man have, say, ten thousand hairs and a half ? "

" That is mere hair-splitting, Willie, and does not come into the question."

" Here is a curious paradox," said George. " If a thousand soldiers are drawn up in battle array on a plane "—they understood him to mean " plain "—" only one man will stand upright."

Nobody could see why. But George explained that, according to Euclid, a plane can touch a sphere only at one point, and that person only who stands at that point, with respect to the centre of the earth, will stand upright. " In the same way," he remarked, " if a billiard-table were quite level—that is, a perfect plane—the balls ought to roll to the centre."

Though he tried to explain this by placing a visiting-card on an orange and expounding the law of gravitation, Mrs. Allgood declined to accept the statement. She could not see that the top of a true billiard-table must, theoretically, be spherical, just like a portion of the orange-peel that George cut out. Of course, the table is so small in proportion to the surface of the earth that the curvature is not appreciable, but it is nevertheless true in theory. A surface that we call level is not the same as our idea of a true geometrical plane.

" Uncle John," broke in Willie Allgood, " there is a certain island situated between England and France, and yet that island is farther from France than England is. What is the island ? "

" That seems absurd, my boy ; because if I place this tumbler, to represent the island, between these two plates, it seems impossible that the tumbler can be farther from either of the plates than they are from each other."

" But isn't Guernsey between England and France ? " asked Willie.

" Yes, certainly."

" Well, then, I think you will find, uncle, that Guernsey is about twenty-six miles from France, and England is only twenty-one miles from France, between Calais and Dover."

" My mathematical master," said George, " has been trying to induce me to accept the axiom that ' if equals be multiplied by equals the products are equal.' "

" It is self-evident," pointed out Mr. Filkins. " For example, if 3 feet equal 1 yard, then twice 3 feet will equal 2 yards. Do you see ? "

" But, Mr. Filkins," asked George, " is this tumbler half full of water equal to a similar glass half empty ? "

" Certainly, George."

" Then it follows from the axiom that a glass full must equal a glass empty. Is that correct ? "

" No, clearly not. I never thought of it in that light."

" Perhaps," suggested Mr. Allgood, " the rule does not apply to liquids."

" Just what I was thinking, Allgood. It would seem that we must make an exception in the case of liquids."

" But it would be awkward," said George, with a smile, " if we also had to except the case of solids. For instance, let us take the solid earth. One mile square equals one square mile. Therefore two miles square must equal two square miles. Is this so ? "

" Well, let me see ! No, of course not," Mr. Filkins replied, " because two miles square is four square miles."

" Then," said George, " if the axiom is not true in these cases, when is it true ? "

Mr. Filkins promised to look into the matter, and perhaps the reader will also like to give it consideration at leisure.

" Look here, George," said his cousin Reginald Woolley : " by what fractional part does four-fourths exceed three-fourths ? "

" By one-fourth ! " shouted everybody at once.

" Try another one," George suggested.

" With pleasure, when you have answered that one correctly," was Reginald's reply.

" Do you mean to say that it isn't one-fourth ? "

" Certainly I do."

Several members of the company failed to see that the correct answer is " one-third," although Reginald tried to explain that three of anything, if increased by one-third, becomes four.

" Uncle John, how do you pronounce ' t-o-o ' ? " asked Willie.

" ' Too,' my boy."

" And how do you pronounce ' t-w-o ' ? "

" That is also ' too.' "

" Then how do you pronounce the second day of the week ? "

" Well, that I should pronounce ' Tuesday,' not ' Toosday.' "

" Would you really ?   I should pronounce it ' Monday.' "

" If you go on like this, Willie," said Uncle John, with mock severity, " you will soon be without a friend in the world."

" Can any of you write down quickly in figures ' twelve thousand twelve hundred and twelve pounds ' ? " asked Mr. Allgood.

His eldest daughter, Miss Mildred, was the only person who happened to have a pencil at hand.

" It can't be done," she declared, after making an attempt on the white table-cloth ; but Mr. Allgood showed her that it should be written, " £13,212."

" Now it is my turn," said Mildred.   " I have been waiting to ask you all a question. In the Massacre of the Innocents under Herod, a number of poor little children were buried in the sand with only their feet sticking out.   How might you distinguish the boys from the girls ? "

" I suppose," said Mrs. Allgood, " it is a conundrum—something to do with their poor little ' souls.' "

But after everybody had given it up, Mildred reminded the company that only boys were put to death.

" Once upon a time," began George, " Achilles had a race with a tortoise——"

" Stop, George ! " interposed Mr. Allgood. " We won't have that one.   I knew two men in my youth who were once the best of friends, but they quarrelled over that infernal thing of Zeno's, and they never spoke to one another again for the rest of their lives.   I draw the line at that, and the other stupid thing by Zeno about the flying arrow.   I don't believe anybody understands them, because I could never do so myself."

" Oh, very well, then, father.   Here is another.   The Post-Office people were about to erect a line of telegraph-posts over a high hill from Turmitville to Wurzleton ; but as it was found that a railway company was making a deep level cutting in the same direction, they arranged to put up the posts beside the line. Now, the posts were to be a hundred yards apart, the length of the road over the hill being five miles, and the length of the level cutting only four and a half miles.   How many posts did they save by erecting them on the level ? "

" That is a very simple matter of calculation," said Mr. Filkins.   " Find how many times one hundred yards will go in five miles, and how many times in four and a half miles.   Then deduct one from the other, and you have the number of posts saved by the shorter route."

" Quite right," confirmed Mr. Allgood. " Nothing could be easier."

" That is just what the Post-Office people said," replied George, " but it is quite wrong. If you look at this sketch that I have just made, you will see that there is no difference whatever. If the posts are a hundred yards apart, just the

same number will be required on the level as over the surface of the hill."

" Surely you must be wrong, George," said Mrs. Allgood, " for if the posts are a hundred yards apart and it is half a mile farther over the hill, you have to put up posts on that extra half-mile."

" Look at the diagram, mother.   You will see that the distance from post to post is not the distance from base to base measured along the ground.   I am just the same distance from you if I stand on this spot on the carpet or stand immediately above it on the chair."

But Mrs. Allgood was not convinced.

Mr. Smoothly, the curate, at the end of the table, said at this point that he had a little question to ask.

" Suppose the earth were a perfect sphere with a smooth surface, and a girdle of steel were placed round the Equator so that it touched at every point."

" ' I'll put a girdle round about the earth in forty minutes,' " muttered George, quoting the words of Puck in A Midsummer Night's Dream.

" Now, if six yards were added to the length of the girdle, what would then be the distance between the girdle and the earth, supposing that distance to be equal all round ? "

" In such a great length," said Mr. Allgood, " I do not suppose the distance would be worth mentioning."

" What do you say, George ? " asked Mr. Smoothly.

" Well, without calculating I should imagine it would be a very minute fraction of an inch."

Reginald and Mr. Filkins were of the same opinion.

" I think it will surprise you all," said the curate, " to learn that those extra six yards would make the distance from the earth all round the girdle very nearly a yard ! "

" Very nearly a yard ! " everybody exclaimed, with astonishment ; but Mr. Smoothly was quite correct.   The increase is independent of the original length of the girdle, which may be round the earth or round an orange ; in any case the additional six yards will give a distance of nearly a yard all round.   This is apt to surprise the non-mathematical mind.

" Did you hear the story of the extraordinary precocity of Mrs. Perkins's baby that died last week ? " asked Mrs. Allgood.   " It was only three months old, and lying at the point of death, when the grief-stricken mother asked the doctor if nothing could save it.   ' Absolutely nothing ! ' said the doctor.   Then the infant looked up pitifully into its mother's face and said —absolutely nothing ! "

" Impossible ! " insisted Mildred.   " And only three months old ! "

"There have been extraordinary cases of infantile precocity," said Mr. Filkins, "the truth of which has often been carefully attested. But are you sure this really happened, Mrs. Allgood?"

"Positive," replied the lady. "But do you really think it astonishing that a child of three months should say absolutely nothing? What would you expect it to say?"

"Speaking of death," said Mr. Smoothly, solemnly, "I knew two men, father and son, who died in the same battle during the South African War. They were both named Andrew Johnson and buried side by side, but there was some difficulty in distinguishing them on the headstones. What would you have done?"

"Quite simple," said Mr. Allgood. "They should have described one as 'Andrew Johnson, Senior,' and the other as 'Andrew Johnson, Junior.'"

"But I forgot to tell you that the father died first."

"What difference can that make?"

"Well, you see, they wanted to be absolutely exact, and that was the difficulty."

"But I don't see any difficulty," said Mr. Allgood, nor could anybody else.

"Well," explained Mr. Smoothly, "it is like this. If the father died first, the son was then no longer 'Junior.' Is that so?"

"To be strictly exact, yes."

"That is just what they wanted—to be strictly exact. Now, if he was no longer 'Junior,' then he did not die 'Junior.' Consequently it must be incorrect so to describe him on the headstone. Do you see the point?"

"Here is a rather curious thing," said Mr. Filkins, "that I have just remembered. A man wrote to me the other day that he had recently discovered two old coins while digging in his garden. One was dated '51 B.C.,' and the other one marked 'George I.' How do I know that he was not writing the truth?"

"Perhaps you know the man to be addicted to lying," said Reginald.

"But that would be no proof that he was not telling the truth in this instance."

"Perhaps," suggested Mildred, "you know that there were no coins made at those dates."

"On the contrary, they were made at both periods."

"Were they silver or copper coins?" asked Willie.

"My friend did not state, and I really cannot see, Willie, that it makes any difference."

"I see it!" shouted Reginald. "The letters 'B.C.' would never be used on a coin made before the birth of Christ. They never anticipated the event in that way. The letters were only adopted later to denote dates previous to those which we call 'A.D.' That is very good; but I cannot see why the other statement could not be correct."

"Reginald is quite right," said Mr. Filkins, "about the first coin. The second one could not exist, because the first George would never be described in his lifetime as 'George I.'"

"Why not?" asked Mrs. Allgood. "He was George I."

"Yes; but they would not know it until there was a George II."

"Then there was no George II. until George III. came to the throne?"

"That does not follow. The second George becomes 'George II.' on account of there having been a 'George I.'"

"Then the first George was 'George I.' on account of there having been no king of that name before him."

"Don't you see, mother," said George Allgood, "we did not call Queen Victoria 'Victoria I.;' but if there is ever a 'Victoria II.,' then she will be known that way."

"But there have been several Georges, and therefore he was 'George I.' There haven't been several Victorias, so the two cases are not similar."

They gave up the attempt to convince Mrs. Allgood, but the reader will, of course, see the point clearly.

"Here is a question," said Mildred Allgood, "that I should like some of you to settle for me. I am accustomed to buy from our greengrocer bundles of asparagus, each 12 inches in circumference. I always put a tape measure round them to make sure I am getting the full quantity. The other day the man had no large bundles in stock, but handed me instead two small ones, each 6 inches in circumference. 'That is the same thing,' I said, 'and, of course, the price will be the same;' but he insisted that the two bundles together contained more than the large one, and charged me a few pence extra. Now, what I want to know is, which of us was correct? Would the two small bundles contain the same quantity as the large one? Or would they contain more?"

"That is the ancient puzzle," said Reginald, laughing, "of the sack of corn that Sempronius borrowed from Caius, which your greengrocer, perhaps, had been reading about somewhere. He caught you beautifully."

"Then they were equal?"

"On the contrary, you were both wrong, and you were badly cheated. You only got half the quantity that would have been contained in a large bundle, and therefore ought to have been charged half the original price, instead of more."

Yes, it was a bad swindle, undoubtedly. A circle with a circumference half that of another must have its area a quarter that of the other. Therefore the two small bundles contained together only half as much asparagus as a large one.

"Mr. Filkins, can you answer this?" asked Willie. "There is a man in the next village who eats two eggs for breakfast every morning."

"Nothing very extraordinary in that," George broke in. "If you told us that the two eggs ate the man it would be interesting."

"Don't interrupt the boy, George," said his mother.

"Well," Willie continued, "this man neither buys, borrows, barters, begs, steals, nor finds the eggs. He doesn't keep hens, and the eggs are not given to him. How does he get the eggs ? "

" Does he take them in exchange for something else ? " asked Mildred.

" That would be bartering them," Willie replied.

" Perhaps some friend sends them to him," suggested Mrs. Allgood.

" I said that they were not given to him."

" I know," said George, with confidence. " A strange hen comes into his place and lays them."

" But that would be finding them, wouldn't it ? "

" Does he hire them ? " asked Reginald.

" If so, he could not return them after they were eaten, so that would be stealing them."

All agreed that Willie's answer was quite satisfactory. Then Uncle John produced a little fallacy that " brought the proceedings to a close," as the newspapers say.

413.—A CHESSBOARD FALLACY.

" HERE is a diagram of a chessboard," he said. " You see there are sixty-four squares—eight by eight. Now I draw a straight line from the top left-hand corner, where the first and second squares meet, to the bottom right-hand corner. I cut along this line with the scissors, slide up the piece that I have marked B, and then clip off the little corner C by a cut along the first upright line. This little piece will exactly fit into its place at the top, and we now have

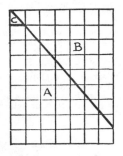

" Perhaps it is a pun on the word ' lay,' " Mr. Filkins said. " Does he lay them on the table ? "

" He would have to get them first, wouldn't he ? The question was, How does he get them ? "

" Give it up ! " said everybody. Then little Willie crept round to the protection of his mother, for George was apt to be rough on such occasions.

" The man keeps ducks ! " he cried, " and his servant collects the eggs every morning."

" But you said he doesn't keep birds ! " George protested.

" I didn't, did I, Mr. Filkins ? I said he doesn't keep hens."

" But he finds them," said Reginald.

" No ; I said his servant finds them."

" Well, then," Mildred interposed, " his servant gives them to him."

" You cannot give a man his own property, can you ? "

an oblong with seven squares on one side and nine squares on the other. There are, therefore, now only sixty-three squares, because seven multiplied by nine makes sixty-three. Where on earth does that lost square go to ? I have tried over and over again to catch the little beggar, but he always eludes me. For the life of me I cannot discover where he hides himself."

" It seems to be like the other old chessboard fallacy, and perhaps the explanation is the same," said Reginald—" that the pieces do not exactly fit."

" But they do fit," said Uncle John. " Try it, and you will see."

Later in the evening Reginald and George were seen in a corner with their heads together, trying to catch that elusive little square, and it is only fair to record that before they retired for the night they succeeded in securing their prey, though some others of the company failed to see it when captured. Can the reader solve the little mystery ?

UNCLASSIFIED PROBLEMS.

"A snapper up of unconsidered trifles."
Winter's Tale, iv. 2.

414.—WHO WAS FIRST?

ANDERSON, Biggs, and Carpenter were staying together at a place by the seaside. One day they went out in a boat and were a mile at sea when a rifle was fired on shore in their direction. Why or by whom the shot was fired fortunately does not concern us, as no information on these points is obtainable, but from the facts I picked up we can get material for a curious little puzzle for the novice.

It seems that Anderson only heard the report of the gun, Biggs only saw the smoke, and Carpenter merely saw the bullet strike the water near them. Now, the question arises : Which of them first knew of the discharge of the rifle?

415.—A WONDERFUL VILLAGE.

THERE is a certain village in Japan, situated in a very low valley, and yet the sun is nearer to the inhabitants every noon, by 3,000 miles and upwards, than when he either rises or sets to these people. In what part of the country is the village situated ?

416.—A CALENDAR PUZZLE.

IF the end of the world should come on the first day of a new century, can you say what are the chances that it will happen on a Sunday ?

417.—THE TIRING IRONS.

The puzzle will be seen to consist of a simple *loop* of wire fixed in a handle to be held in the left hand, and a certain number of *rings* secured by *wires* which pass through holes in the *bar* and are kept there by their blunted ends. The wires work freely in the bar, but cannot come apart from it, nor can the wires be removed from the rings. The general puzzle is to detach the loop completely from all the rings, and then to put them all on again.

Now, it will be seen at a glance that the first ring (to the right) can be taken off at any time by sliding it over the end and dropping it through the loop ; or it may be put on by reversing the operation. With this exception, the only ring that can ever be removed is the one that happens to be a contiguous second on the loop at the right-hand end. Thus, with all the rings on, the second can be dropped at once ; with the first ring down, you cannot drop the second, but may remove the third ; with the first three rings down, you cannot drop the fourth, but may remove the fifth ; and so on. It will be found that the first and second rings can be dropped together or put on together ; but to prevent confusion we will throughout disallow this exceptional double move, and say that only one ring may be put on or removed at a time.

We can thus take off one ring in 1 move ; two rings in 2 moves ; three rings in 5 moves ; four rings in 10 moves ; five rings in 21 moves ; and if we keep on doubling (and adding one where the number of rings is odd) we may easily ascertain the number of moves for completely removing any number of rings. To get off all

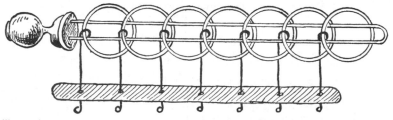

THE illustration represents one of the most ancient of all mechanical puzzles. Its origin is unknown. Cardan, the mathematician, wrote about it in 1550, and Wallis in 1693 ; while it is said still to be found in obscure English villages (sometimes deposited in strange places, such as a church belfry), made of iron, and appropriately called "tiring-irons," and to be used by the Norwegians to-day as a lock for boxes and bags. In the toyshops it is sometimes called the "Chinese rings," though there seems to be no authority for the description, and it more frequently goes by the unsatisfactory name of "the puzzling rings." The French call it "Baguenaudier."

the seven rings requires 85 moves. Let us look at the five moves made in removing the first three rings, the circles above the line standing for rings on the loop and those under for rings off the loop.

Drop the first ring ; drop the third ; put up the first ; drop the second ; and drop the first— 5 moves, as shown clearly in the diagrams. The dark circles show at each stage, from the starting position to the finish, which rings it is possible to drop. After move 2 it will be noticed that no ring can be dropped until one has been put on, because the first and second rings from the right now on the loop are not together. After the fifth move, if we wish to remove all seven rings we must now drop the fifth. But before we can then remove the fourth it is necessary to put on the first three and remove the first

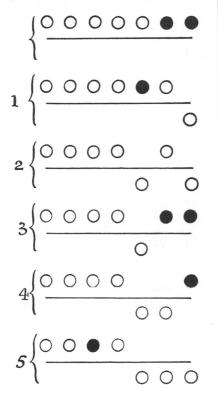

two. We shall then have 7, 6, 4, 3 on the loop, and may therefore drop the fourth. When we have put on 2 and 1 and removed 3, 2, 1, we may drop the seventh ring. The next operation then will be to get 6, 5, 4, 3, 2, 1 on the loop and remove 4, 3, 2, 1, when 6 will come off; then get 5, 4, 3, 2, 1 on the loop, and remove 3, 2, 1, when 5 will come off; then get 4, 3, 2, 1 on the loop and remove 2, 1, when 4 will come off; then get 3, 2, 1 on the loop and remove 1, when 3 will come off; then get 2, 1 on the loop, when 2 will come off; and 1 will fall through on the 85th move, leaving the loop quite free. The reader should now be able to understand the puzzle, whether or not he has it in his hand in a practical form.

The particular problem I propose is simply this. Suppose there are altogether fourteen rings on the tiring-irons, and we proceed to take them all off in the correct way so as not to waste any moves. What will be the position of the rings after the 9,999th move has been made?

418.—SUCH A GETTING UPSTAIRS.

In a suburban villa there is a small staircase with eight steps, not counting the landing. The little puzzle with which Tommy Smart perplexed his family is this. You are required to start from the bottom and land twice on the floor above (stopping there at the finish), having returned once to the ground floor. But you must be careful to use every tread the same number of times. In how few steps can you make the ascent? It seems a very simple matter, but it is more than likely that at your first attempt you will make a great many more steps than are necessary. Of course you must not go more than one riser at a time.

Tommy knows the trick, and has shown it to his father, who professes to have a contempt for such things; but when the children are in bed the pater will often take friends out into the hall and enjoy a good laugh at their bewilderment. And yet it is all so very simple when you know how it is done.

419.—THE FIVE PENNIES.

Here is a really hard puzzle, and yet its conditions are so absurdly simple. Every reader knows how to place four pennies so that they are equidistant from each other. All you have to do is to arrange three of them flat on the table so that they touch one another in the form of a triangle, and lay the fourth penny on top in the centre. Then, as every penny touches every other penny, they are all at equal distances from one another. Now try to do the same thing with five pennies—place them so that every penny shall touch every other penny—and you will find it a different matter altogether.

420.—THE INDUSTRIOUS BOOKWORM.

Our friend Professor Rackbrane is seen in the illustration to be propounding another of his little posers. He is explaining that since he last had occasion to take down those three volumes of a learned book from their place on his shelves a bookworm has actually bored a hole straight through from the first page to the last. He says that the leaves are together three inches thick in each volume, and that every cover is exactly one-eighth of an inch thick, and he asks how long a tunnel had the industrious worm to bore in preparing his new tube railway. Can you tell him?

421.—A CHAIN PUZZLE.

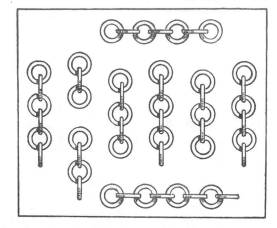

THIS is a puzzle based on a pretty little idea first dealt with by the late Mr. Sam Loyd. A man had nine pieces of chain, as shown in the illustration. He wanted to join these fifty links into one endless chain. It will cost a penny to open any link and twopence to weld a link together again, but he could buy a new endless chain of the same character and quality for 2s. 2d. What was the cheapest course for him to adopt? Unless the reader is cunning he may find himself a good way out in his answer.

422.—THE SABBATH PUZZLE.

I HAVE come across the following little poser in an old book. I wonder how many readers will see the author's intended solution to the riddle.

Christians the week's *first* day for Sabbath hold;
The Jews the *seventh*, as they did of old;
The Turks the *sixth*, as we have oft been told.
How can these three, in the same place and day,
Have each his own true Sabbath? tell, I pray.

423.—THE RUBY BROOCH.

THE annals of Scotland Yard contain some remarkable cases of jewel robberies, but one of the most perplexing was the theft of Lady Little-

wood's rubies. There have, of course, been many greater robberies in point of value, but few so artfully conceived. Lady Littlewood, of Romley Manor, had a beautiful but rather eccentric heirloom in the form of a ruby brooch. While staying at her town house early in the eighties she took the jewel to a shop in Brompton for some slight repairs.

"A fine collection of rubies, madam," said the shopkeeper, to whom her ladyship was a stranger.

"Yes," she replied; "but curiously enough I have never actually counted them. My mother once pointed out to me that if you start from the centre and count up one line, along the outside and down the next line, there are

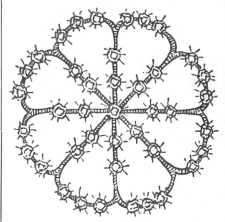

always eight rubies. So I should always know if a stone were missing."

Six months later a brother of Lady Littlewood's, who had returned from his regiment in India, noticed that his sister was wearing the ruby brooch one night at a county ball, and on their return home asked to look at it more closely. He immediately detected the fact that four of the stones were gone.

"How can that possibly be?" said Lady Littlewood. "If you count up one line from the centre, along the edge, and down the next line, in any direction, there are always eight stones. This was always so and is so now. How, therefore, would it be possible to remove a stone without my detecting it?"

"Nothing could be simpler," replied the brother. "I know the brooch well. It originally contained forty-five stones, and there are now only forty-one. Somebody has stolen four rubies, and then reset as small a number of the others as possible in such a way that there shall always be eight in any of the directions you have mentioned."

There was not the slightest doubt that the Brompton jeweller was the thief, and the matter was placed in the hands of the police. But the man was wanted for other robberies. and had left the neighbourhood some time before. To this day he has never been found.

The interesting little point that at first baffled the police, and which forms the subject of our puzzle, is this: How were the forty-five rubies originally arranged on the brooch? The illustration shows exactly how the forty-one were arranged after it came back from the jeweller; but although they count eight correctly in any of the directions mentioned, there are four stones missing.

424.—THE DOVETAILED BLOCK.

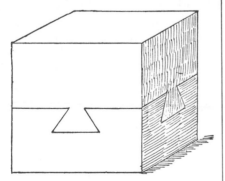

HERE is a curious mechanical puzzle that was given to me some years ago, but I cannot say who first invented it. It consists of two solid blocks of wood securely dovetailed together. On the other two vertical sides that are not visible the appearance is precisely the same as on those shown. How were the pieces put together? When I published this little puzzle in a London newspaper I received (though they were unsolicited) quite a stack of models,

in oak, in teak, in mahogany, rosewood, satinwood, elm, and deal; some half a foot in length, and others varying in size right down to a delicate little model about half an inch square. It seemed to create considerable interest.

425.—JACK AND THE BEANSTALK.

THE illustration, by a British artist, is a sketch of Jack climbing the beanstalk. Now, the artist has made a serious blunder in this drawing. Can you find out what it is?

426.—THE HYMN-BOARD POSER.

THE worthy vicar of Chumpley St. Winifred is in great distress. A little church difficulty has arisen that all the combined intelligence of the parish seems unable to surmount. What this difficulty is I will state hereafter, but it may add to the interest of the problem if I first give a short account of the curious position that has been brought about. It all has to do with the church hymn-boards, the plates of which have become so damaged that they have ceased to fulfil the purpose for which they were devised. A generous parishioner has promised to pay for a new set of plates at a certain rate of cost; but strange as it may seem, no agreement can be come to as to what that cost should be.

The proposed maker of the plates has named

a price which the donor declares to be absurd. The good vicar thinks they are both wrong, so he asks the schoolmaster to work out the little sum. But this individual declares that he can find no rule bearing on the subject in any of his

arithmetic books. An application having been made to the local medical practitioner, as a man of more than average intellect at Chumpley, he has assured the vicar that his practice is so heavy that he has not had time even to look at it, though his assistant whispers that the doctor has been sitting up unusually late for several nights past. Widow Wilson has a smart son, who is reputed to have once won a prize for puzzle-solving. He asserts that as he cannot find any solution to the problem it must have something to do with the squaring of the circle, the duplication of the cube, or the trisection of an angle ; at any rate, he has never before seen a puzzle on the principle, and he gives it up.

This was the state of affairs when the assistant curate (who, I should say, had frankly confessed from the first that a profound study of theology had knocked out of his head all the knowledge of mathematics he ever possessed) kindly sent me the puzzle.

A church has three hymn-boards, each to indicate the numbers of five different hymns to be sung at a service. All the boards are in use at the same service. The hymn-book contains 700 hymns. A new set of numbers is required, and a kind parishioner offers to present a set

painted on metal plates, but stipulates that only the smallest number of plates necessary shall be purchased. The cost of each plate is to be 6d., and for the painting of each plate the charges are to be : For one plate, 1s. ; for two plates alike, 11¾d. each ; for three plates alike, 11½d. each, and so on, the charge being one farthing less per plate for each similarly painted plate. Now, what should be the lowest cost ?

Readers will note that they are required to use every legitimate and practical method of economy. The illustration will make clear the nature of the three hymn-boards and plates. The five hymns are here indicated by means of twelve plates. These plates slide in separately at the back, and in the illustration there is room, of course, for three more plates.

427.—PHEASANT-SHOOTING.

A COCKNEY friend, who is very apt to draw the long bow, and is evidently less of a sportsman than he pretends to be, relates to me the following not very credible yarn :—

"I've just been pheasant-shooting with my friend the duke. We had splendid sport, and I made some wonderful shots. What do you think of this, for instance ? Perhaps you can twist it into a puzzle. The duke and I were crossing a field when suddenly twenty-four pheasants rose on the wing right in front of us. I fired, and two-thirds of them dropped dead at my feet. Then the duke had a shot at what were left, and brought down three-twenty-fourths of them, wounded in the wing. Now, out of those twenty-four birds, how many still remained ? "

It seems a simple enough question, but can the reader give a correct answer ?

428.—THE GARDENER AND THE COOK.

A CORRESPONDENT, signing himself " Simple Simon," suggested that I should give a special catch puzzle in the issue of *The Weekly Dispatch* for All Fools' Day, 1900. So I gave the following, and it caused considerable amusement ; for out of a very large body of competitors, many quite expert, not a single person solved it, though it ran for nearly a month.

" The illustration is a fancy sketch of my correspondent, 'Simple Simon,' in the act of trying to solve the following innocent little arithmetical puzzle. A race between a man and a woman that I happened to witness one All Fools' Day has fixed itself indelibly on my memory. It happened at a country-house, where the gardener and the cook decided to run a race to a point 100 feet straight away and return. I found that the gardener ran 3 feet at every bound and the cook only 2 feet, but then she made three bounds to his two. Now, what was the result of the race?"

A fortnight after publication I added the your second coin at exactly the distance of an inch from the first, the third an inch distance from the second, and so on. No halfpenny may touch another halfpenny or cross the boundary. Our illustration will make the matter perfectly clear. No. 2 coin is an inch from No. 1; No. 3 an inch from No. 2; No. 4 an inch from No. 3; but after No. 10 is placed we can go no further in this attempt. Yet several more halfpennies might have been got in. How many can the reader place?

following note: "It has been suggested that perhaps there is a catch in the 'return,' but there is not. The race is to a point 100 feet away and home again—that is, a distance of 200 feet. One correspondent asks whether they take exactly the same time in turning, to which I reply that they do. Another seems to suspect that it is really a conundrum, and that the answer is that 'the result of the race was a (matrimonial) tie.' But I had no such intention. The puzzle is an arithmetical one, as it purports to be."

429.—PLACING HALFPENNIES.

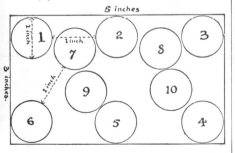

HERE is an interesting little puzzle suggested to me by Mr. W. T. Whyte. Mark off on a sheet of paper a rectangular space 5 inches by 3 inches, and then find the greatest number of halfpennies that can be placed within the enclosure under the following conditions. A halfpenny is exactly an inch in diameter. Place your first halfpenny where you like, then place

430.—FIND THE MAN'S WIFE.

ONE summer day in 1903 I was loitering on the Brighton front, watching the people strolling about on the beach, when the friend who was with me suddenly drew my attention to an individual who was standing alone, and said, "Can you point out that man's wife? They are stopping at the same hotel as I am, and the lady is one of those in view." After a few minutes' observation, I was successful in indicating the lady correctly. My friend was curious to know by what method of reasoning I had arrived at the result. This was my answer:—

"We may at once exclude that Sister of Mercy and the girl in the short frock; also the woman selling oranges. It cannot be the lady in widows' weeds. It is not the lady in the bath chair, because she is not staying at your hotel, for I happened to see her come out of a private house this morning assisted by her maid. The two ladies in red breakfasted at my hotel this morning, and as they were not wearing outdoor dress I conclude they are staying there. It therefore rests between the lady in blue and the one with the green parasol. But the left hand that holds the parasol is, you see, ungloved and bears no wedding-ring. Consequently I am driven to the conclusion that the lady in blue is the man's wife—and you say this is correct."

Now, as my friend was an artist, and as I thought an amusing puzzle might be devised on the lines of his question, I asked him to make me a drawing according to some directions that I gave him, and I have pleasure in presenting his production to my readers. It will be seen that the picture shows six men and six ladies: Nos. 1, 3, 5, 7, 9, and 11 are ladies, and Nos. 2,

4, 6, 8, 10, and 12 are men. These twelve individuals represent six married couples, all strangers to one another, who, in walking aimlessly about, have got mixed up. But we are only concerned with the man that is wearing a straw hat—Number 10. The puzzle is to find this man's wife. Examine the six ladies carefully, and see if you can determine which one of them it is.

I showed the picture at the time to a few friends, and they expressed very different opinions on the matter. One said, " I don't believe he would marry a girl like Number 7." Another said, " I am sure a nice girl like Number 3 would not marry such a fellow ! " Another said, " It must be Number 1, because she has

got as far away as possible from the brute ! " It was suggested, again, that it must be Number 11, because " he seems to be looking towards her ; " but a cynic retorted, " For that very reason, if he is really looking at her, I should say that she is not his wife ! "

I now leave the question in the hands of my readers. Which is really Number 10's wife ?

The illustration is of necessity considerably reduced from the large scale on which it originally appeared in *The Weekly Dispatch* (24th May 1903), but it is hoped that the details will be sufficiently clear to allow the reader to derive entertainment from its examination. In any case the solution given will enable him to follow the points with interest.

SOLUTIONS.

1.—A POST-OFFICE PERPLEXITY.

THE young lady supplied 5 twopenny stamps, 30 penny stamps, and 8 twopence-halfpenny stamps, which delivery exactly fulfils the conditions and represents a cost of five shillings.

2.—YOUTHFUL PRECOCITY.

THE price of the banana must have been one penny farthing. Thus, 960 bananas would cost £5, and 480 sixpences would buy 2,304 bananas.

3.—AT A CATTLE MARKET.

JAKES must have taken 7 animals to market, Hodge must have taken 11, and Durrant must have taken 21. There were thus 39 animals altogether.

4.—THE BEANFEAST PUZZLE.

THE cobblers spent 35s., the tailors spent also 35s., the hatters spent 42s., and the glovers spent 21s. Thus, they spent altogether £6, 13s., while it will be found that the five cobblers spent as much as four tailors, twelve tailors as much as nine hatters, and six hatters as much as eight glovers.

5.—A QUEER COINCIDENCE.

PUZZLES of this class are generally solved in the old books by the tedious process of " working backwards." But a simple general solution is as follows : If there are n players, the amount held by every player at the end will be $m(2^n)$, the last winner must have held $m(n+1)$ at the start, the next $m(2n+1)$, the next $m(4n+1)$, the next $m(8n+1)$, and so on to the first player, who must have held $m(2^{n-1}n+1)$.

Thus, in this case, $n=7$, and the amount held by every player at the end was 2^7 farthings. Therefore $m=1$, and G started with 8 farthings, F with 15, E with 29, D with 57, C with 113, B with 225, and A with 449 farthings.

6.—A CHARITABLE BEQUEST.

THERE are seven different ways in which the money may be distributed : 5 women and 19

men, 10 women and 16 men, 15 women and 13 men, 20 women and 10 men, 25 women and 7 men, 30 women and 4 men, and 35 women and 1 man. But the last case must not be counted, because the condition was that there should be " men," and a single man is not men. Therefore the answer is six years.

7.—THE WIDOW'S LEGACY.

THE widow's share of the legacy must be £205, 2s. 6d. and $\frac{10}{13}$ of a penny.

8.—INDISCRIMINATE CHARITY

THE gentleman must have had 3s. 6d. in his pocket when he set out for home.

9.—THE TWO AEROPLANES.

THE man must have paid £500 and £750 for the two machines, making together £1,250 ; but as he sold them for only £1,200, he lost £50 by the transaction.

10.—BUYING PRESENTS.

JORKINS had originally £19, 18s. in his pocket, and spent £9, 19s.

11.—THE CYCLISTS' FEAST.

THERE were ten cyclists at the feast. They should have paid 8s. each ; but, owing to the departure of two persons, the remaining eight would pay 10s. each.

12.—A QUEER THING IN MONEY.

THE answer is as follows : £44,444, 4s. 4d.= 28, and, reduced to pence, 10,666,612=28.

It is a curious little coincidence that in the answer 10,666,612 the four central figures indicate the only other answer, £66, 6s. 6d.

13.—A NEW MONEY PUZZLE.

THE smallest sum of money, in pounds, shillings, pence, and farthings, containing all the nine digits once, and once only, is £2,567, 18s. 9¾d.

14.—SQUARE MONEY.

THE answer is 1½d. and 3d. Added together they make 4½d., and 1½d. multiplied by 3 is also 4½d.

15.—POCKET MONEY.

THE largest possible sum is 15s. 9d., composed of a crown and a half-crown (or three half-crowns), four florins, and a threepenny piece.

16.—THE MILLIONAIRE'S PERPLEXITY.

THE answer to this quite easy puzzle may, of course, be readily obtained by trial, deducting the largest power of 7 that is contained in one million dollars, then the next largest power from the remainder, and so on. But the little problem is intended to illustrate a simple direct method. The answer is given at once by converting 1,000,000 to the septenary scale, and it is on this subject of scales of notation that I propose to write a few words for the benefit of those who have never sufficiently considered the matter.

Our manner of figuring is a sort of perfected arithmetical shorthand, a system devised to enable us to manipulate numbers as rapidly and correctly as possible by means of symbols. If we write the number 2,341 to represent two thousand three hundred and forty-one dollars, we wish to imply 1 dollar, added to four times 10 dollars, added to three times 100 dollars, added to two times 1,000 dollars. From the number in the units place on the right, every figure to the left is understood to represent a multiple of the particular power of 10 that its position indicates, while a cipher (o) must be inserted where necessary in order to prevent confusion, for if instead of 207 we wrote 27 it would be obviously misleading. We thus only require ten figures, because directly a number exceeds 9 we put a second figure to the left, directly it exceeds 99 we put a third figure to the left, and so on. It will be seen that this is a purely arbitrary method. It is working in the denary (or ten) scale of notation, a system undoubtedly derived from the fact that our forefathers who devised it had ten fingers upon which they were accustomed to count, like our children of to-day. It is unnecessary for us ordinarily to state that we are using the denary scale, because this is always understood in the common affairs of life.

But if a man said that he had 6,553 dollars in the septenary (or seven) scale of notation, you will find that this is precisely the same amount as 2,341 in our ordinary denary scale. Instead of using powers of ten, he uses powers of 7, so that he never needs any figure higher than 6, and 6,553 really stands for 3, added to five times 7, added to five times 49, added to six times 343 (in the ordinary notation), or 2,341. To reverse the operation, and convert 2,341 from the denary to the septenary scale, we divide it by 7, and get 334 and remainder 3; divide 334 by 7, and get 47 and remainder 5; and so keep on dividing by 7 as long as there is anything to divide. The remainders, read backwards, 6, 5, 5, 3, give us the answer, 6,553.

Now, as I have said, our puzzle may be solved at once by merely converting 1,000,000 dollars to the septenary scale. Keep on dividing this number by 7 until there is nothing more left to divide, and the remainders will be found to be 11333311, which is 1,000,000 expressed in the septenary scale. Therefore, 1 gift of 1 dollar, 1 gift of 7 dollars, 3 gifts of 49 dollars, 3 gifts of 343 dollars, 3 gifts of 2,401 dollars, 3 gifts of 16,807 dollars, 1 gift of 117,649 dollars, and one substantial gift of 823,543 dollars, satisfactorily solves our problem. And it is the only possible solution. It is thus seen that no " trials " are necessary ; by converting to the septenary scale of notation we go direct to the answer.

17.—THE PUZZLING MONEY BOXES.

THE correct answer to this puzzle is as follows : John put into his money-box two double florins (8s.), William a half-sovereign and a florin (12s.), Charles a crown (5s.), and Thomas a sovereign (20s.). There are six coins in all, of a total value of 45s. If John had 2s. more, William 2s. less, Charles twice as much, and Thomas half as much as they really possessed, they would have each have had exactly 10s.

18.—THE MARKET WOMEN.

THE price received was in every case 105 farthings. Therefore the greatest number of women is eight, as the goods could only be sold at the following rates : 105 lbs. at 1 farthing, 35 at 3, 21 at 5, 15 at 7, 7 at 15, 5 at 21, 3 at 35, and 1 lb. at 105 farthings.

19.—THE NEW YEAR'S EVE SUPPERS.

THE company present on the occasion must have consisted of seven pairs, ten single men, and one single lady. Thus, there were twenty-five persons in all, and at the prices stated they would pay exactly £5 together.

20.—BEEF AND SAUSAGES.

THE lady bought 48 lbs. of beef at 2s., and the same quantity of sausages at 1s. 6d., thus spending £8, 8s. Had she bought 42 lbs. of beef and 56 lbs. of sausages she would have spent £4, 4s. on each, and have obtained 98 lbs. instead of 96 lbs.—a gain in weight of 2 lbs.

21.—A DEAL IN APPLES.

I WAS first offered sixteen apples for my shilling, which would be at the rate of ninepence a dozen. The two extra apples gave me eighteen for a shilling, which is at the rate of eightpence a dozen, or one penny a dozen less than the first price asked.

22.—A DEAL IN EGGS.

THE man must have bought ten eggs at five-pence, ten eggs at one penny, and eighty eggs at a halfpenny. He would then have one hundred eggs at a cost of eight shillings and fourpence, and the same number of eggs of two of the qualities.

23.—THE CHRISTMAS-BOXES.

THE distribution took place " some years ago," when the fourpenny-piece was in circulation. Nineteen persons must each have received nineteen pence. There are five different ways in which this sum may have been paid in silver coins. We need only use two of these ways. Thus if fourteen men each received four fourpenny-pieces and one threepenny-piece, and five men each received five threepenny-pieces and one fourpenny-piece, each man would receive nineteen pence, and there would be exactly one hundred coins of a total value of £1, 10s. 1d.

24.—A SHOPPING PERPLEXITY.

THE first purchase amounted to 1s. 5¾d., the second to 1s. 11½d., and together they make 3s. 5¼d. Not one of these three amounts can be paid in fewer than six current coins of the realm.

25.—CHINESE MONEY.

As a ching-chang is worth twopence and four-fifteenths of a ching-chang, the remaining eleven-fifteenths of a ching-chang must be worth twopence. Therefore eleven ching-changs are worth exactly thirty pence, or half a crown. Now, the exchange must be made with seven round-holed coins and one square-holed coin. Thus it will be seen that 7 round-holed coins are worth seven-elevenths of 15 ching-changs, and 1 square-holed coin is worth one-eleventh of 16 ching-changs—that is, 77 rounds equal 105 ching-changs and 11 squares equal 16 ching-changs. Therefore 77 rounds added to 11 squares equal 121 ching-changs ; or 7 rounds and 1 square equal 11 ching-changs, or its equivalent, half a crown. This is more simple in practice than it looks here.

26.—THE JUNIOR CLERKS' PUZZLE.

ALTHOUGH Snoggs's *reason* for wishing to take his rise at £2, 10s. half-yearly did not concern our puzzle, the *fact* that he was duping his employer into paying him more than was intended did concern it. Many readers will be surprised to find that, although Moggs only received £350 in five years, the artful Snoggs actually obtained £362, 10s. in the same time. The rest is simplicity itself. It is evident that if Moggs saved £87, 10s. and Snoggs £181, 5s., the latter would be saving twice as great a proportion of his salary as the former (namely, one-half as against one-quarter), and the two sums added together make £268, 15s.

27.—GIVING CHANGE.

THE way to help the American tradesman out of his dilemma is this. Describing the coins by the number of cents that they represent, the tradesman puts on the counter 50 and 25 ; the buyer puts down 100, 3, and 2 ; the stranger adds his 10, 10, 5, 2, and 1. Now, considering that the cost of the purchase amounted to 34 cents, it is clear that out of this pooled money the tradesman has to receive 109, the buyer 71, and the stranger his 28 cents. Therefore it is obvious at a glance that the 100-piece must go to the tradesman, and it then follows that the 50-piece must go to the buyer, and then the 25-piece can only go to the stranger. Another glance will now make it clear that the two 10-cent pieces must go to the buyer, because the tradesman now only wants 9 and the stranger 3. Then it becomes obvious that the buyer must take the 1 cent, that the stranger must take the 3 cents, and the tradesman the 5, 2, and 2. To sum up, the tradesman takes 100, 5, 2, and 2 ; the buyer, 50, 10, 10, and 1 ; the stranger, 25 and 3. It will be seen that not one of the three persons retains any one of his own coins.

28.—DEFECTIVE OBSERVATION.

OF course the date on a penny is on the same side as Britannia—the " tail " side. Six pennies may be laid around another penny, all flat on the table, so that every one of them touches the central one. The number of threepenny-pieces that may be laid on the surface of a half-crown, so that no piece lies on another or overlaps the edge of the half-crown, is one. A second threepenny-piece will overlap the edge of the larger coin. Few people guess fewer than three, and many persons give an absurdly high number.

29.—THE BROKEN COINS.

IF the three broken coins when perfect were worth 253 pence, and are now in their broken condition worth 240 pence, it should be obvious that $\frac{13}{253}$ of the original value has been lost. And as the same fraction of each coin has been broken away, each coin has lost $\frac{13}{253}$ of its original bulk.

30.—TWO QUESTIONS IN PROBABILITIES.

IN tossing with the five pennies all at the same time, it is obvious that there are 32 different ways in which the coins may fall, because the first coin may fall in either of two ways, then the second coin may also fall in either of two ways, and so on. Therefore five 2's multiplied together make 32. Now, how are these 32 ways made up ? Here they are :—

(a) 5 heads		1 way
(b) 5 tails		1 way
(c) 4 heads and 1 tail . . .		5 ways
(d) 4 tails and 1 head . . .		5 ways
(e) 3 heads and 2 tails . . .		10 ways
(f) 3 tails and 2 heads . . .		10 ways

Now, it will be seen that the only favourable cases are a, b, c, and d—12 cases. The remaining 20 cases are unfavourable, because they do not give at least four heads or four tails. Therefore the chances are only 12 to 20 in your favour, or (which is the same thing) 3 to 5. Put another way, you have only 3 chances out of 8.

The amount that should be paid for a draw from the bag that contains three sovereigns and one shilling is 15s. 3d. Many persons will say that, as one's chances of drawing a sovereign were 3 out of 4, one should pay three-fourths of a pound, or 15s., overlooking the fact that one must draw at least a shilling—there being no blanks.

31.—DOMESTIC ECONOMY.

WITHOUT the hint that I gave, my readers would probably have been unanimous in deciding that Mr. Perkins's income must have been £1,710. But this is quite wrong. Mrs. Perkins says, " We have spent a third of his yearly income in rent," etc., etc.—that is, in two years they have spent an amount in rent, etc., equal to one-third of his yearly income. Note that she does *not* say that they have spent *each year* this sum, whatever it is, but that *during the two years* that amount has been spent. The only possible answer, according to the exact reading of her words, is, therefore, that his income was £180 per annum. Thus the amount spent in two years, during which his income has amounted to £360, will be £60 in rent, etc., £90 in domestic expenses, £20 in other ways, leaving the balance of £190 in the bank as stated.

32.—THE EXCURSION TICKET PUZZLE.

NINETEEN shillings and ninepence may be paid in 458,908,622 different ways.

I do not propose to give my method of solution. Any such explanation would occupy an amount of space out of proportion to its interest or value. If I could give within reasonable limits a general solution for all money payments, I would strain a point to find room; but such a solution would be extremely complex and cumbersome, and I do not consider it worth the labour of working out.

Just to give an idea of what such a solution would involve, I will merely say that I find that, dealing only with those sums of money that are multiples of threepence, if we only use bronze coins any sum can be paid in $(n+1)^2$ ways where n always represents the number of pence. If threepenny-pieces are admitted, there are $\frac{2n^3+15n^2+33n}{18}+1$ ways. If sixpences are also used there are $\frac{n^4+22n^3+159n^2+414n+216}{216}$ ways, when the sum is a multiple of sixpence, and the constant, 216, changes to 324 when the money is not such a multiple. And so the formulas increase in complexity in an accelerating ratio as we go on to the other coins.

I will, however, add an interesting little table of the possible ways of changing our current coins which I believe has never been given in a book before. Change may be given for a

Farthing in	.	.	.	0 way.
Halfpenny in	.	.	.	1 way.
Penny in	.	.	.	3 ways.
Threepenny-piece in	.	16 ways.
Sixpence in	.	.	.	66 ways.
Shilling in	.	.	.	402 ways.
Florin in	.	.	.	3,818 ways.
Half-crown in	.	.	.	8,709 ways.
Double florin in	.	.	60,239 ways.
Crown in	.	.	.	166,651 ways.
Half-sovereign in	.	.	6,261,622 ways.
Sovereign in	.	.	.	500,291,833 ways.

It is a little surprising to find that a sovereign may be changed in over five hundred million different ways. But I have no doubt as to the correctness of my figures.

33.—A PUZZLE IN REVERSALS.

(1) £13. (2) £23, 19s. 11d. The words " the number of pounds exceeds that of the pence " exclude such sums of money as £2, 16s. 2d. and all sums under £1.

34.—THE GROCER AND DRAPER.

THE grocer was delayed half a minute and the draper eight minutes and a half (seventeen times as long as the grocer), making together nine minutes. Now, the grocer took twenty-four minutes to weigh out the sugar, and, with the half-minute delay, spent 24 min. 30 sec. over the task ; but the draper had only to make *forty-seven* cuts to divide the roll of cloth, containing forty-eight yards, into yard pieces! This took him 15 min. 40 sec., and when we add the eight minutes and a half delay we get 24 min. 10 sec., from which it is clear that the draper won the race by twenty seconds. The majority of solvers make forty-eight cuts to divide the roll into forty-eight pieces !

35.—JUDKINS'S CATTLE.

As there were five droves with an equal number of animals in each drove, the number must be divisible by 5 ; and as every one of the eight dealers bought the same number of animals, the number must be divisible by 8. Therefore the number must be a multiple of 40. The highest possible multiple of 40 that will work will be found to be 120, and this number could be made up in one of two ways—1 ox, 23 pigs, and 96 sheep, or 3 oxen, 8 pigs, and 109 sheep. But the first is excluded by the statement that the animals consisted of " oxen, pigs, and sheep," because a single ox is not oxen. Therefore the second grouping is the correct answer.

36.—BUYING APPLES.

As there were the same number of boys as girls, it is clear that the number of children must be even, and, apart from a careful and exact reading of the question, there would be three different answers. There might be two, six, or fourteen children. In the first of these cases there are ten different ways in which the apples could be bought. But we were told there was an equal number of " boys and girls," and one boy and one girl are not boys and girls, so this case has to be excluded. In the case of fourteen children, the only possible distribution is that each child receives one halfpenny apple. But we were told that each child was to receive an equal distribution of " apples," and one apple is not apples, so this case has also to be excluded. We are therefore driven back on our third case, which exactly fits in with all the conditions. Three boys and three girls each receive 1 halfpenny apple and 2 third-penny apples. The value of these 3 apples is one penny and one-sixth, which multiplied by six makes seven-pence. Consequently, the correct answer is that there were six children—three girls and three boys.

37.—BUYING CHESTNUTS.

In solving this little puzzle we are concerned with the exact interpretation of the words used by the buyer and seller. I will give the question again, this time adding a few words to make the matter more clear. The added words are printed in italics.

" A man went into a shop to buy chestnuts. He said he wanted a pennyworth, and was given five chestnuts. ' It is not enough ; I ought to have a sixth of *a chestnut more*,' he remarked. ' But if I give you one chestnut more,' the shopman replied, " you will have five-*sixths* too many.' Now, strange to say, they were both right. How many chestnuts should the buyer receive for half a crown ? "

The answer is that the price was 155 chestnuts for half a crown. Divide this number by 30, and we find that the buyer was entitled to 5⅙ chestnuts in exchange for his penny. He was, therefore, right when he said, after receiving five only, that he still wanted a sixth. And the salesman was also correct in saying that if he gave one chestnut more (that is, six chestnuts in all) he would be giving five-sixths of a chestnut in excess.

38.—THE BICYCLE THIEF.

People give all sorts of absurd answers to this question, and yet it is perfectly simple if one just considers that the salesman cannot possibly have lost more than the cyclist actually stole. The latter rode away with a bicycle which cost the salesman eleven pounds, and the ten pounds " change ; " he thus made off with twenty-one pounds, in exchange for a worthless bit of paper. This is the exact amount of the salesman's loss, and the other operations of changing the cheque and borrowing from a friend do not affect the question in the slightest. The loss of prospective profit on the sale of the bicycle is, of course, not direct loss of money out of pocket.

39.—THE COSTERMONGER'S PUZZLE.

Bill must have paid 8s. per hundred for his oranges—that is, 125 for 10s. At 8s. 4d. per

hundred, he would only have received 120 oranges for 10s. This exactly agrees with Bill's statement.

40.—MAMMA'S AGE.

The age of Mamma must have been 29 years 2 months ; that of Papa, 35 years ; and that of the child, Tommy, 5 years 10 months. Added together, these make seventy years. The father is six times the age of the son, and, after 23 years 4 months have elapsed, their united ages will amount to 140 years, and Tommy will be just half the age of his father.

41.—THEIR AGES.

The gentleman's age must have been 54 years and that of his wife 45 years.

42.—THE FAMILY AGES.

The ages were as follows : Billie, 3½ years ; Gertrude, 1¾ year ; Henrietta, 5½ years ; Charlie, 10½ years ; and Janet, 21 years.

43.—MRS. TIMPKINS'S AGE.

The age of the younger at marriage is always the same as the number of years that expire before the elder becomes twice her age, if he was three times as old at marriage. In our case it was eighteen years afterwards ; therefore Mrs. Timpkins was eighteen years of age on the wedding-day, and her husband fifty-four.

44.—A CENSUS PUZZLE.

Miss Ada Jorkins must have been twenty-four and her little brother Johnnie three years of age, with thirteen brothers and sisters between. There was a trap for the solver in the words " seven times older than little Johnnie." Of course, " seven times older " is equal to eight times as old. It is surprising how many people hastily assume that it is the same as " seven times as old." Some of the best writers have committed this blunder. Probably many of my readers thought that the ages 24½ and 3½ were correct.

45.—MOTHER AND DAUGHTER.

In four and a half years, when the daughter will be sixteen years and a half and the mother forty-nine and a half years of age.

46.—MARY AND MARMADUKE.

Marmaduke's age must have been twenty-nine years and two-fifths, and Mary's nineteen years and three-fifths. When Marmaduke was aged nineteen and three-fifths, Mary was only nine and four-fifths ; so Marmaduke was at that time twice her age.

47.—ROVER'S AGE.

Rover's present age is ten years and Mildred's thirty years. Five years ago their respective ages were five and twenty-five. Remember that we said " four times older than the dog," which is the same as " five times as old." (See answer to No. 44.)

48.—CONCERNING TOMMY'S AGE.

TOMMY SMART'S age must have been nine years and three-fifths. Ann's age was sixteen and four-fifths, the mother's thirty-eight and two-fifths, and the father's fifty and two-fifths.

49.—NEXT-DOOR NEIGHBOURS.

MR. JUPP 39, Mrs. Jupp 34, Julia 14, and Joe 13; Mr. Simkin 42; Mrs. Simkin 40; Sophy 10; and Sammy 8.

50.—THE BAG OF NUTS.

IT will be found that when Herbert takes twelve, Robert and Christopher will take nine and fourteen respectively, and that they will have together taken thirty-five nuts. As 35 is contained in 770 twenty-two times, we have merely to multiply 12, 9, and 14 by 22 to discover that Herbert's share was 264, Robert's 198, and Christopher's 308. Then, as the total of their ages is 17½ years or half the sum of 12, 9, and 14, their respective ages must be 6, 4½, and 7 years.

51.—HOW OLD WAS MARY?

THE age of Mary to that of Ann must be as 5 to 3. And as the sum of their ages was 44, Mary was 27½ and Ann 16½. One is exactly 11 years older than the other. I will now insert in brackets in the original statement the various ages specified : " Mary is (27½) twice as old as Ann was (13¾) when Mary was half as old (24¾) as Ann will be (49½) when Ann is three times as old (49½) as Mary was (16½) when Mary was (16½) three times as old as Ann (5½)." Now, check this backwards. When Mary was three times as old as Ann, Mary was 16½ and Ann 5½ (11 years younger). Then we get 49½ for the age Ann will be when she is three times as old as Mary was then. When Mary was half this she was 24¾. And at that time Ann must have been 13¾ (11 years younger). Therefore Mary is now twice as old—27½, and Ann 11 years younger—16½.

52.—QUEER RELATIONSHIPS.

IF a man marries a woman, who dies, and he then marries his deceased wife's sister and himself dies, it may be correctly said that he had (previously) married the sister of his widow.

The youth was not the nephew of Jane Brown, because he happened to be her son. Her surname was the same as that of her brother, because she had married a man of the same name as herself.

53.—HEARD ON THE TUBE RAILWAY.

THE gentleman was the second lady's uncle.

54.—A FAMILY PARTY.

THE party consisted of two little girls and a boy, their father and mother, and their father's father and mother.

55.—A MIXED PEDIGREE.

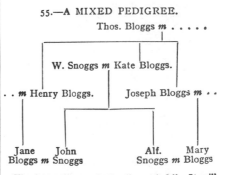

The letter m stands for " married." It will be seen that John Snoggs can say to Joseph Bloggs, " You are my *father's brother-in-law*, because my father married your sister Kate ; you are my *brother's father-in-law*, because my brother Alfred married your daughter Mary ; and you are my *father-in-law's brother*, because my wife Jane was your brother Henry's daughter."

56.—WILSON'S POSER.

IF there are two men, each of whom marries the mother of the other, and there is a son of each marriage, then each of such sons will be at the same time uncle and nephew of the other. There are other ways in which the relationship may be brought about, but this is the simplest.

57.—WHAT WAS THE TIME?

THE time must have been 9.36 p.m. A quarter of the time since noon is 2 hr. 24 min., and a half of the time till noon next day is 7 hr. 12 min. These added together make 9 hr. 36 min.

58.—A TIME PUZZLE.

TWENTY-SIX minutes.

59.—A PUZZLING WATCH.

IF the 65 minutes be counted on the face of the same watch, then the problem would be impossible : for the hands must coincide every 65⁵⁄₁₁ minutes as shown by its face, and it matters not whether it runs fast or slow ; but if it is measured by true time, it gains ₁⁄₁₁ of a minute in 65 minutes, or ⁶⁰⁄₁₄₃ of a minute per hour.

60.—THE WAPSHAW'S WHARF MYSTERY.

THERE are eleven different times in twelve hours when the hour and minute hands of a clock are exactly one above the other. If we divide 12 hours by 11 we get 1 hr. 5 min. $27\frac{3}{11}$ sec., and this is the time after twelve o'clock when they are first together, and also the time that elapses between one occasion of the hands being together and the next. They are together for the second time at 2 hr. 10 min. $54\frac{6}{11}$ sec. (twice the above time); next at 3 hr. 16 min. $21\frac{9}{11}$ sec.; next at 4 hr. 21 min. $49\frac{1}{11}$ sec. This last is the only occasion on which the two hands are together with the second hand "just past the forty-ninth second." This, then, is the time at which the watch must have stopped. Guy Boothby, in the opening sentence of his *Across the World for a Wife*, says, " It was a cold, dreary winter's afternoon, and by the time the hands of the clock on my mantelpiece joined forces and stood at twenty minutes past four, my chambers were well-nigh as dark as midnight." It is evident that the author here made a slip, for, as we have seen above, he is 1 min. $49\frac{1}{11}$ sec. out in his reckoning.

61.—CHANGING PLACES.

THERE are thirty-six pairs of times when the hands exactly change places between three p.m. and midnight. The number of pairs of times from any hour (n) to midnight is the sum of $12 - (n+1)$ natural numbers. In the case of the puzzle $n = 3$; therefore $12 - (3+1) = 8$ and $1 + 2 + 3 + 4 + 5 + 6 + 7 + 8 = 36$, the required answer. The first pair of times is 3 hr. $21\frac{57}{143}$ min. and 4 hr. $16\frac{112}{143}$ min., and the last pair is 10 hr. $59\frac{83}{143}$ min. and 11 hr. $54\frac{138}{143}$ min. I will not give all the remainder of the thirty-six pairs of times, but supply a formula by which any of the sixty-six pairs that occur from midday to midnight may be at once found :—

$$a \text{ hr. } \frac{720b + 60a}{143} \text{ min. and } b \text{ hr. } \frac{720a + 60b}{143} \text{ min.}$$

For the letter a may be substituted any hour from 0, 1, 2, 3 up to 10 (where nought stands for 12 o'clock midday) ; and b may represent any hour, later than a, up to 11.

By the aid of this formula there is no difficulty in discovering the answer to the second question : $a = 8$ and $b = 11$ will give the pair 8 hr. $58\frac{106}{143}$ min. and 11 hr. $44\frac{28}{143}$ min., the latter being the time when the minute hand is nearest of all to the point IX—in fact, it is only $\frac{16}{143}$ of a minute distant.

Readers may find it instructive to make a table of all the sixty-six pairs of times when the hands of a clock change places. An easy way is as follows : Make a column for the first times and a second column for the second times of the pairs. By making $a = 0$ and $b = 1$ in the above expressions we find the first case, and enter 0 hr. $5\frac{5}{143}$ min. at the head of the first column, and 1 hr. $0\frac{60}{143}$ min. at the head of the second column. Now, by successively adding $5\frac{5}{143}$ min. in the first, and 1 hr. $0\frac{60}{143}$ min. in the second column, we get all the *eleven* pairs in which the first time is a certain number of minutes after nought, or mid-day. Then there is a "jump" in the times, but you can find the next pair by making $a = 1$ and $b = 2$, and then by successively adding these two times as before you will get all the *ten* pairs

after 1 o'clock. Then there is another "jump," and you will be able to get by addition all the *nine* pairs after 2 o'clock. And so on to the end. I will leave readers to investigate for themselves the nature and cause of the "jumps." In this way we get under the successive hours, $11 + 10 + 9 + 8 + 7 + 6 + 5 + 4 + 3 + 2 + 1 = 66$ pairs of times, which result agrees with the formula in the first paragraph of this article.

Some time ago the principal of a Civil Service Training College, who conducts a " Civil Service Column " in one of the periodicals, had the query addressed to him, " How soon after XII o'clock will a clock with both hands of the same length be ambiguous ? " His first answer was, " Some time past one o'clock," but he varied the answer from issue to issue. At length some of his readers convinced him that the answer is, " At $5\frac{5}{143}$ min. past XII ; " and this he finally gave as correct, together with the reason for it that at that time *the time indicated is the same whichever hand you may assume as hour hand !*

62.—THE CLUB CLOCK.

THE positions of the hands shown in the illustration could only indicate that the clock stopped at 44 min. $51\frac{143}{1427}$ sec. after eleven o'clock. The second hand would next be " exactly midway between the other two hands " at 45 min. $52\frac{496}{1427}$ sec. after eleven o'clock. If we had been dealing with the points on the circle to which the three hands are directed, the answer would be 45 min. $22\frac{106}{1427}$ sec. after eleven ; but the question applied to the hands, and the second hand would not be between the others at that time, but outside them.

63.—THE STOP-WATCH.

THE time indicated on the watch was $5\frac{5}{11}$ min. past 9, when the second hand would be at $27\frac{3}{11}$ sec. The next time the hands would be similar distances apart would be $54\frac{6}{11}$ min. past 2, when the second hand would be at $32\frac{8}{11}$ sec. But you need only hold the watch (or our previous illustration of it) in front of a mirror, when you will see the second time reflected in it ! Of course, when reflected, you will read XI as I, X as II, and so on.

64.—THE THREE CLOCKS.

As a mere arithmetical problem this question presents no difficulty. In order that the hands shall all point to twelve o'clock at the same time, it is necessary that B shall gain at least twelve hours and that C shall lose twelve hours. As B gains a minute in a day of twenty-four hours, and C loses a minute in precisely the same time, it is evident that one will have gained 720 minutes (just twelve hours) in 720 days, and the other will have lost 720 minutes in 720 days. Clock A keeping perfect time, all three clocks must indicate twelve o'clock simultaneously at noon on the 720th day from April 1, 1898. What day of the month will that be ?

I published this little puzzle in 1898 to see

how many people were aware of the fact that 1900 would not be a leap year. It was surprising how many were then ignorant on the point. Every year that can be divided by four without a remainder is bissextile or leap year, with the exception that one leap year is cut off in the century. 1800 was not a leap year, nor was 1900. On the other hand, however, to make the calendar more nearly agree with the sun's course, every fourth hundred year is still considered bissextile. Consequently, 2000, 2400, 2800, 3200, etc., will all be leap years. May my readers live to see them. We therefore find that 720 days from noon of April 1, 1898, brings us to noon of March 22, 1900.

65.—THE RAILWAY STATION CLOCK.

THE time must have been $43\frac{7}{11}$ min. past two o'clock.

66.—THE VILLAGE SIMPLETON.

THE day of the week on which the conversation took place was Sunday. For when the day after to-morrow (Tuesday) is "yesterday," "to-day" will be Wednesday; and when the day before yesterday (Friday) was "to-morrow," "to-day" was Thursday. There are two days between Thursday and Sunday, and between Sunday and Wednesday.

67.—AVERAGE SPEED.

THE average speed is twelve miles an hour, not twelve and a half, as most people will hastily declare. Take any distance you like, say sixty miles. This would have taken six hours going and four hours returning. The double journey of 120 miles would thus take ten hours, and the average speed is clearly twelve miles an hour.

68.—THE TWO TRAINS.

ONE train was running just twice as fast as the other.

69.—THE THREE VILLAGES.

CALLING the three villages by their initial letters, it is clear that the three roads form a triangle, A, B, C, with a perpendicular, measuring twelve miles, dropped from C to the base A, B. This divides our triangle into two right-angled triangles with a twelve-mile side in common. It is then found that the distance from A to C is 15 miles, from C to B 20 miles, and from A to B 25 (that is 9 and 16) miles. These figures are easily proved, for the square of 12 added to the square of 9 equals the square of 15, and the square of 12 added to the square of 16 equals the square of 20.

70.—DRAWING HER PENSION.

THE distance must be $6\frac{3}{4}$ miles.

71.—SIR EDWYN DE TUDOR.

THE distance must have been sixty miles. If Sir Edwyn left at noon and rode 15 miles an hour, he would arrive at four o'clock—an hour too soon. If he rode 10 miles an hour, he would arrive at six o'clock—an hour too late. But if he went at 12 miles an hour, he would reach the castle of the wicked baron exactly at five o'clock—the time appointed.

72.—THE HYDROPLANE QUESTION.

THE machine must have gone at the rate of seven-twenty-fourths of a mile per minute and the wind travelled five-twenty-fourths of a mile per minute. Thus, going, the wind would help, and the machine would do twelve-twenty-fourths, or half a mile a minute, and returning only two-twenty-fourths, or one-twelfth of a mile per minute, the wind being against it. The machine without any wind could therefore do the ten miles in thirty-four and two-sevenths minutes, since it could do seven miles in twenty-four minutes.

73.—DONKEY RIDING.

THE complete mile was run in nine minutes. From the facts stated we cannot determine the time taken over the first and second quarter-miles separately, but together they, of course, took four and a half minutes. The last two quarters were run in two and a quarter minutes each.

74.—THE BASKET OF POTATOES.

MULTIPLY together the number of potatoes, the number less one, and twice the number less one, then divide by 3. Thus 50, 49, and 99 multiplied together make 242,550, which, divided by 3, gives us 80,850 yards as the correct answer. The boy would thus have to travel 45 miles and fifteen-sixteenths—a nice little recreation after a day's work.

75.—THE PASSENGER'S FARE.

MR. TOMPKINS should have paid fifteen shillings as his correct share of the motor-car fare. He only shared half the distance travelled for £3, and therefore should pay half of thirty shillings, or fifteen shillings.

76.—THE BARREL OF BEER.

HERE the digital roots of the six numbers are 6, 4, 1, 2, 7, 9, which together sum to 29, whose digital root is 2. As the contents of the barrels sold must be a number divisible by 3, if one buyer purchased twice as much as the other, we must find a barrel with root 2, 5, or 8 to set on one side. There is only one barrel, that containing 20 gallons, that fulfils these conditions. So the man must have kept these 20 gallons of beer for his own use and sold one man 33 gallons (the 18-gallon and 15-gallon barrels) and sold the other man 66 gallons (the 16, 19, and 31 gallon barrels).

77.—DIGITS AND SQUARES.

THE top row must be one of the four following numbers: 192, 219, 273, 327. The first was the example given.

78.—ODD AND EVEN DIGITS.

As we have to exclude complex and improper fractions and recurring decimals, the simplest solution is this : $79+5\frac{1}{4}$ and $84+\frac{3}{4}$, both equal $84\frac{3}{4}$. Without any use of fractions it is obviously impossible.

79.—THE LOCKERS PUZZLE.

THE smallest possible total is $356=107+249$, and the largest sum possible is $981=235+746$, or $657+324$. The middle sum may be either $720=134+586$, or $702=134+568$, or $407=138+269$. The total in this case must be made up of three of the figures 0, 2, 4, 7, but no sum other than the three given can possibly be obtained. We have therefore no choice in the case of the first locker, an alternative in the case of the third, and any one of three arrangements in the case of the middle locker. Here is one solution :—

107	134	235
249	586	746
356	720	981

Of course, in each case figures in the first two lines may be exchanged vertically without altering the total, and as a result there are just 3,072 different ways in which the figures might be actually placed on the locker doors. I must content myself with showing one little principle involved in this puzzle. The sum of the digits in the total is always governed by the digit omitted. $\frac{8}{9}-\frac{7}{10}-\frac{6}{11}-\frac{5}{12}-\frac{4}{13}-\frac{3}{14}-\frac{2}{15}-\frac{1}{16}-\frac{0}{17}-\frac{9}{18}$. Whichever digit shown here in the upper line we omit, the sum of the digits in the total will be found beneath it. Thus in the case of locker A we omitted 8, and the figures in the total sum up to 14. If, therefore, we wanted to get 356, we may know at once to a certainty that it can only be obtained (if at all) by dropping the 8.

80.—THE THREE GROUPS.

THERE are nine solutions to this puzzle, as follows, and no more :—

$12\times483=5,796$	$27\times198=5,346$
$42\times138=5,796$	$39\times186=7,254$
$18\times297=5,346$	$48\times159=7,632$

$$28\times157=4,396$$
$$4\times1,738=6,952$$
$$4\times1,963=7,852$$

The seventh answer is the one that is most likely to be overlooked by solvers of the puzzle.

81.—THE NINE COUNTERS.

IN this case a certain amount of mere " trial " is unavoidable. But there are two kinds of " trials "—those that are purely haphazard, and those that are methodical. The true puzzle lover is never satisfied with mere haphazard trials. The reader will find that by just reversing the figures in 23 and 46 (making the multipliers 32 and 64) both products will be

5,056. This is an improvement, but it is not the correct answer. We can get as large a product as 7,008 if we multiply 584 by 12 and 96 by 73, but this solution is not to be found without the exercise of some judgment and patience.

82.—THE TEN COUNTERS.

As I pointed out, it is quite easy so to arrange the counters that they shall form a pair of simple multiplication sums, each of which will give the same product—in fact, this can be done by anybody in five minutes with a little patience. But it is quite another matter to find that pair which gives the largest product and that which gives the smallest product.

Now, in order to get the smallest product, it is necessary to select as multipliers the two smallest possible numbers. If, therefore, we place 1 and 2 as multipliers, all we have to do is to arrange the remaining eight counters in such a way that they shall form two numbers, one of which is just double the other ; and in doing this we must, of course, try to make the smaller number as low as possible. Of course the lowest number we could get would be 3,045 ; but this will not work, neither will 3,405, 3,450, etc., and it may be ascertained that 3,485 is the lowest product. One of the required answers is $3,485\times2=6,970$, and $6,970\times1=6,970$.

The other part of the puzzle (finding the pair with the highest product) is, however, the real knotty point, for it is not at all easy to discover whether we should let the multiplier consist of one or of two figures, though it is clear that we must keep, so far as we can, the largest figures to the left in both multiplier and multiplicand. It will be seen that by the following arrangement so high a number as 58,560 may be obtained. Thus, $915\times64=58,560$, and $732\times80=58,560$.

83.—DIGITAL MULTIPLICATION.

THE solution that gives the smallest possible sum of digits in the common product is $23\times174=58\times69=4,002$, and the solution that gives the largest possible sum of digits, $9\times654=18\times327=5,886$. In the first case the digits sum to 6 and in the second case to 27. There is no way of obtaining the solution but by actual trial.

84.—THE PIERROT'S PUZZLE.

THERE are just six different solutions to this puzzle, as follows :—

8 multiplied by	473	equals	3784
9	,,	351	,,	3159
15	,,	93	,,	1395
21	,,	87	,,	1827
27	,,	81	,,	2187
35	,,	41	,,	1435

It will be seen that in every case the two multipliers contain exactly the same figures as the product.

85.—THE CAB NUMBERS.

THE highest product is, I think, obtained by multiplying 8,745,231 by 96—namely, 839,542,176.

Dealing here with the problem generally, I have shown in the last puzzle that with three digits there are only two possible solutions, and with four digits only six different solutions. These cases have all been given. With five digits there are just twenty-two solutions, as follows :—

3	×	4128	=	12384
3	×	4281	=	12843
3	×	7125	=	21375
3	×	7251	=	21753
2541	×	6	=	15246
651	×	24	=	15624
678	×	42	=	28476
246	×	51	=	12546
57	×	834	=	47538
75	×	231	=	17325
624	×	78	=	48672
435	×	87	=	37845

9	×	7461	=	67149
72	×	936	=	67392

2	×	8714	=	17428
2	×	8741	=	17482
65	×	281	=	18265
65	×	983	=	63895

4973	×	8	=	39784
6521	×	8	=	52168
14	×	926	=	12964
86	×	251	=	21586

Now, if we took every possible combination and tested it by multiplication, we should need to make no fewer than 30,240 trials, or, if we at once rejected the number 1 as a multiplier, 28,560 trials—a task that I think most people would be inclined to shirk. But let us consider whether there be no shorter way of getting at the results required. I have already explained that if you add together the digits of any number and then, as often as necessary, add the digits of the result, you must ultimately get a number composed of one figure. This last number I call the "digital root." It is necessary in every solution of our problem that the root of the sum of the digital roots of our multipliers shall be the same as the root of their product. There are only four ways in which this can happen : when the digital roots of the multipliers are 3 and 6, or 9 and 9, or 2 and 2, or 5 and 8. I have divided the twenty-two answers above into these four classes. It is thus evident that the digital root of any product in the first two classes must be 9, and in the second two classes 4.

Owing to the fact that no number of five figures can have a digital sum less than 15 or more than 35, we find that the figures of our product must sum to either 18 or 27 to produce the root 9, and to either 22 or 31 to produce the root 4. There are 3 ways of selecting five

different figures that add up to 18, there are 11 ways of selecting five figures that add up to 27, there are 9 ways of selecting five figures that add up to 22, and 5 ways of selecting five figures that add up to 31. There are, therefore, 28 different groups, and no more, from any one of which a product may be formed.

We next write out in a column these 28 sets of five figures, and proceed to tabulate the possible factors, or multipliers, into which they may be split. Roughly speaking, there would now appear to be about 2,000 possible cases to be tried, instead of the 30,240 mentioned above; but the process of elimination now begins, and if the reader has a quick eye and a clear head he can rapidly dispose of the large bulk of these cases, and there will be comparatively few test multiplications necessary. It would take far too much space to explain my own method in detail, but I will take the first set of figures in my table and show how easily it is done by the aid of little tricks and dodges that should occur to everybody as he goes along.

My first product group of five figures is 84,321. Here, as we have seen, the root of each factor must be 3 or a multiple of 3. As there is no 6 or 9, the only single multiplier is 3. Now, the remaining four figures can be arranged in 24 different ways, but there is no need to make 24 multiplications. We see at a glance that, in order to get a five-figure product, either the 8 or the 4 must be the first figure to the left. But unless the 2 is preceded on the right by the 8, it will produce when multiplied either a 6 or a 7, which must not occur. We are, therefore, reduced at once to the two cases, 3 × 4,128 and 3 × 4,281, both of which give correct solutions. Suppose next that we are trying the two-figure factor, 21. Here we see that if the number to be multiplied is under 500 the product will either have only four figures or begin with 10. Therefore we have only to examine the cases 21 × 843 and 21 × 834. But we know that the first figure will be repeated, and that the second figure will be twice the first figure added to the second. Consequently, as twice 3 added to 4 produces a nought in our product, the first case is at once rejected. It only remains to try the remaining case by multiplication, when we find it does not give a correct answer. If we are next trying the factor 12, we see at the start that neither the 8 nor the 3 can be in the units place, because they would produce a 6, and so on. A sharp eye and an alert judgment will enable us thus to run through our table in a much shorter time than would be expected. The process took me a little more than three hours.

I have not attempted to enumerate the solutions in the cases of six, seven, eight, and nine digits, but I have recorded nearly fifty examples with nine digits alone.

86.—QUEER MULTIPLICATION.

IF we multiply 32547891 by 6, we get the product, 195287346. In both cases all the nine digits are used once and once only.

87.—THE NUMBER CHECKS PUZZLE.

DIVIDE the ten checks into the following three groups : 7 1 5—4 6—3 2 8 9 0, and the first multiplied by the second produces the third.

88.—DIGITAL DIVISION.

IT is convenient to consider the digits as arranged to form fractions of the respective values, one-half, one-third, one-fourth, one-fifth, one-sixth, one-seventh, one-eighth, and one-ninth. I will first give the eight answers, as follows :—

$$\frac{6729}{13458}=\tfrac{1}{2}, \quad \frac{5823}{17469}=\tfrac{1}{3}, \quad \frac{3942}{15768}=\tfrac{1}{4},$$
$$\frac{2697}{13485}=\tfrac{1}{5}, \quad \frac{2943}{17658}=\tfrac{1}{6}, \quad \frac{2394}{16758}=\tfrac{1}{7},$$
$$\frac{3187}{25496}=\tfrac{1}{8}, \quad \frac{6381}{57429}=\tfrac{1}{9}.$$

The sum of the numerator digits and the denominator digits will, of course, always be 45, and the "digital root" is 9. Now, if we separate the nine digits into any two groups, the sum of the two digital roots will always be 9. In fact, the two digital roots must be either 9—9, 8—1, 7—2, 6—3, or 5—4. In the first case the actual sum is 18, but then the digital root of this number is itself 9. The solutions in the cases of one-third, one-fourth, one-sixth, one-seventh, and one-ninth must be of the form 9—9 ; that is to say, the digital roots of both numerator and denominator will be 9. In the cases of one-half and one-fifth, however, the digital roots are 6—3, but of course the higher root may occur either in the numerator or in the denominator ; thus $\frac{2697}{13485}$, $\frac{2769}{13845}$, $\frac{2973}{14865}$, $\frac{3729}{18645}$, where, in the first two arrangements, the roots of the numerator and denominator are respectively 6—3, and in the last two 3—6. The most curious case of all is, perhaps, one-eighth, for here the digital roots may be of any one of the five forms given above.

The denominators of the fractions being regarded as the numerators multiplied by 2, 3, 4, 5, 6, 7, 8, and 9 respectively, we must pay attention to the "carryings over." In order to get five figures in the product there will, of course, always be a carry-over after multiplying the last figure to the left, and in every case higher than 4 we must carry over at least three times. Consequently in cases from one-fifth to one-ninth we cannot produce different solutions by a mere change of position of pairs of figures, as, for example, we may with $\frac{5823}{17469}$ and $\frac{5832}{17496}$, where the 6 and 3 change places. It is true that the same figures may often be differently arranged, as shown in the two pairs of values for one-fifth that I have given in the last paragraph, but here it will be found there is a general readjustment of figures and not a simple changing of the positions of pairs. There are other little points that would occur to every solver—such as that the figure 5 cannot ever appear to the extreme right of the numerator, as this would result in our getting either a nought or a second 5 in the denominator. Similarly 1 cannot ever appear in the same position, nor 6 in the fraction one-sixth, nor an even figure in the fraction one-fifth, and so on. The pre-

liminary consideration of such points as I have touched upon will not only prevent our wasting a lot of time in trying to produce impossible forms, but will lead us more or less directly to the desired solutions.

89.—ADDING THE DIGITS.

THE smallest possible sum of money is £1, 8s. 9¾d., the digits of which add to 25.

90.—THE CENTURY PUZZLE.

THE problem of expressing the number 100 as a mixed number or fraction, using all the nine digits once, and once only, has, like all these digital puzzles, a fascinating side to it. The merest tyro can by patient trial obtain correct results, and there is a singular pleasure in discovering and recording each new arrangement akin to the delight of the botanist in finding some long-sought plant. It is simply a matter of arranging those nine figures correctly, and yet with the thousands of possible combinations that confront us the task is not so easy as might at first appear, if we are to get a considerable number of results. Here are eleven answers, including the one I gave as a specimen :—

$$96\tfrac{2148}{537}, \ 96\tfrac{1752}{438}, \ 96\tfrac{1428}{357}, \ 94\tfrac{1578}{263}, \ 91\tfrac{7524}{836},$$
$$91\tfrac{5823}{647}, \ 91\tfrac{5742}{638}, \ 82\tfrac{3546}{197}, \ 81\tfrac{7524}{396},$$
$$81\tfrac{5643}{297}, \ 3\tfrac{69258}{714}.$$

Now, as all the fractions necessarily represent whole numbers, it will be convenient to deal with them in the following form : 96+4, 94+6, 91+9, 82+18, 81+19, and 3+97.

With any whole number the digital roots of the fraction that brings it up to 100 will always be of one particular form. Thus, in the case of 96+4, one can say at once that if any answers are obtainable, then the roots of both the numerator and the denominator of the fraction will be 6. Examine the first three arrangements given above, and you will find that this is so. In the case of 94+6 the roots of the numerator and denominator will be respectively 3—2, in the case of 91+9 they will be 9—8, in the case of 82+18 they will be 9—9, in the case of 81+19 they will be 9—9, and in the case of 3+97 they will be 3—3. Every fraction that can be employed has, therefore, its particular digital root form, and you are only wasting your time in unconsciously attempting to break through this law.

Every reader will have perceived that certain whole numbers are evidently impossible. Thus, if there is a 5 in the whole number, there will also be a nought or a second 5 in the fraction, which are barred by the conditions. Then multiples of 10, such as 90 and 80, cannot of course occur, nor can the whole number conclude with a 9, like 89 and 79, because the fraction, equal to 11 or 21, will have 1 in the last place, and will therefore repeat a figure. Whole numbers that repeat a figure, such as 88 and 77, are also clearly useless. These cases, as I have said, are all obvious to every reader. But when I declare that such combinations as $98+2$, $92+8$, $86+14$, $83+17$, $74+26$, etc., etc., are to be at once dismissed as impossible, the reason is not so evident, and I unfortunately cannot spare space to explain it.

But when all those combinations have been struck out that are known to be impossible, it does not follow that all the remaining " possible forms " will actually work. The elemental form may be right enough, but there are other and deeper considerations that creep in to defeat our attempts. For example, $98+2$ is an impossible combination, because we are able to say at once that there is no possible form for the digital roots of the fraction equal to 2. But in the case of $97+3$ there is a possible form for the digital roots of the fraction, namely, $6-5$, and it is only on further investigation that we are able to determine that this form cannot in practice be obtained, owing to curious considerations. The working is greatly simplified by a process of elimination, based on such considerations as that certain multiplications produce a repetition of figures, and that the whole number cannot be from 12 to 23 inclusive, since in every such case sufficiently small denominators are not available for forming the fractional part.

91.—MORE MIXED FRACTIONS.

THE point of the present puzzle lies in the fact that the numbers 15 and 18 are not capable of solution. There is no way of determining this without trial. Here are answers for the ten possible numbers :—

$$9\frac{5472}{1368} = 13 \; ; \quad 9\frac{8235}{1647} = 14 \; ;$$
$$12\frac{3576}{894} = 16 \; ; \quad 6\frac{13258}{947} = 20 \; ;$$
$$15\frac{9432}{786} = 27 \; ; \quad 24\frac{9756}{813} = 36 \; ;$$
$$27\frac{5148}{396} = 40 \; ; \quad 65\frac{1892}{473} = 69 \; ;$$
$$59\frac{3614}{278} = 72 \; ; \quad 75\frac{3648}{192} = 94.$$

I have only found the one arrangement for each of the numbers 16, 20, and 27; but the other numbers are all capable of being solved in more than one way. As for 15 and 18, though these may be easily solved as a simple fraction, yet a "mixed fraction" assumes the presence of a whole number ; and though my own idea for dodging the conditions is the following, where the fraction is both complex and mixed, it will be fairer to keep exactly to the form indicated :—

$$3\frac{8952}{746} = 15 \; ; \quad 9\frac{5742}{638} = 18.$$

I have proved the possibility of solution for all numbers up to 100, except 1, 2, 3, 4, 15, and 18. The first three are easily shown to be impossible. I have also noticed that numbers whose digital root is 8—such as 26, 35, 44, 53, etc.—seem to lend themselves to the greatest number of answers. For the number 26 alone I have recorded no fewer than twenty-five different arrangements, and I have no doubt that there are many more.

92.—DIGITAL SQUARE NUMBERS.

So far as I know, there are no published tables of square numbers that go sufficiently high to be available for the purposes of this puzzle. The lowest square number containing all the nine digits once, and once only, is 139,854,276, the square of 11,826. The highest square number under the same conditions is, 923,187,456, the square of 30,384.

93.—THE MYSTIC ELEVEN.

MOST people know that if the sum of the digits in the odd places of any number is the same as the sum of the digits in the even places, then the number is divisible by 11 without remainder. Thus in 896743012 the odd digits, 20468, add up 20, and the even digits, 1379, also add up 20. Therefore the number may be divided by 11. But few seem to know that if the difference between the sum of the odd and the even digits is 11, or a multiple of 11, the rule equally applies. This law enables us to find, with a very little trial, that the smallest number containing nine of the ten digits (calling nought a digit) that is divisible by 11 is 102,347,586, and the highest number possible, 987,652,413.

94.—THE DIGITAL CENTURY.

THERE is a very large number of different ways in which arithmetical signs may be placed between the nine digits, arranged in numerical order, so as to give an expression equal to 100. In fact, unless the reader investigated the matter very closely, he might not suspect that so many ways are possible. It was for this reason that I added the condition that not only must the fewest possible signs be used, but also the fewest possible strokes. In this way we limit the problem to a single solution, and arrive at the simplest and therefore (in this case) the best result.

Just as in the case of magic squares there are methods by which we may write down with the greatest ease a large number of solutions, but not all the solutions, so there are several ways in which we may quickly arrive at dozens of arrangements of the " Digital Century," without finding all the possible arrangements. There is, in fact, very little principle in the thing, and there is no certain way of demonstrating that we have got the best possible solution. All I can say is that the arrangement I shall give as the best is the best I have up to the present succeeded in discovering. I will give the reader a few interesting specimens, the first being the solution usually published, and the last the best solution that I know.

	Signs.	Strokes.
$1+2+3+4+5+6+7+(8×9)=100$	(9	.. 18)
$\begin{aligned}-(1×2)-3-4-5+(6×7)+(8×9)\\=100 \quad . \quad . \quad . \quad . \quad . \quad . \quad .\end{aligned}$	(12	.. 20)
$\begin{aligned}1+(2×3)+(4×5)-6+7+(8×9)\\=100 \quad . \quad . \quad . \quad . \quad . \quad . \quad .\end{aligned}$	(11	.. 21)
$(1+2-3-4)(5-6-7-8-9)=100$	(9	.. 12)

	Signs.	Strokes.
$1 + (2 \times 3) + 4 + 5 + 67 + 8 + 9 = 100$.	(8	.. 16)
$(1 \times 2) + 34 + 56 + 7 - 8 + 9 = 100$	(7	.. 13)
$12 + 3 - 4 + 5 + 67 + 8 + 9 = 100$.	(6	.. 11)
$123 - 4 - 5 - 6 - 7 + 8 - 9 = 100$.	(6	.. 7)
$123 + 4 - 5 + 67 - 8 - 9 = 100$	(4	.. 6)
$123 + 45 - 67 + 8 - 9 = 100$	(4	.. 6)
$123 - 45 - 67 + 89 = 100$	(3 ..	4)

It will be noticed that in the above I have counted the bracket as one sign and two strokes. The last solution is singularly simple, and I do not think it will ever be beaten.

95.—THE FOUR SEVENS.

THE way to write four sevens with simple arithmetical signs so that they represent 100 is as follows :—

$$\frac{7}{.7} \times \frac{7}{.7} = 100.$$

Of course the fraction, 7 over decimal 7, equals 7 divided by $\frac{7}{10}$, which is the same as 70 divided by 7, or 10. Then 10 multiplied by 10 is 100, and there you are ! It will be seen that this solution applies equally to any number whatever that you may substitute for 7.

96.—THE DICE NUMBERS.

THE sum of all the numbers that can be formed with any given set of four different figures is always 6,666 multiplied by the sum of the four figures. Thus, 1, 2, 3, 4 add up 10, and ten times 6,666 is 66,660. Now, there are thirty-five different ways of selecting four figures from the seven on the dice—remembering the 6 and 9 trick. The figures of all these thirty-five groups add up to 600. Therefore 6,666 multiplied by 600 gives us 3,999,600 as the correct answer.

Let us discard the dice and deal with the problem generally, using the nine digits, but excluding nought. Now, if you were given simply the sum of the digits—that is, if the condition were that you could use any four figures so long as they summed to a given amount—then we have to remember that several combinations of four digits will, in many cases, make the same sum.

10	11	12	13	14	15	16	17	18	19	20
1	1	2	3	5	6	8	9	11	11	12

21	22	23	24	25	26	27	28	29	30
11	11	9	8	6	5	3	2	1	1

Here the top row of numbers gives all the possible sums of four different figures, and the bottom row the number of different ways in which each sum may be made. For example, 13 may be made in three ways : 1237, 1246, and 1345. It will be found that the numbers in the bottom row add up to 126, which is the number of combinations of nine figures taken four at a time. From this table we may at once calculate the answer to such a question as this :

What is the sum of all the numbers composed of four different digits (nought excluded) that add up to 14 ? Multiply 14 by the number beneath it in the table, 5, and multiply the result by 6,666, and you will have the answer. It follows that, to know the sum of all the numbers composed of four different digits, if you multiply all the pairs in the two rows and then add the results together, you will get 2,520, which, multiplied by 6,666, gives the answer 16,798,320.

The following general solution for any number of digits will doubtless interest readers. Let n represent number of digits, then 5 $(10^n - 1)$ |8 divided by |9—n equals the required sum. Note that |0 equals 1. This may be reduced to the following practical rule : Multiply together $4 \times 7 \times 6 \times 5 \dots$ to $(n - 1)$ factors ; now add $(n + 1)$ ciphers to the right, and from this result subtract the same set of figures with a single cipher to the right. Thus for $n = 4$ (as in the case last mentioned), $4 \times 7 \times 6 = 168$. Therefore 16,800,000 less 1,680 gives us 16,798,320 in another way.

97.—THE SPOT ON THE TABLE.

THE ordinary schoolboy would correctly treat this as a quadratic equation. Here is the actual arithmetic. Double the product of the two distances from the walls. This gives us 144, which is the square of 12. The sum of the two distances is 17. If we add these two numbers, 12 and 17, together, and also subtract one from the other, we get the two answers that 29 or 5 was the radius, or half-diameter, of the table. Consequently, the full diameter was 58 in. or 10 in. But a table of the latter dimensions would be absurd, and not at all in accordance with the illustration. Therefore the table must have been 58 in. in diameter. In this case the spot was on the edge nearest to the corner of the room—to which the boy was pointing. If the other answer were admissible, the spot would be on the edge farthest from the corner of the room.

98.—ACADEMIC COURTESIES.

THERE must have been ten boys and twenty girls. The number of bows girl to girl was therefore 380, of boy to boy 90, of girl with boy 400, and of boys and girls to teacher 30, making together 900, as stated. It will be remembered that it was not said that the teacher himself returned the bows of any child.

99.—THE THIRTY-THREE PEARLS.

THE value of the large central pearl must have been £3,000. The pearl at one end (from which they increased in value by £100) was £1,400 ; the pearl at the other end, £600.

100.—THE LABOURER'S PUZZLE.

THE man said, " I am going twice as deep," not " as deep again." That is to say, he was still going twice as deep as he had gone already, so that when finished the hole would be three times its present depth. Then the answer is that at present the hole is 3 ft. 6 in. deep and the man 2 ft. 4 in. above ground. When completed the hole will be 10 ft. 6 in. deep, and therefore the man will then be 4 ft. 8 in. below the surface, or twice the distance that he is now above ground.

101.—THE TRUSSES OF HAY.

ADD together the ten weights and divide by 4, and we get 289 lbs. as the weight of the five trusses together. If we call the five trusses in the order of weight A, B, C, D, and E, the lightest being A and the heaviest E, then the lightest, 110 lbs., must be the weight of A and B ; and the next lightest, 112 lbs., must be the weight of A and C. Then the two heaviest, D and E, must weigh 121 lbs., and C and E must weigh 120 lbs. We thus know that A, B, D, and E weigh together 231 lbs., which, deducted from 289 lbs. (the weight of the five trusses), gives us the weight of C as 58 lbs. Now, by mere subtraction, we find the weight of each of the five trusses—54 lbs., 56 lbs., 58 lbs., 59 lbs., and 62 lbs. respectively.

102.—MR. GUBBINS IN A FOG.

THE candles must have burnt for three hours and three-quarters. One candle had one-sixteenth of its total length left and the other four-sixteenths.

103.—PAINTING THE LAMP-POSTS.

PAT must have painted six more posts than Tim, no matter how many lamp-posts there were. For example, suppose twelve on each side ; then Pat painted fifteen and Tim nine. If a hundred on each side, Pat painted one hundred and three, and Tim only ninety-seven

104.—CATCHING THE THIEF.

THE constable took thirty steps. In the same time the thief would take forty-eight, which, added to his start of twenty-seven, carried him seventy-five steps. This distance would be exactly equal to thirty steps of the constable.

105.—THE PARISH COUNCIL ELECTION.

THE voter can vote for one candidate in 23 ways, for two in 253 ways, for three in 1,771, for four in 8,855, for five in 33,649, for six in 100,947, for seven in 245,157, for eight in 490,314, and for nine candidates in 817,190 different ways. Add these together, and we get the total of 1,698,159 ways of voting.

106.—THE MUDDLETOWN ELECTION.

THE numbers of votes polled respectively by the Liberal, the Conservative, the Independent, and the Socialist were 1,553, 1,535, 1,407, and 978. All that was necessary was to add the sum of the three majorities (739) to the total poll of 5,473 (making 6,212) and divide by 4, which gives us 1,553 as the poll of the Liberal. Then the polls of the other three candidates can, of course, be found by deducting the successive majorities from the last-mentioned number.

107.—THE SUFFRAGISTS' MEETING.

EIGHTEEN were present at the meeting and eleven left. If twelve had gone, two-thirds would have retired. If only nine had gone, the meeting would have lost half its members.

108.—THE LEAP-YEAR LADIES.

THE correct and only answer is that 11,616 ladies made proposals of marriage. Here are all the details, which the reader can check for himself with the original statements. Of 10,164 spinsters, 8,085 married bachelors, 627 married widowers, 1,221 were declined by bachelors, and 231 declined by widowers. Of the 1,452 widows, 1,155 married bachelors, and 297 married widowers. No widows were declined. The problem is not difficult, by algebra, when once we have succeeded in correctly stating it.

109.—THE GREAT SCRAMBLE.

THE smallest number of sugar plums that will fulfil the conditions is 26,880. The five boys obtained respectively : Andrew, 2,863 ; Bob, 6,335 ; Charlie, 2,438 ; David, 10,294 ; Edgar, 4,950. There is a little trap concealed in the words near the end, " one-fifth of the same," that seems at first sight to upset the whole account of the affair. But a little thought will show that the words could only mean " one-fifth of five-eighths, the fraction last mentioned " — that is, one-eighth of the three-quarters that Bob and Andrew had last acquired.

110.—THE ABBOT'S PUZZLE.

THE only answer is that there were 5 men, 25 women, and 70 children. There were thus 100 persons in all, 5 times as many women as men, and as the men would together receive 15 bushels, the women 50 bushels, and the children 35 bushels, exactly 100 bushels would be distributed.

111.—REAPING THE CORN.

THE whole field must have contained 46.626 square rods. The side of the central square, left by the farmer, is 4.8 284 rods, so it contains 23.313 square rods. The area of the field was thus something more than a quarter of an acre and less than one-third ; to be more precise, .2 914 of an acre.

112.—A PUZZLING LEGACY.

As the share of Charles falls in through his death, we have merely to divide the whole hundred acres between Alfred and Benjamin in the proportion of one-third to one-fourth—that is in the proportion of four-twelfths to three-twelfths, which is the same as four to three. Therefore Alfred takes four-sevenths of the hundred acres and Benjamin three-sevenths.

113.—THE TORN NUMBER.

THE other number that answers all the requirements of the puzzle is 9,801. If we divide this in the middle into two numbers and add them together we get 99, which, multiplied by itself, produces 9,801. It is true that 2,025 may be treated in the same way, only this number is excluded by the condition which requires that no two figures should be alike.

The general solution is curious. Call the number of figures in each half of the torn label n. Then, if we add 1 to each of the exponents of the prime factors (other than 3) of $10^n - 1$ (1 being regarded as a factor with the constant exponent, 1), their product will be the number of solutions. Thus, for a label of six figures, $n=3$. The factors of $10^3 - 1$ are $1^1 \times 37^1$ (not considering the 3^3), and the product of $2 \times 2 = 4$, the number of solutions. This always includes the special cases $98 - 01$, $00 - 01$, $998 - 01$, $000 - 001$, etc. The solutions are obtained as follows:—Factorize $10^3 - 1$ in all possible ways, always keeping the powers of 3 together; thus, 37×27, 999×1. Then solve the equation $37x = 27y + 1$. Here $x = 19$ and $y = 26$. Therefore, $19 \times 37 = 703$, the square of which gives one label, 494,209. A complementary solution (through $27x = 37y + 1$) can at once be found by $10^n - 703 = 297$, the square of which gives 088,209 for second label. (These non-significant noughts to the left must be included, though they lead to peculiar cases like $00238 - 04641 = 4879^2$, where $0238 - 4641$ would not work.) The special case 999×1 we can write at once $998,001$, according to the law shown above, by adding nines on one half and noughts on the other, and its complementary will be 1 preceded by five noughts, or 000001. Thus we get the squares of 999 and 1. These are the four solutions.

114.—CURIOUS NUMBERS.

THE three smallest numbers, in addition to 48, are 1,680, 57,120, and 1,940,448. It will be found that 1,681 and 841, 57,121 and 28,561, 1,940,449 and 970,225, are respectively the squares of 41 and 29, 239 and 169, 1,393 and 985.

115.—A PRINTER'S ERROR.

THE answer is that $2^5 . 9^2$ is the same as 2592, and this is the only possible solution to the puzzle.

116.—THE CONVERTED MISER.

As we are not told in what year Mr. Jasper Bullyon made the generous distribution of his accumulated wealth, but are required to find the lowest possible amount of money, it is clear that we must look for a year of the most favourable form.

There are four cases to be considered—an ordinary year with fifty-two Sundays and with fifty-three Sundays, and a leap-year with fifty-

two and fifty-three Sundays respectively. Here are the lowest possible amounts in each case :—

313 weekdays, 52 Sundays	. . .	£112,055
312 weekdays, 53 Sundays	. . .	19,345
314 weekdays, 52 Sundays	.	{ No solution possible.
313 weekdays, 53 Sundays	. . .	£69,174

The lowest possible amount, and therefore the correct answer, is £19,345, distributed in an ordinary year that began on a Sunday. The last year of this kind was 1911. He would have paid £53 on every day of the year, or £62 on every weekday, with £1 left over, as required, in the latter event.

117.—A FENCE PROBLEM.

THOUGH this puzzle presents no great difficulty to any one possessing a knowledge of algebra, it has perhaps rather interesting features.

Seeing, as one does in the illustration, just one corner of the proposed square, one is scarcely prepared for the fact that the field, in order to comply with the conditions, must contain exactly 501,760 acres, the fence requiring the same number of rails. Yet this is the correct answer, and the only answer, and if that gentleman in Iowa carries out his intention, his field will be twenty-eight miles long on each side, and a little larger than the county of Westmorland. I am not aware that any limit has ever been fixed to the size of a "field," though they do not run so large as this in Great Britain. Still, out in Iowa, where my correspondent resides, they do these things on a very big scale. I have, however, reason to believe that when he finds the sort of task he has set himself, he will decide to abandon it; for if that cow decides to roam to fresh woods and pastures new, the milkmaid may have to start out a week in advance in order to obtain the morning's milk.

Here is a little rule that will always apply where the length of the rail is half a pole. Multiply the number of rails in a hurdle by four, and the result is the exact number of miles in the side of a square field containing the same number of acres as there are rails in the complete fence. Thus, with a one-rail fence the field is four miles square; a two-rail fence gives eight miles square; a three-rail fence, twelve miles square; and so on, until we find that a seven-rail fence multiplied by four gives a field of twenty-eight miles square. In the case of our present problem, if the field be made smaller, then the number of rails will exceed the number of acres; while if the field be made larger, the number of rails will be less than the acres of the field.

118.—CIRCLING THE SQUARES.

THOUGH this problem might strike the novice as being rather difficult, it is, as a matter of fact, quite easy, and is made still easier by inserting four out of the ten numbers.

First, it will be found that squares that are diametrically opposite have a common difference. For example, the difference between the square of 14 and the square of 2, in the diagram, is 192; and the difference between the square of 16 and the square of 8 is also 192. This must be so in every case. Then it should be remembered that the difference between squares of two consecutive numbers is always twice the smaller number plus 1, and that the difference between the squares of any two numbers can always be expressed as the difference of the numbers multiplied by their sum. Thus the square of 5 (25) less the square of 4 (16) equals $(2 \times 4) + 1$, or 9; also, the square of 7 (49) less the square of 3 (9) equals $(7 + 3) \times (7 - 3)$, or 40.

Now, the number 192, referred to above, may be divided into five different pairs of even factors: 2×96, 4×48, 6×32, 8×24, and 12×16, and these divided by 2 give us, 1×48, 2×24, 3×16, 4×12, and 6×8. The difference and sum respectively of each of these pairs in turn produce 47, 49; 22, 26; 13, 19; 8, 16; and 2, 14. These are the required numbers, four of which are already placed. The six numbers that have to be added may be placed in just six different ways, one of which is as follows, reading round the circle clockwise: 16, 2, 49, 22, 19, 8, 14, 47, 26, 13.

I will just draw the reader's attention to one other little point. In all circles of this kind, the difference between diametrically opposite numbers increases by a certain ratio, the first numbers (with the exception of a circle of 6) being 4 and 6, and the others formed by doubling the next preceding but one. Thus, in the above case, the first difference is 2, and then the numbers increase by 4, 6, 8, and 12. Of course, an infinite number of solutions may be found if we admit fractions. The number of squares in a circle of this kind must, however, be of the form $4n + 6$; that is, it must be a number composed of 6 plus a multiple of 4.

119.—RACKBRANE'S LITTLE LOSS.

THE professor must have started the game with thirteen shillings, Mr. Potts with four shillings, and Mrs. Potts with seven shillings.

120.—THE FARMER AND HIS SHEEP.

THE farmer had one sheep only! If he divided this sheep (which is best done by weight) into two parts, making one part two-thirds and the other part one-third, then the difference between these two numbers is the same as the difference between their squares—that is, one-third. Any two fractions will do if the denominator equals the sum of the two numerators.

121.—HEADS OR TAILS.

CROOKS must have lost, and the longer he went on the more he would lose. In two tosses he would be left with three-quarters of his money, in four tosses with nine-sixteenths of his money,

in six tosses with twenty-seven sixty-fourths of his money, and so on. The order of the wins and losses makes no difference, so long as their number is in the end equal.

122.—THE SEE-SAW PUZZLE.

THE boy's weight must have been about 39.79 lbs. A brick weighed 3 lbs. Therefore 16 bricks weighed 48 lbs. and 11 bricks 33 lbs. Multiply 48 by 33 and take the square root.

123.—A LEGAL DIFFICULTY.

IT was clearly the intention of the deceased to give the son twice as much as the mother, or the daughter half as much as the mother. Therefore the most equitable division would be that the mother should take two-sevenths, the son four-sevenths, and the daughter one-seventh.

124.—A QUESTION OF DEFINITION.

THERE is, of course, no difference in *area* between a mile square and a square mile. But there may be considerable difference in *shape*. A mile square can be no other shape than square; the expression describes a surface of a certain specific size and shape. A square mile may be of any shape; the expression names a unit of area, but does not prescribe any particular shape.

125.—THE MINERS' HOLIDAY.

BILL HARRIS must have spent thirteen shillings and sixpence, which would be three shillings more than the average for the seven men—half a guinea.

126.—SIMPLE MULTIPLICATION.

THE number required is 3,529,411,764,705,882, which may be multiplied by 3 and divided by 2, by the simple expedient of removing the 3 from one end of the row to the other. If you want a longer number, you can increase this one to any extent by repeating the sixteen figures in the same order.

127.—SIMPLE DIVISION.

SUBTRACT every number in turn from every other number, and we get 358 (twice), 716, 1,611, 1,253, and 895. Now, we see at a glance that, as 358 equals 2×179, the only number that can divide in every case without a remainder will be 179. On trial we find that this is such a divisor. Therefore, 179 is the divisor we want, which always leaves a remainder 164 in the case of the original numbers given.

128.—A PROBLEM IN SQUARES.

THE sides of the three boards measure 31 in., 41 in., and 49 in. The common difference of area is exactly five square feet. Three numbers whose squares are in A.P., with a common difference of 7, are $1\frac{13}{24}$, $2\frac{17}{24}$, $3\frac{23}{24}$; and with a common difference of 13 are $\frac{80809}{1680}$, $\frac{104921}{1680}$, and $\frac{127729}{1680}$. In the case of whole square numbers the common difference will always be divisible by 24, so it is obvious that our squares must be fractional. Readers should now try to solve the case where the common difference is 23. It is rather a hard nut.

129.—THE BATTLE OF HASTINGS.

ANY number (not itself a square number) may be multiplied by a square that will give a product 1 less than another square. The given number must not itself be a square, because a square multiplied by a square produces a square, and no square plus 1 can be a square. My remarks throughout must be understood to apply to whole numbers, because fractional soldiers are not of much use in war.

Now, of all the numbers from 2 to 99 inclusive, 61 happens to be the most awkward one to work, and the lowest possible answer to our puzzle is that Harold's army consisted of 3,119,882,982,860,264,400 men. That is, there would be 51,145,622,669,840,400 men (the square of 226,153,980) in each of the sixty-one squares. Add one man (Harold), and they could then form one large square with 1,766,319,049 men on every side. The general problem, of which this is a particular case, is known as the "Pellian Equation"—apparently because Pell neither first propounded the question nor first solved it! It was issued as a challenge by Fermat to the English mathematicians of his day. It is readily solved by the use of continued fractions.

Next to 61, the most difficult number under 100 is 97, where $97 \times 6,377,352^2 + 1 =$ a square.

The reason why I assumed that there must be something wrong with the figures in the chronicle is that we can confidently say that Harold's army did not contain over three trillion men! If this army (not to mention the Normans) had had the whole surface of the earth (sea included) on which to encamp, each man would have had slightly more than a quarter of a square inch of space in which to move about! Put another way: Allowing one square foot of standing-room per man, each small square would have required all the space allowed by a globe three times the diameter of the earth.

130.—THE SCULPTOR'S PROBLEM.

A LITTLE thought will make it clear that the answer must be fractional, and that in one case the numerator will be greater and in the other case less than the denominator. As a matter of fact, the height of the larger cube must be $\frac{8}{5}$ ft., and of the smaller $\frac{3}{5}$ ft., if we are to have the answer in the smallest possible figures. Here the lineal measurement is $\frac{11}{5}$ ft.—that is, $1\frac{1}{5}$ ft. What are the cubic contents of the two cubes? First $\frac{8}{5} \times \frac{8}{5} \times \frac{8}{5} = \frac{512}{125}$, and secondly $\frac{3}{5} \times \frac{3}{5} \times \frac{3}{5} = \frac{27}{125}$. Add these together and the result is $\frac{539}{125}$, which reduces to $\frac{11}{5}$, or $1\frac{1}{5}$ ft. We thus see that the answers in cubic feet and lineal feet are precisely the same.

The germ of the idea is to be found in the works of Diophantus of Alexandria, who wrote about the beginning of the fourth century. These fractional numbers appear in triads, and are obtained from three generators, a, b, c, where a is the largest and c the smallest.

Then $ab + c^2 =$ denominator, and $a^2 - c^2$, $b^2 - c^2$, and $a^2 - b^2$ will be the three numerators. Thus, using the generators 3, 2, 1, we get $\frac{5}{7}$, $\frac{3}{7}$, $\frac{4}{7}$ and we can pair the first and second, as in the above solution, or the first and third for a second solution. The denominator must always be a prime number of the form $6n + 1$, or composed of such primes. Thus you can have 13, 19, etc., as denominators, but not 25, 55, 187, etc.

When the principle is understood there is no difficulty in writing down the dimensions of as many sets of cubes as the most exacting collector may require. If the reader would like one, for example, with plenty of nines, perhaps the following would satisfy him : $\frac{98910}{99000001}$ and $\frac{10999}{99000001}$.

131.—THE SPANISH MISER.

THERE must have been 386 doubloons in one box, 8,450 in another, and 16,514 in the third, because 386 is the smallest number that can occur. If I had asked for the smallest aggregate number of coins, the answer would have been 482, 3,362, and 6,242. It will be found in either case that if the contents of any two of the three boxes be combined, they form a square number of coins. It is a curious coincidence (nothing more, for it will not always happen) that in the first solution the digits of the three numbers add to 17 in every case, and in the second solution to 14. It should be noted that the middle one of the three numbers will always be half a square.

132.—THE NINE TREASURE BOXES.

HERE is the answer that fulfils the conditions :—

A=4	B=3,364	C=6,724
D=2,116	E=5,476	F=8,836
G=9,409	H=12,769	I=16,129

Each of these is a square number, the roots, taken in alphabetical order, being 2, 58, 82, 46, 74, 94, 97, 113, and 127, while the required difference between A and B, B and C, D and E, etc., is in every case 3,360.

133.—THE FIVE BRIGANDS.

THE sum of 200 doubloons might have been held by the five brigands in any one of 6,627 different ways. Alfonso may have held any number from 1 to 11. If he held 1 doubloon, there are 1,005 different ways of distributing the remainder ; if he held 2, there are 985 ways ; if 3, there are 977 ways ; if 4, there are 903 ways ; if 5 doubloons, 832 ways ; if 6 doubloons, 704 ways ; if 7 doubloons, 570 ways ; if 8 doubloons, 388 ways ; if 9 doubloons, 200 ways ; if 10 doubloons, 60 ways ; and if Alfonso held 11 doubloons, the remainder could be distributed in 3 different **ways**. More than 11 doubloons he could not possibly have had. It will scarcely be expected that I shall give all these 6,627 ways at length. What I propose to do is to enable the reader, if he should feel so disposed, to write out all the answers where Alfonso has one and the same amount. Let us take the cases where Alfonso has 6 doubloons, and see how we may obtain all the 704 different ways indicated above. Here are two tables that will serve as keys to all these answers :—

Table I.	Table II.
A=6.	A=6.
B=n.	B=n
C=(63−5n)+m.	C=1+m.
D=(128+4n)−4m.	D=(376−16n)−4m.
E=3+3m.	E=(15n−183)+3m.

In the first table we may substitute for n any whole number from 1 to 12 inclusive, and m may be nought or any whole number from 1 to $(31+n)$ inclusive. In the second table n may have the value of any whole number from 13 to 23 inclusive, and m may be nought or any whole number from 1 to $(93−4n)$ inclusive. The first table thus gives $(32+n)$ answers for every value of n; and the second table gives $(94−4n)$ answers for every value of n. The former, therefore, produces 462 and the latter 242 answers, which together make 704, as already stated. Let us take Table I., and say $n=5$ and $m=2$; also in Table II. take $n=13$ and $m=0$. Then we at once get these two answers :—

A=	6	A=	6
B=	5	B=	13
C=	40	C=	1
D=	140	D=	168
E=	9	E=	12
	200 doubloons		200 doubloons

These will be found to work correctly. All the rest of the 704 answers, where Alfonso always holds six doubloons, may be obtained in this way from the two tables by substituting the different numbers for the letters m and n.

Put in another way, for every holding of Alfonso the number of answers is the sum of two arithmetical progressions, the common difference in one case being 1 and in the other −4. Thus in the case where Alfonso holds 6 doubloons one progression is 33+34+35+36 + ... +43+44, and the other 42+38+34 +30+ .... +6+2. The sum of the first series is 462, and of the second 242—results which again agree with the figures already given. The problem may be said to consist in finding the first and last terms of these progressions. I should remark that where Alfonso holds 9, 10, or 11 there is only one progression, of the second form.

134.—THE BANKER'S PUZZLE.

In order that a number of sixpences may not be divisible into a number of equal piles, it is necessary that the number should be a prime.

If the banker can bring about a prime number, he will win ; and I will show how he can always do this, whatever the customer may put in the box, and that therefore the banker will win to a certainty. The banker must first deposit forty sixpences, and then, no matter how many the customer may add, he will desire the latter to transfer from the counter the square of the number next below what the customer put in. Thus, banker puts 40, customer, we will say, adds 6, then transfers from the counter 25 (the square of 5), which leaves 71 in all, a prime number. Try again. Banker puts 40, customer adds 12, then transfers 121 (the square of 11), as desired, which leaves 173, a prime number. The key to the puzzle is the curious fact that any number up to 39, if added to its square and the sum increased by 41, makes a prime number. This was first discovered by Euler, the great mathematician. It has been suggested that the banker might desire the customer to transfer sufficient to raise the contents of the box to a given number ; but this would not only make the thing an absurdity, but breaks the rule that neither knows what the other puts in.

135.—THE STONEMASON'S PROBLEM.

The puzzle amounts to this. Find the smallest square number that may be expressed as the sum of more than three consecutive cubes, the cube 1 being barred. As more than three heaps were to be supplied, this condition shuts out the otherwise smallest answer, $23^3+24^3+25^3=204^2$. But it admits the answer, $25^3+26^3+27^3+28^3 +29^3=315^2$. The correct answer, however, requires more heaps, but a smaller aggregate number of blocks. Here it is : $14^3+15^3+ ....$ up to 25^3 inclusive, or twelve heaps in all, which, added together, make 97,344 blocks of stone that may be laid out to form a square 312 × 312. I will just remark that one key to the solution lies in what are called triangular numbers. (See pp. 13, 25, and 166.)

136.—THE SULTAN'S ARMY.

The smallest primes of the form $4n+1$ are 5, 13, 17, 29, and 37, and the smallest of the form $4n−1$ are 3, 7, 11, 19, and 23. Now, primes of the first form can always be expressed as the sum of two squares, and in only one way. Thus, $5=4+1$; $13=9+4$; $17=16+1$; $29=25+4$; $37=36+1$. But primes of the second form can never be expressed as the sum of two squares in any way whatever.

In order that a number may be expressed as the sum of two squares in several different ways, it is necessary that it shall be a composite number containing a certain number of primes of our first form. Thus, 5 or 13 alone can only be so expressed in one way ; but 65, (5×13), can be expressed in two ways, 1,105, (5×13 ×17), in four ways, 32,045, (5×13×17×29), in eight ways. We thus get double as many ways for every new factor of this form that we introduce. Note, however, that I say *new* factor, for the *repetition* of factors is subject to another law. We cannot express 25, (5 × 5), in two ways, but only in one; yet 125, (5 × 5 × 5), can be given in two ways, and so can 625, (5 × 5 × 5 × 5) ; while if we take in yet another 5 we can express the number as the sum of two squares in three different ways.

If a prime of the second form gets into your composite number, then that number cannot be the sum of two squares. Thus 15, (3 × 5), will not work, nor will 135, (3 × 3 × 3 × 5) ; but if we take in an even number of 3's it will work, because these 3's will themselves form a square number, but you will only get one solution. Thus, 45, (3 × 3 × 5, or 9 × 5)=36+9. Similarly, the factor 2 may always occur, or any power of 2, such as 4, 8, 16, 32 ; but its introduction or omission will never affect the number of your solutions, except in such a case as 50, where it doubles a square and therefore gives you the two answers, 49+1 and 25+25.

Now, directly a number is decomposed into its prime factors, it is possible to tell at a glance whether or not it can be split into two squares ; and if it can be, the process of discovery in how many ways is so simple that it can be done in the head without any effort. The number 1 gave was 130. I at once saw that this was 2 × 5 × 13, and consequently that, as 65 can be expressed in two ways (64+1 and 49+16), 130 can also be expressed in two ways, the factor 2 not affecting the question.

The smallest number that can be expressed as the sum of two squares in twelve different ways is 160,225, and this is therefore the smallest army that would answer the Sultan's purpose. The number is composed of the factors 5 × 5 × 13 × 17 × 29, each of which is of the required form. If they were all different factors, there would be sixteen ways ; but as one of the factors is repeated, there are just twelve ways. Here are the sides of the twelve pairs of squares : (400 and 15), (399 and 32), (393 and 76), (392 and 81), (384 and 113), (375 and 140), (360 and 175), (356 and 183), (337 and 216), (329 and 228), (311 and 252), (265 and 300). Square the two numbers in each pair, add them together, and their sum will in every case be 160,225.

137.—A STUDY IN THRIFT.

Mrs. Sandy McAllister will have to save a tremendous sum out of her housekeeping allowance if she is to win that sixth present that her canny husband promised her. And the allowance must be a very liberal one if it is to admit of such savings. The problem required that we should find five numbers higher than 36 the units of which may be displayed so as to form a square, a triangle, two triangles, and three triangles, using the complete number in every one of the four cases.

Every triangular number is such that if we multiply it by 8 and add 1 the result is an odd square number. For example, multiply 1, 3, 6, 10, 15 respectively by 8 and add 1, and we get 9, 25, 49, 81, 121, which are the squares of the odd numbers 3, 5, 7, 9, 11. Therefore in every

case where $8x^2+1$=a square number, x^2 is also a triangular. This point is dealt with in our puzzle, "The Battle of Hastings." I will now merely show again how, when the first solution is found, the others may be discovered without any difficulty. First of all, here are the figures :—

$$8 \times 1^2 + 1 = 3^2$$
$$8 \times 6^2 + 1 = 17^2$$
$$8 \times 35^2 + 1 = 99^2$$
$$8 \times 204^2 + 1 = 577^2$$
$$8 \times 1189^2 + 1 = 3363^2$$
$$8 \times 6930^2 + 1 = 19601^2$$
$$8 \times 40391^2 + 1 = 114243^2$$

The successive pairs of numbers are found in this way :—

$(1 \times 3)+ (3 \times 1)= 6 \qquad (8 \times 1)+ (3 \times 3)= 17$
$(1 \times 17)+ (3 \times 6)= 35 \qquad (8 \times 6)+(3 \times 17)= 99$
$(1 \times 99)+(3 \times 35)=204 \qquad (8 \times 35)+(3 \times 99)=577$

and so on. Look for the numbers in the table above, and the method will explain itself.

Thus we find that the numbers 36, 1225, 41616, 1413721, 48024900, and 1631432881 will form squares with sides of 6, 35, 204, 1189, 6930, and 40391 ; and they will also form single triangles with sides of 8, 49, 288, 1681, 9800, and 57121. These numbers may be obtained from the last column in the first table above in this way : simply divide the numbers by 2 and reject the remainder. Thus the integral halves of 17, 99, and 577 are 8, 49, and 288.

All the numbers we have found will form either two or three triangles at will. The following little diagram will show you graphically at a glance that every square number must necessarily be the sum of two triangulars, and that the side of one triangle will be the same as the side of the corresponding square, while the other will be just 1 less.

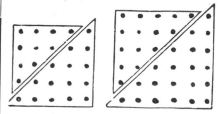

Thus a square may always be divided easily into two triangles, and the sum of two consecutive triangulars will always make a square. In numbers it is equally clear, for if we examine the first triangulars—1, 3, 6, 10, 15, 21, 28— we find that by adding all the consecutive pairs in turn we get the series of square numbers— 4, 9, 16, 25, 36, 49, etc.

The method of forming three triangles from our numbers is equally direct, and not at all a matter of trial. But I must content myself with giving actual figures, and just stating that every triangular higher than 6 will form three triangulars. I give the sides of the triangles, and readers will know from my remarks when stating the puzzle how to find from these sides the number of counters or coins in each, and so check the results if they so wish.

writing out the series until we come to a square number does not appeal to the mathematical mind, but it serves to show how the answer to

Number.	Side of Square.	Side of Triangle.	Sides of Two Triangles.	Sides of Three Triangles.
36	6	8	6+5	5+5+3
1225	35	49	36+34	33+32+16
41616	204	288	204+203	192+192+95
1413721	1189	1681	1189+1188	1121+1120+560
48024900	6930	9800	6930+6929	6533+6533+3267
1631432881	40391	57121	40391+40390	38081+38080+19040

I should perhaps explain that the arrangements given in the last two columns are not the only ways of forming two and three triangles. There are others, but one set of figures will fully serve our purpose. We thus see that before Mrs. McAllister can claim her sixth £5 present she must save the respectable sum of £1,631,432,881.

138.—THE ARTILLERYMEN'S DILEMMA.

WE were required to find the smallest number of cannon balls that we could lay on the ground to form a perfect square, and could pile into a square pyramid. I will try to make the matter clear to the merest novice.

I	2	3	4	5	6	7
I	3	6	10	15	21	28
I	4	10	20	35	56	84
I	5	14	30	55	91	140

Here in the first row we place in regular order the natural numbers. Each number in the second row represents the sum of the numbers in the row above, from the beginning to the number just over it. Thus 1, 2, 3, 4, added together, make 10. The third row is formed in exactly the same way as the second. In the fourth row every number is formed by adding together the number just above it and the preceding number. Thus 4 and 10 make 14, 20 and 35 make 55. Now, all the numbers in the second row are triangular numbers, which means that these numbers of cannon balls may be laid out on the ground so as to form equilateral triangles. The numbers in the third row will all form our triangular pyramids, while the numbers in the fourth row will all form square pyramids.

Thus the very process of forming the above numbers shows us that every square pyramid is the sum of two triangular pyramids, one of which has the same number of balls in the side at the base, and the other one ball fewer. If we continue the above table to twenty-four places, we shall reach the number 4,900 in the fourth row. As this number is the square of 70, we can lay out the balls in a square, and can form a square pyramid with them. This manner of

the particular puzzle may be easily arrived at by anybody. As a matter of fact, I confess my failure to discover any number other than 4,900 that fulfils the conditions, nor have I found any rigid proof that this is the only answer. The problem is a difficult one, and the second answer, if it exists (which I do not believe), certainly runs into big figures.

For the benefit of more advanced mathematicians I will add that the general expression for square pyramid numbers is $\frac{2n^3+3n^2+n}{6}$.

For this expression to be also a square number (the special case of 1 excepted) it is necessary that $n=p^2-1=6t^2$, where $2p^2-1=q^2$ (the "Pellian Equation"). In the case of our solution above, $n=24$, $p=5$, $t=2$, $q=7$.

139.—THE DUTCHMEN'S WIVES.

THE money paid in every case was a square number of shillings, because they bought 1 at 1s., 2 at 2s., 3 at 3s., and so on. But every husband pays altogether 63s. more than his wife, so we have to find in how many ways 63 may be the difference between two square numbers. These are the three only possible ways : the square of 8 less the square of 1, the square of 12 less the square of 9, and the square of 32 less the square of 31. Here 1, 9, and 31 represent the number of pigs bought and the number of shillings per pig paid by each woman, and 8, 12, and 32 the same in the case of their respective husbands. From the further information given as to their purchases, we can now pair them off as follows : Cornelius and Gurtrün bought 8 and 1 ; Elas and Katrün bought 12 and 9 ; Hendrick and Anna bought 32 and 31. And these pairs represent correctly the three married couples.

The reader may here desire to know how we may determine the maximum number of ways in which a number may be expressed as the difference between two squares, and how we are to find the actual squares. Any integer except 1, 4, and twice any odd number, may be expressed as the difference of two integral squares in as many ways as it can be split up into pairs of factors, counting 1 as a factor. Suppose the number to be 5,940. The factors are

$2^2.3^3.5.11$. Here the exponents are 2, 3, 1, 1. Always deduct 1 from the exponents of 2 and add 1 to all the other exponents; then we get 1, 4, 2, 2, and half the product of these four numbers will be the required number of ways in which 5,940 may be the difference of two squares—that is, 8. To find these eight squares, as it is an *even* number, we first divide by 4 and get 1485, the eight pairs of factors of which are 1×1485, 3×495, 5×297, 9×165, 11×135, 15×99, 27×55, and 33×45. The sum and difference of any one of these pairs will give the required numbers. Thus, the square of 1,486 less the square of 1,484 is 5,940, the square of 498 less the square of 492 is the same, and so on. In the case of 63 above, the number is *odd*; so we factorize at once, 1×63, 3×21, 7×9. Then we find that *half* the sum and difference will give us the numbers 32 and 31, 12 and 9, and 8 and 1, as shown in the solution to the puzzle.

The reverse problem, to find the factors of a number when you have expressed it as the difference of two squares, is obvious. For example, the sum and difference of any pair of numbers in the last sentence will give us the factors of 63. Every prime number (except 1 and 2) may be expressed as the difference of two squares in one way, and in one way only. If a number can be expressed as the difference of two squares in more than one way, it is composite; and having so expressed it, we may at once obtain the factors, as we have seen. Fermat showed in a letter to Mersenne or Frénicle, in 1643, how we may discover whether a number may be expressed as the difference of two squares in more than one way, or proved to be a prime. But the method, when dealing with large numbers, is necessarily tedious, though in practice it may be considerably shortened. In many cases it is the shortest method known for factorizing large numbers, and I have always held the opinion that Fermat used it in performing a certain feat in factorizing that is historical and wrapped in mystery.

140.—FIND ADA'S SURNAME.

The girls' names were Ada Smith, Annie Brown, Emily Jones, Mary Robinson, and Bessie Evans.

141.—SATURDAY MARKETING.

As every person's purchase was of the value of an exact number of shillings, and as the party possessed when they started out forty shilling coins altogether, there was no necessity for any lady to have any smaller change, or any evidence that they actually had such change. This being so, the only answer possible is that the women were named respectively Anne Jones, Mary Robinson, Jane Smith, and Kate Brown. It will now be found that there would be exactly eight shillings left, which may be divided equally among the eight persons in coin without any change being required.

142.—THE SILK PATCHWORK.

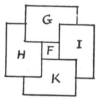

Our illustration will show how to cut the stitches of the patchwork so as to get the square F entire, and four equal pieces, G, H, I, K, that will form a perfect Greek cross. The reader will know how to assemble these four pieces from Fig. 13 in the article.

143.—TWO CROSSES FROM ONE.

It will be seen that one cross is cut out entire, as A in Fig. 1, while the four pieces marked

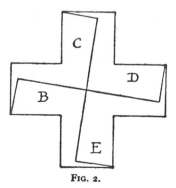

FIG. 1.   FIG. 2.

B, C, D and E form the second cross, as in Fig. 2, which will be exactly the same size as the other. I will leave the reader the pleasant task of discovering for himself the best way of finding the direction of the cuts. Note that the Swastika again appears.

The difficult question now presents itself: How are we to cut three Greek crosses from one in the fewest possible pieces? As a matter of fact, this problem may be solved in as few as thirteen pieces; but as I know many of my readers, advanced geometricians, will be glad to have something to work on of which they are not shown the solution, I leave the mystery for the present undisclosed.

144.—THE CROSS AND THE TRIANGLE.

THE line A B in the following diagram represents the side of a square having the same area as the cross. I have shown elsewhere, as stated, how to make a square and equilateral triangle of equal area. I need not go, therefore, into the preliminary question of finding the dimensions of the triangle that is to equal our cross. We will assume that we have already found this, and the question then becomes, How are we to cut up one of these into pieces that will form the other?

First draw the line A B where A and B are midway between the extremities of the two side arms. Next make the lines D C and E F equal in length to half the side of the triangle. Now from E and F describe with the same radius the intersecting arcs at G and draw F G. Finally make I K equal to H C and L B equal to A D. If we now draw I L, it should be parallel to F G, and all the six pieces are marked out. These fit together and form a perfect equilateral triangle, as shown in the second diagram. Or we might have first found the direction of the line M N in our triangle, then placed the point O over the point E in the cross and turned round the triangle over the cross until the line

I have seen many attempts at a solution involving the assumption that the height of the triangle is exactly the same as the height of the cross. This is a fallacy: the cross will always be higher than the triangle of equal area.

145.—THE FOLDED CROSS.

FIRST fold the cross along the dotted line A B in Fig. 1. You then have it in the form shown

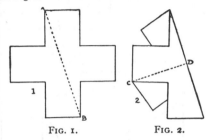

FIG. 1.　　FIG. 2.

in Fig. 2. Next fold it along the dotted line C D (where D is, of course, the centre of the cross). and you get the form shown in Fig. 3.

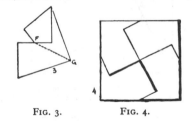

FIG. 3.　　FIG. 4.

Now take your scissors and cut from G to F, and the four pieces, all of the same size and

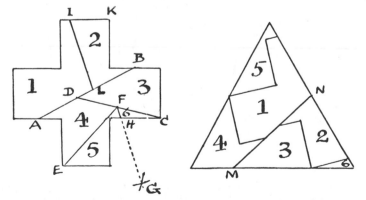

M N was parallel to A B. The piece 5 can then be marked off and the other pieces in succession. | shape, will fit together and form a square, as shown in Fig. 4.

146.—AN EASY DISSECTION PUZZLE.

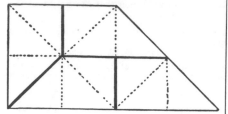

THE solution to this puzzle is shown in the illustration. Divide the figure up into twelve equal triangles, and it is easy to discover the directions of the cuts, as indicated by the dark lines.

147.—AN EASY SQUARE PUZZLE.

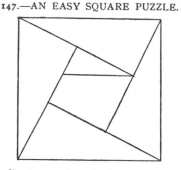

THE diagram explains itself, one of the five pieces having been cut in two to form a square.

148.—THE BUN PUZZLE.

THE secret of the bun puzzle lies in the fact that, with the relative dimensions of the circles as given, the three diameters will form a right-angled triangle, as shown by A, B, C. It follows that the two smaller buns are exactly equal to the large bun. Therefore, if we give David and Edgar the two halves marked D and E, they will have their fair shares—one quarter of the confectionery each. Then if we place the small bun, H, on the top of the remaining one and trace its circumference in the manner shown, Fred's piece, F, will exactly equal Harry's small bun, H, with the addition of the piece marked G—half the rim of the other. Thus each boy gets an exactly equal share, and there are only five pieces necessary.

149.—THE CHOCOLATE SQUARES.

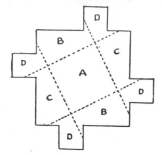

SQUARE A is left entire ; the two pieces marked B fit together and make a second square ; the two pieces C make a third square ; and the four pieces marked D will form the fourth square.

150.—DISSECTING A MITRE.

THE diagram on the next page shows how to cut into five pieces to form a square. The dotted lines are intended to show how to find the points C and F—the only difficulty. A B is half B D, and A E is parallel to B H. With the point of the compasses at B describe the arc H E, and A E will be the distance of C from B. Then F G equals B C less A B.

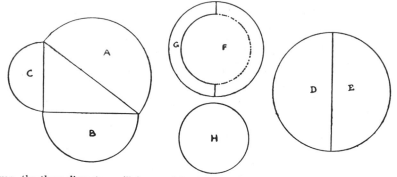

This puzzle—with the added condition that it shall be cut into four parts of the same size and shape—I have not been able to trace to an earlier date than 1835. Strictly speaking,

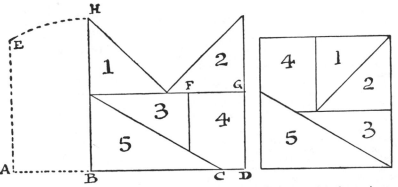

it is, in that form, impossible of solution; but I give the answer that is always presented, and that seems to satisfy most people.

We are asked to assume that the two portions containing the same letter—AA, BB, CC, DD—are joined by "a mere hair," and are, therefore, only one piece. To the geometrician this is

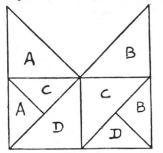

absurd, and the four shares are not equal in area unless they consist of two pieces each. If you make them equal in area, they will not be exactly alike in shape.

151.—THE JOINER'S PROBLEM.

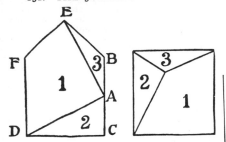

NOTHING could be easier than the solution of this puzzle—when you know how to do it.

And yet it is apt to perplex the novice a good deal if he wants to do it in the fewest possible pieces—three. All you have to do is to find the point A, midway between B and C, and then cut from A to D and from A to E. The three pieces then form a square in the manner shown. Of course, the proportions of the original figure must be correct; thus the triangle BEF is just a quarter of the square BCDF. Draw lines from B to D and from C to F, and this will be clear.

152.—ANOTHER JOINER'S PROBLEM.

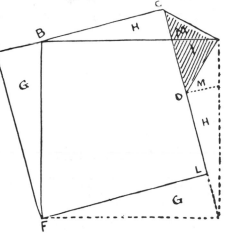

THE point was to find a general rule for forming a perfect square out of another square combined with a "right-angled isosceles triangle." The triangle to which geometricians give this high-sounding name is, of course, nothing more or less than half a square that has been divided from corner to corner.

The precise relative proportions of the square and triangle are of no consequence whatever.

It is only necessary to cut the wood or material into five pieces.

Suppose our original square to be A C L F in the above diagram and our triangle to be the shaded portion C E D. Now, we first find half the length of the long side of the triangle (C D) and measure off this length at A B. Then we place the triangle in its present position against the square and make two cuts—one from B to F, and the other from B to E. Strange as it may seem, that is all that is necessary ! If we now remove the pieces G, H, and M to their new places, as shown in the diagram, we get the perfect square B E K F.

Take any two square pieces of paper, of different sizes but perfect squares, and cut the smaller one in half from corner to corner. Now proceed in the manner shown, and you will find that the two pieces may be combined to form a larger square by making these two simple cuts, and that no piece will be required to be turned over.

The remark that the triangle might be " a little larger or a good deal smaller in proportion " was intended to bar cases where area of triangle is greater than area of square. In such cases six pieces are necessary, and if triangle and square are of equal area there is an obvious solution in three pieces, by simply cutting the square in half diagonally.

153.—A CUTTING-OUT PUZZLE.

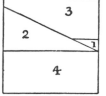

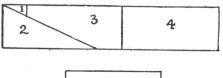

THE illustration shows how to cut the four pieces and form with them a square. First find the side of the square (the mean proportional between the length and height of the rectangle), and the method is obvious. If our strip is exactly in the proportions 9×1, or 16×1, or 25×1, we can clearly cut it in 3, 4, or 5 rectangular pieces respectively to form a square. Excluding these special cases, the general law is that for a strip in length more than n^2 times the breadth, and not more than $(n+1)^2$ times the breadth, it may be cut in $n+2$ pieces to form a square, and there will be $n-1$ rectangular pieces like piece 4 in the diagram. Thus, for example, with a strip 24×1, the length is more than 16 and less than 25 times the breadth. Therefore it can be done in 6 pieces (n here being 4), 3 of which will be rectangular. In the case where n equals 1, the rectangle disappears and we get a solution in three pieces. Within these limits, of course, the sides need not be rational : the solution is purely geometrical.

154.—MRS. HOBSON'S HEARTHRUG.

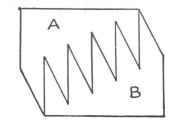

As I gave full measurements of the mutilated rug, it was quite an easy matter to find the precise dimensions for the square. The two pieces cut off would, if placed together, make an oblong piece 12×6, giving an area of 72 (inches or yards, as we please), and as the original complete rug measured 36×27, it had an area of 972. If, therefore, we deduct the pieces that have been cut away, we find that our new rug will contain 972 less 72, or 900; and as 900 is the square of 30, we know that the new rug must measure 30×30 to be a perfect square. This is a great help towards the solution, because we may safely conclude that the two horizontal sides measuring 30 each may be left intact.

There is a very easy way of solving the puzzle in four pieces, and also a way in three pieces that can scarcely be called difficult, but the correct answer is in only two pieces.

It will be seen that if, after the cuts are made, we insert the teeth of the piece B one tooth lower down, the two portions will fit together and form a square.

155.—THE PENTAGON AND SQUARE.

A REGULAR pentagon may be cut into as few as six pieces that will fit together without any turning over and form a square, as I shall show below. Hitherto the best answer has been in seven pieces—the solution produced some years ago by a foreign mathematician, Paul Busschop. We first form a parallelogram, and from that the square. The process will be seen in the diagram on the next page.

The pentagon is A B C D E. By the cut A C and the cut F M (F being the middle point between A and C, and M being the same distance from A as F) we get two pieces that may be placed in position at G H E A and form the parallelogram G H D C. We then find the mean proportional between the length H D and the *height* of the parallelogram. This distance we mark off from C at K, then draw C K, and from G drop the line G L, perpendicular to K C. The rest is easy and rather obvious. It will be seen that the six pieces will form either the pentagon or the square.

I have received what purported to be a solution in five pieces, but the method was based on the rather subtle fallacy that half the diagonal plus half the side of a pentagon equals the side of a square of the same area. I say subtle, because it is an extremely close approximation that will deceive the eye, and is quite difficult to prove inexact. I am not aware that attention has before been drawn to this curious approximation.

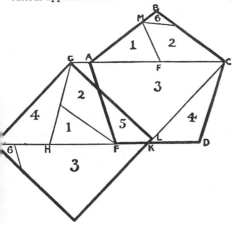

Another correspondent made the side of his square 1¼ of the side of the pentagon. As a matter of fact, the ratio is irrational. I calculate that if the side of the pentagon is 1—inch, foot, or anything else—the side of the square of equal area is 1.3117 nearly, or say roughly 1 $\frac{3}{10}$. So we can only hope to solve the puzzle by geometrical methods.

156.—THE DISSECTED TRIANGLE.

DIAGRAM A is our original triangle. We will say it measures 5 inches (or 5 feet) on each side. If we take off a slice at the bottom of any equilateral triangle by a cut parallel with the base, the portion that remains will always be an equilateral triangle; so we first cut off piece 1 and get a triangle 3 inches on every side. The manner of finding directions of the other cuts in A is obvious from the diagram.

Now, if we want two triangles, 1 will be one of them, and 2, 3, 4, and 5 will fit together, as in B, to form the other. If we want three equilateral triangles, 1 will be one, 4 and 5 will form the second, as in C, and 2 and 3 will form the third, as in D. In B and C the piece 5 is turned over; but there can be no objection to this, as it is not forbidden, and is in no way opposed to the nature of the puzzle.

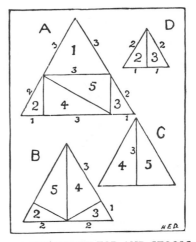

157.—THE TABLE-TOP AND STOOLS.

ONE object that I had in view when presenting this little puzzle was to point out the uncertainty of the meaning conveyed by the word "oval." Though originally derived from the Latin word *ovum*, an egg, yet what we understand as the egg-shape (with one end smaller than the other) is only one of many forms of the oval; while some eggs are spherical in shape, and a sphere or circle is most certainly not an oval. If we speak of an ellipse—a conical ellipse—we are on safer ground, but here we must be careful of error. I recollect a Liverpool town councillor, many years ago, whose ignorance of the poultry-yard led him to substitute the word "hen" for "fowl," remarking,

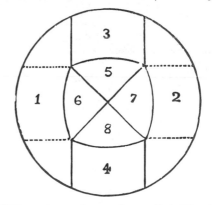

"We must remember, gentlemen, that although every cock is a hen, every hen is not a cock!" Similarly, we must always note that although every ellipse is an oval, every oval is not an

ellipse. It is correct to say that an oval is an oblong curvilinear figure, having two unequal

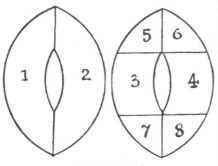

THE TWO STOOLS.

diameters, and bounded by a curve line returning into itself; and this includes the ellipse, but all other figures which in any way approach towards the form of an oval without necessarily having the properties above described are included in the term "oval." Thus the following solution that I give to our puzzle involves the pointed "oval," known among architects as the "vesica piscis."

The dotted lines in the table are given for greater clearness, the cuts being made along the other lines. It will be seen that the eight pieces form two stools of exactly the same size and shape with similar hand-holes. These holes are a trifle longer than those in the schoolmaster's stools, but they are much narrower and of considerably smaller area. Of course 5 and 6 can be cut out in one piece—also 7 and 8—making only *six pieces* in all. But I wished to keep the same number as in the original story.

When I first gave the above puzzle in a London newspaper, in competition, no correct solution was received, but an ingenious and neatly executed attempt by a man lying in a London infirmary was accompanied by the following note: "Having no compasses here, I was compelled to improvise a pair with the aid of a small penknife, a bit of firewood from a bundle, a piece of tin from a toy engine, a tin tack, and two portions of a hairpin, for points. They are a fairly serviceable pair of compasses, and I shall keep them as a memento of your puzzle."

158.—THE GREAT MONAD.

THE areas of circles are to each other as the squares of their diameters. If you have a circle 2 in. in diameter and another 4 in. in diameter, then one circle will be four times as great in area as the other, because the square of 4 is four times as great as the square of 2. Now, if we refer to Diagram 1, we see how two equal squares may be cut into four pieces that will form one larger square; from which it is self-evident that any square has just half the area of the square of its diagonal. In Diagram 2 I have introduced a square as it often occurs in ancient drawings of the Monad; which was my reason for believing that the symbol had mathematical meanings, since it will be found to demonstrate the fact that the area of the outer ring or annulus is exactly equal to the area of the inner circle. Compare Diagram 2 with Diagram 1, and you will see that as the square of the diameter C D is double the square of the diameter of the inner circle, or C E, therefore the area of the larger circle is double the area of the smaller one, and consequently the area of the annulus is exactly equal to that of the inner circle. This answers our first question.

In Diagram 3 I show the simple solution to the second question. It is obviously correct, and may be proved by the cutting and superposition of parts. The dotted lines will also

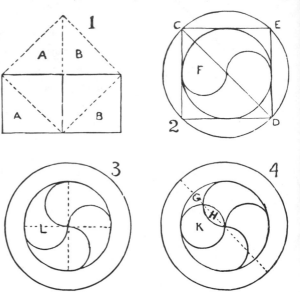

serve to make it evident. The third question is solved by the cut C D in Diagram 2, but it remains to be proved that the piece F is really one-half of the Yin or the Yan. This we will do in Diagram 4. The circle K has one-quarter the area of the circle containing Yin and Yan, because its diameter is just one-half the length. Also L in Diagram 3 is, we know, one-quarter the area. It is therefore evident that G is exactly equal to H, and therefore half G is equal to half H. So that what F loses from L it gains from K, and F must be half of Yin or Yan.

159.—THE SQUARE OF VENEER.

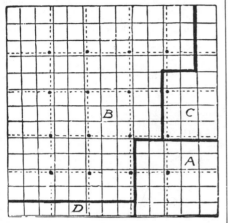

ANY square number may be expressed as the sum of two squares in an infinite number of different ways. The solution of the present puzzle forms a simple demonstration of this rule. It is a condition that we give actual dimensions.

wards determine. Divide the square as shown (where the dotted lines indicate the original markings) into 169 squares. As 169 is the sum of the two squares 144 and 25, we will proceed to divide the veneer into two squares, measuring respectively 12 × 12 and 5 × 5 ; and as we know that two squares may be formed from one square by dissection in four pieces, we seek a solution in this number. The dark lines in the diagram show where the cuts are to be made. The square 5 × 5 is cut out whole, and the larger square is formed from the remaining three pieces, B, C, and D, which the reader can easily fit together.

Now, n is clearly $\frac{1}{13}$ of an inch. Consequently our larger square must be $\frac{60}{13}$ in. × $\frac{60}{13}$ in., and our smaller square $\frac{25}{13}$ in. × $\frac{25}{13}$ in. The square of $\frac{60}{13}$ added to the square of $\frac{25}{13}$ is 25. The square is thus divided into as few as four pieces that form two squares of known dimensions, and all the sixteen nails are avoided.

Here is a general formula for finding two squares whose sum shall equal a given square, say a^2. In the case of the solution of our puzzle $p=3$, $q=2$, and $a=5$.

$$\frac{2pqa}{p^2+q^2} = x \; ; \; \frac{\sqrt{a^2(p^2+q^2)^2-(2pqa)^2}}{p^2+q^2} = y$$

Here $x^2+y^2=a^2$.

160.—THE TWO HORSESHOES.

THE puzzle was to cut the two shoes (including the hoof contained within the outlines) into four pieces, two pieces each, that would fit together and form a perfect circle. It was also stipulated that all four pieces should be different in shape. As a matter of fact, it is a puzzle based on the principle contained in that curious Chinese symbol the Monad. (See No. 158.)

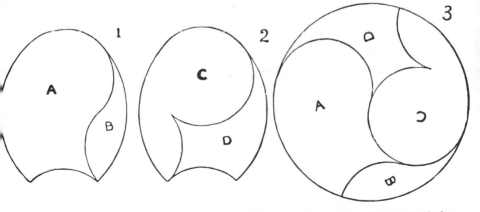

In this puzzle I ignore the known dimensions of our square and work on the assumption that it is 13n by 13n. The value of n we can after-

The above diagrams give the correct solution to the problem. It will be noticed that 1 and 2 are cut into the required four pieces, all different in shape, that fit together and form the perfect circle shown in Diagram 3. It will further be observed that the two pieces A and B of one shoe and the two pieces C and D of the other form two exactly similar halves of the circle—the Yin and the Yan of the great Monad. It will be seen that the shape of the horseshoe is more easily determined from the circle than the dimensions of the circle from the horseshoe, though the latter presents no difficulty when you know that the curve of the long side of the shoe is part of the circumference of your circle. The difference between B and D is instructive, and the idea is useful in all such cases where it is a condition that the pieces must be different in shape. In forming D we simply add on a symmetrical piece, a curvilinear square, to the piece B. Therefore, in giving either B or D a quarter turn before placing in the new position, a precisely similar effect must be produced.

161.—THE BETSY ROSS PUZZLE.

FOLD the circular piece of paper in half along the dotted line shown in Fig. 1, and divide the

162.—THE CARDBOARD CHAIN.

THE reader will probably feel rewarded for any care and patience that he may bestow on cutting out the cardboard chain. We will suppose that he has a piece of cardboard measuring 8 in. by 2½ in., though the dimensions are of no importance. Yet if you want a long chain you must, of course, take a long strip of cardboard. First rule pencil lines B B and C C, half an inch from the edges, and also the short perpendicular lines half an inch apart. (See next page.) Rule lines on the other side in just the same way, and in order that they shall coincide it is well to prick through the card with a needle the points where the short lines end. Now take your penknife and split the card from A A down to B B, and from D D up to C C. Then cut right through the card along all the short perpendicular lines, and half through the card along the short portions of B B and C C that are not dotted. Next turn the card over and cut half through along the short lines on B B and C C at the places that are immediately beneath the dotted lines on the upper side. With a little careful separation of the parts with the penknife, the cardboard may

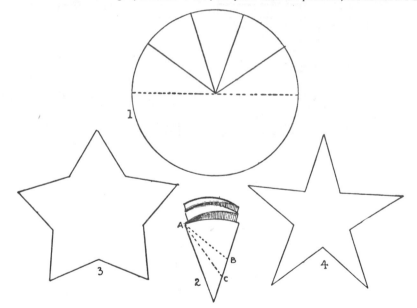

upper half into five equal parts as indicated. Now fold the paper along the lines, and it will have the appearance shown in Fig. 2. If you want a star like Fig. 3, cut from A to B ; if you wish one like Fig. 4, cut from A to C. Thus, the nearer you cut to the point at the bottom the longer will be the points of the star, and the farther off from the point that you cut the shorter will be the points of the star.

now be divided into two interlacing ladder-like portions, as shown in Fig. 2; and if you cut away all the shaded parts you will get the chain, cut solidly out of the cardboard, without any join, as shown in the illustrations on page 40.

It is an interesting variant of the puzzle to cut out two keys on a ring—in the same manner without join.

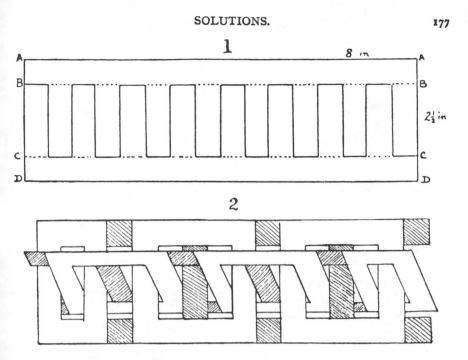

164.—THE POTATO PUZZLE.

As many as twenty-two pieces may be obtained by the six cuts. The illustration shows a pretty symmetrical solution. The rule in such cases

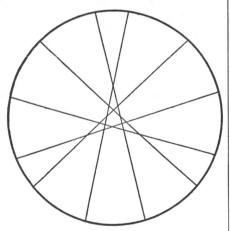

is that every cut shall intersect every other cut and no two intersections coincide; that is to say, every line passes through every other line, but more than two lines do not cross at the same point anywhere. There are other ways of making the cuts, but this rule must always be observed if we are to get the full number of pieces.

The general formula is that with n cuts we can always produce $\dfrac{n(n+1)}{2}+1$ pieces. One of the problems proposed by the late Sam Loyd was to produce the maximum number of pieces by n straight cuts through a solid cheese. Of course, again, the pieces cut off may not be moved or piled. Here we have to deal with the intersection of planes (instead of lines), and the general formula is that with n cuts we may produce $\dfrac{(n-1)n(n+1)}{6}+n+1$ pieces. It is extremely difficult to " see " the direction and effects of the successive cuts for more than a few of the lowest values of n.

165.—THE SEVEN PIGS.

THE illustration shows the direction for placing the three fences so as to enclose every pig in a separate sty. The greatest number of spaces that can be enclosed with three straight lines in a square is seven, as shown in the last puzzle. Bearing this fact in mind, the puzzle must be solved by trial.

THE SEVEN PIGS.

166.—THE LANDOWNER'S FENCES.

FOUR fences only are necessary, as follows :—

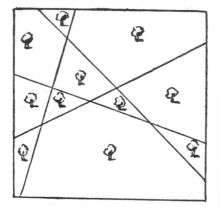

167.—THE WIZARD'S CATS.

THE illustration requires no explanation. It shows clearly how the three circles may be drawn so that every cat has a separate enclosure, and cannot approach another cat without crossing a line.

168.—THE CHRISTMAS PUDDING.

THE illustration shows how the pudding may be cut into two parts of exactly the same size and shape. The lines must necessarily pass through the points A, B, C, D, and E. But, subject to this condition, they may be varied in an infinite number of ways. For example, at a point midway between A and the edge, the

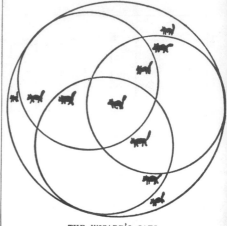

THE WIZARD'S CATS.

line may be completed in an unlimited number of ways (straight or crooked), provided it be

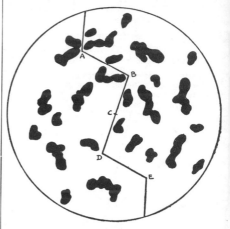

exactly reflected from E to the opposite edge. And similar variations may be introduced at other places.

169.—A TANGRAM PARADOX.

THE diagrams will show how the figures are constructed—each with the seven Tangrams. It will be noticed that in both cases the head, hat, and arm are precisely alike, and the width at the base of the body the same. But this body contains four pieces in the first case, and in the second design only three. The first is larger than the second by exactly that narrow strip indicated by the dotted line between A and B.

This strip is therefore exactly equal in area to the piece forming the foot in the other design,

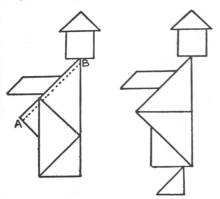

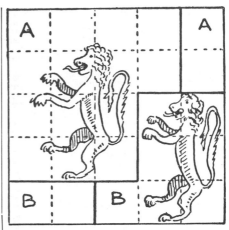

though when thus distributed along the side of the body the increased dimension is not easily apparent to the eye.

170.—THE CUSHION COVERS.

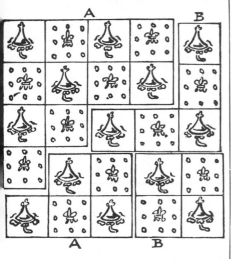

THE two pieces of brocade marked A will fit together and form one perfect square cushion top, and the two pieces marked B will form the other.

171.—THE BANNER PUZZLE.

THE illustration explains itself. Divide the bunting into 25 squares (because this number is the sum of two other squares—16 and 9), and then cut along the thick lines. The two pieces marked A form one square, and the two pieces marked B form the other.

172.—MRS. SMILEY'S CHRISTMAS PRESENT.

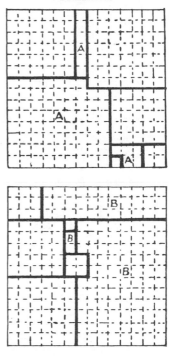

THE first step is to find six different square numbers that sum to 196. For example, $1+4+25+36+49+81 = 196$; $1+4+9+25+36+121 = 196$; $1+9+16+25+64+81 = 196$. The rest calls for individual judgment and ingenuity, and no definite rules can be given for procedure. The annexed diagrams will show solutions for the first two cases stated. Of course the three pieces marked A and those marked B will fit together and form a square in each case. The assembling of the parts may be slightly varied, and the reader may be interested in finding a solution for the third set of squares I have given.

173.—MRS. PERKINS'S QUILT.

THE following diagram shows how the quilt should be constructed.

There is, I believe, practically only one solution to this puzzle. The fewest separate squares must be eleven. The portions must be of the sizes given, the three largest pieces must be arranged as shown, and the remaining group of eight squares may be "reflected," but cannot be differently arranged.

174.—THE SQUARES OF BROCADE.

So far as I have been able to discover, there is only one possible solution to fulfil the conditions.

The pieces fit together as in Diagram 1, Diagrams 2 and 3 showing how the two original

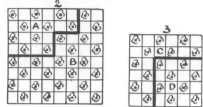

squares are to be cut. It will be seen that the pieces A and C have each twenty chequers, and are therefore of equal area. Diagram 4 (built up with the dissected square No. 5) solves the puzzle, except for the small condition contained in the words, " I cut the *two* squares in the manner desired." In this case the smaller square is preserved intact. Still I give it as

an illustration of a feature of the puzzle. It is impossible in a problem of this kind to give a *quarter-turn* to any of the pieces if the pattern is to properly match, but (as in the case of F, in Diagram 4) we may give a symmetrical piece a *half-turn* — that is, turn it upside down. Whether or not a piece may be given a quarter-turn, a half-turn, or no turn at all in these chequered problems, depends on the character of the design, on the material employed, and also on the form of the piece itself.

175.—ANOTHER PATCHWORK PUZZLE.

THE lady need only unpick the stitches along the dark lines in the larger portion of patchwork,

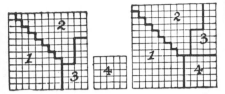

when the four pieces will fit together and form a square, as shown in our illustration.

176.—LINOLEUM CUTTING.

THERE is only one solution that will enable us to retain the larger of the two pieces with as little as possible cut from it. Fig. 1 in the following diagram shows how the smaller piece is to be cut, and Fig. 2 how we should dissect

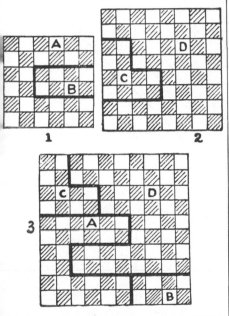

the larger piece, while in Fig. 3 we have the new square 10 × 10 formed by the four pieces with all the chequers properly matched. It will be seen that the piece D contains fifty-two chequers, and this is the largest piece that it is possible to preserve under the conditions.

177.—ANOTHER LINOLEUM PUZZLE.

CUT along the thick lines, and the four pieces will fit together and form a perfect square in the manner shown in the smaller diagram.

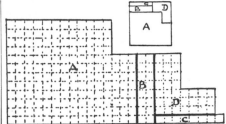

ANOTHER LINOLEUM PUZZLE.

178.—THE CARDBOARD BOX.

THE areas of the top and side multiplied together and divided by the area of the end give the square of the length. Similarly, the product of top and end divided by side gives the square of the breadth ; and the product of side and end divided by the top gives the square of the depth. But we only need one of these operations. Let us take the first. Thus, 120 × 96 divided by 80 equals 144, the square of 12. Therefore the length is 12 inches, from which we can, of course, at once get the breadth and depth—10 in. and 8 in. respectively.

179.—STEALING THE BELL-ROPES.

WHENEVER we have one side (a) of a right-angled triangle, and know the difference between the second side and the hypotenuse (which difference we will call b), then the length of the hypotenuse will be $\dfrac{a^2}{2b}+\dfrac{b}{2}$. In the case of our puzzle this will be $\dfrac{48 \times 48}{6}+1\frac{1}{2}$ in. = 32 ft. $1\frac{1}{2}$ in., which is the length of the rope.

180.—THE FOUR SONS.

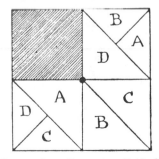

THE diagram shows the most equitable division of the land possible, " so that each son shall receive land of exactly the same area and exactly similar in shape," and so that each shall have access to the well in the centre without trespass on another's land. The conditions do not require that each son's land shall be in one piece, but it is necessary that the two portions assigned to an individual should be kept apart, or two adjoining portions might be held to be one piece, in which case the condition as to shape would have to be broken. At present there is only one shape for each

183.—DRAWING A SPIRAL.

MAKE a fold in the paper, as shown by the dotted line in the illustration. Then, taking any two points, as A and B, describe semicircles on the line alternately from the centres B and A, being careful to make the ends join,

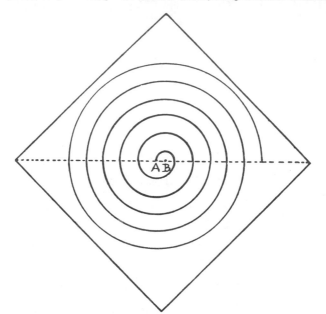

piece of land—half a square divided diagonally. And A, B, C, and D can each reach their land from the outside, and have each equal access to the well in the centre.

181.—THE THREE RAILWAY STATIONS.

THE three stations form a triangle, with sides 13, 14, and 15 miles. Make the 14 side the base ; then the height of the triangle is 12 and the area 84. Multiply the three sides together and divide by four times the area. The result is eight miles and one-eighth, the distance required.

182.—THE GARDEN PUZZLE.

HALF the sum of the four sides is 144. From this deduct in turn the four sides, and we get 64, 99, 44, and 81. Multiply these together, and we have as the result the square of 4,752. Therefore the garden contained 4,752 square yards. Of course the tree being equidistant from the four corners shows that the garden is a quadrilateral that may be inscribed in a circle.

and the thing is done. Of course this is not a *true* spiral, but the puzzle was to produce the *particular* spiral that was shown, and that was drawn in this simple manner.

184.—HOW TO DRAW AN OVAL.

IF you place your sheet of paper round the surface of a cylindrical bottle or canister, the oval can be drawn with one sweep of the compasses.

185.—ST. GEORGE'S BANNER.

As the flag measures 4 ft. by 3 ft., the length of the diagonal (from corner to corner) is 5 ft. All you need do is to deduct half the length of this diagonal (2½ ft.) from a quarter of the distance all round the edge of the flag (3½ ft.)—a quarter of 14 ft. The difference (1 ft.) is the required width of the arm of the red cross. The area of the cross will then be the same as that of the white ground.

186.—THE CLOTHES LINE PUZZLE.

MULTIPLY together, and also add together, the heights of the two poles and divide one result by the other. That is, if the two heights are a and b respectively, then $\dfrac{ab}{a+b}$ will give the height of the intersection. In the particular case of our puzzle, the intersection was therefore 2 ft. 11 in. from the ground. The distance that the poles are apart does not affect the answer. The reader who may have imagined that this was an accidental omission will perhaps be interested in discovering the reason why the distance between the poles may be ignored.

187.—THE MILKMAID PUZZLE.

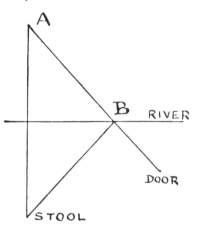

DRAW a straight line, as shown in the diagram, from the milking-stool perpendicular to the near bank of the river, and continue it to the point A, which is the same distance from that bank as the stool. If you now draw the straight line from A to the door of the dairy, it will cut the river at B. Then the shortest route will be from the stool to B and thence to the door. Obviously the shortest distance from A to the door is the straight line, and as the distance from the stool to any point of the river is the same as from A to that point, the correctness of the solution will probably appeal to every reader without any acquaintance with geometry.

188.—THE BALL PROBLEM.

IF a round ball is placed on the level ground, six similar balls may be placed round it (all on the ground), so that they shall all touch the central ball.

As for the second question, the ratio of the diameter of a circle to its circumference we call pi, and though we cannot express this ratio in exact numbers, we can get sufficiently near to it for all practical purposes. However, in this case it is not necessary to know the value of pi at all. Because, to find the area of the surface of a sphere we multiply the square of the diameter by pi; to find the volume of a sphere

we multiply the cube of the diameter by one-sixth of pi. Therefore we may ignore pi, and have merely to seek a number whose square shall equal one-sixth of its cube. This number is obviously 6. Therefore the ball was 6 ft. in diameter, the area of its surface will be 36 times pi in square feet, and its volume also 36 times pi in cubic feet.

189.—THE YORKSHIRE ESTATES.

THE triangular piece of land that was not for sale contains exactly eleven acres. Of course it is not difficult to find the answer if we follow the eccentric and tricky tracks of intricate trigonometry ; or I might say that the application of a well-known formula reduces the problem to finding one-quarter of the square root of $(4 \times 370 \times 116) - (370 + 116 - 74)^2$—that is a quarter of the square root of 1936, which is one-quarter of 44, or 11 acres. But all that the reader really requires to know is the Pythagorean law on which many puzzles have been built, that in any right-angled triangle the square of the hypotenuse is equal to the sum of the squares of the other two sides. I shall dispense with all " surds " and similar absurdities, notwithstanding the fact that the sides of our triangle are clearly incommensurate, since we cannot exactly extract the square roots of the three square areas.

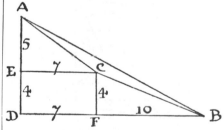

In the above diagram A B C represents our triangle. A D B is a right-angled triangle, A D measuring 9 and B D measuring 17, because the square of 9 added to the square of 17 equals 370, the known area of the square on A B. Also A E C is a right-angled triangle, and the square of 5 added to the square of 7 equals 74, the square estate on A C. Similarly, C F B is a right-angled triangle, for the square of 4 added to the square of 10 equals 116, the square estate on B C. Now, although the sides of our triangular estate are incommensurate, we have in this diagram all the exact figures that we need to discover the area with precision.

The area of our triangle A D B is clearly half of 9×17, or $76\frac{1}{2}$ acres. The area of A E C is half of 5×7, or $17\frac{1}{2}$ acres ; the area of C F B is half of 4×10, or 20 acres ; and the area of the oblong E D F C is obviously 4×7, or 28 acres. Now, if we add together $17\frac{1}{2}$, 20, and

28=65½, and deduct this sum from the area of the large triangle A D B (which we have found to be 76½ acres), what remains must clearly be the area of A B C. That is to say, the area we want must be 76½ − 65½=11 acres exactly.

190.—FARMER WURZEL'S ESTATE.

THE area of the complete estate is exactly one hundred acres. To find this answer I use the following little formula, $\dfrac{\sqrt{4ab-(a+b-c)^2}}{4}$ where a, b, c represent the three square areas, in any order. The expression gives the area of the triangle A. This will be found to be 9 acres. It can be easily proved that A, B, C, and D are all equal in area ; so the answer is 26+20+ 18+9+9+9+9=100 acres.

Here is the proof. If every little dotted square in the diagram represents an acre, this must be

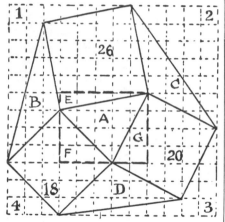

a correct plan of the estate, for the squares of 5 and 1 together equal 26 ; the squares of 4 and 2 equal 20 ; and the squares of 3 and 3 added together equal 18. Now we see at once that the area of the triangle E is 2½, F is 4½, and G is 4. These added together make 11 acres, which we deduct from the area of the rectangle, 20 acres, and we find that the field A contains exactly 9 acres. If you want to prove that B, C, and D are equal in size to A, divide them in two by a line from the middle of the longest side to the opposite angle, and you will find that the two pieces in every case, if cut out, will exactly fit together and form A.

Or we can get our proof in a still easier way. The complete area of the squared diagram is 12×12=144 acres, and the portions 1, 2, 3, 4, not included in the estate, have the respective areas of 12½, 17½, 9½, and 4½. These added together make 44, which, deducted from 144, leaves 100 as the required area of the complete estate.

191.—THE CRESCENT PUZZLE.

REFERRING to the original diagram, let A C be x, let C D be $x−9$, and let E C be $x−5$. Then $x−5$ is a mean proportional between $x−9$ and x, from which we find that x equals 25. Therefore the diameters are 50 in. and 41 in. respectively.

192.—THE PUZZLE WALL.

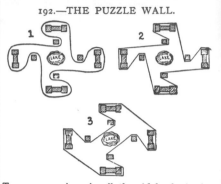

THE answer given in all the old books is that shown in Fig. 1, where the curved wall shuts out the cottages from access to the lake. But in seeking the direction for the "shortest possible" wall most readers to-day, remembering that the shortest distance between two points is a straight line, will adopt the method shown in Fig. 2. This is certainly an improvement, yet the correct answer is really that indicated in Fig. 3. A measurement of the lines will show that there is a considerable saving of length in this wall.

193.—THE SHEEP-FOLD.

THIS is the answer that is always given and accepted as correct : Two more hurdles would be necessary, for the pen was twenty-four by one (as in Fig. A) and on next page, and by moving one of the sides and placing an extra hurdle at each end (as in Fig. B) the area would be doubled. The diagrams are not to scale. Now there is no condition in the puzzle that requires the sheep-fold to be of any particular form. But even if we accept the point that the pen was twenty-four by one, the answer utterly fails, for two extra hurdles are certainly not at all necessary. For example, I arrange the fifty hurdles as in Fig. C, and as the area is increased from twenty-four "square hurdles" to 156, there is now accommodation for 650 sheep. If it be held that the area must be exactly double that of the original pen, then I construct it (as in Fig. D) with twenty-eight hurdles only, and have twenty-two in hand for other purposes on the farm. Even if it were insisted that all the original hurdles must be used, then I should construct it as in Fig. E, where I can get the area as exact as any farmer could possibly require, even if we have to allow for the fact that the sheep might not be able to graze at the extreme ends. Thus we see that, from any

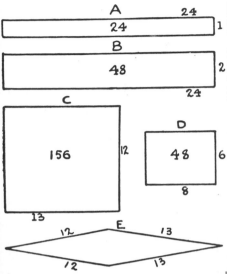

point of view, the accepted answer to this ancient little puzzle breaks down. And yet attention has never before been drawn to the absurdity.

194.—THE GARDEN WALLS.

THE puzzle was to divide the circular field into four equal parts by three walls, each wall being of exactly the same length. There are two essential difficulties in this problem. These are : (1) the thickness of the walls, and (2) the condition that these walls are three in number. As to the first point, since we are told that the walls are brick walls, we clearly cannot ignore their thickness, while we have to find a solution that will equally work, whether the walls be of a thickness of one, two, three, or more bricks.

The second point requires a little more consideration. How are we to distinguish between a wall and walls ? A straight wall without any bend in it, no matter how long, cannot ever become "walls," if it is neither broken nor intersected in any way. Also our circular field is clearly enclosed by one wall. But if it had happened to be a square or a triangular enclosure, would there be respectively four and three walls or only one enclosing wall in each case ? It is true that we speak of "the four walls " of a square building or garden, but this is only a conventional way of saying "the four sides." If you were speaking of the actual brick-work, you would say, "I am going to enclose this square garden with a wall." Angles clearly do not affect the question, for we may have a zig-zag wall just as well as a straight one, and the Great Wall of China is a good example of a wall with plenty of angles. Now, if you look at Diagrams 1, 2, and 3, you may be puzzled to

declare whether there are in each case two or four new walls ; but you cannot call them three, as required in our puzzle. The intersection either affects the question or it does not affect it.

If you tie two pieces of string firmly together, or splice them in a nautical manner, they become " one piece of string." If you simply let them lie across one another or overlap, they remain " two pieces of string." It is all a question of joining and welding. It may similarly be held that if two walls be built into one another—I might almost say, if they be made homogeneous—they become one wall, in which case Diagrams 1, 2, and 3 might each be said to show one wall or two, if it be indicated that the four ends only touch, and are not really built into the outer circular wall.

The objection to Diagram 4 is that although it shows the three required walls (assuming the ends are not built into the outer circular wall), yet it is only absolutely correct when we assume the walls to have no thickness. A brick has thickness, and therefore the fact throws the whole method out and renders it only approximately correct.

Diagram 5 shows, perhaps, the only correct and perfectly satisfactory solution. It will be noticed that, in addition to the circular wall, there are three new walls, which touch (and so enclose) but are not built into one another. This solution may be adapted to any desired thickness of wall, and its correctness as to area and length of wall space is so obvious that it is unnecessary to explain it. I will,

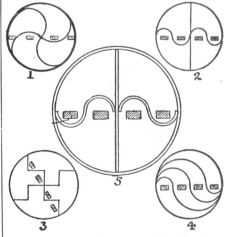

however, just say that the semicircular piece of ground that each tenant gives to his neighbour is exactly equal to the semicircular piece that his neighbour gives to him, while any section of wall space found in one garden is precisely repeated in all the others. Of course there is an infinite number of ways in which this solution may be correctly varied.

195.—LADY BELINDA'S GARDEN.

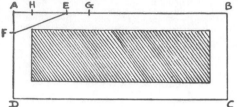

ALL that Lady Belinda need do was this : She should measure from A to B, fold her tape in four and mark off the point E, which is thus one quarter of the side. Then, in the same way, mark off the point F, one-fourth of the side A D Now, if she makes E G equal to A F, and G H equal to E F, then A H is the required width for the path in order that the bed shall be exactly half the area of the garden. An exact numerical measurement can only be obtained when the sum of the squares of the two sides is a square number. Thus, if the garden measured 12 poles by 5 poles (where the squares of 12 and 5, 144 and 25, sum to 169, the square of 13), then 12 added to 5, less 13, would equal four, and a quarter of this, 1 pole, would be the width of the path.

196.—THE TETHERED GOAT.

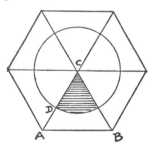

THIS problem is quite simple if properly attacked. Let us suppose the triangle A B C to represent our half-acre field, and the shaded portion to be the quarter-acre over which the goat will graze when tethered to the corner C. Now, as six equal equilateral triangles placed together will form a regular hexagon, as shown, it is evident that the shaded pasture is just one-sixth of the complete area of a circle. Therefore all we require is the radius (CD) of a circle containing six quarter-acres or 1½ acres, which is equal to 9,408,960 square inches. As we only want our answer " to the nearest inch," it is sufficiently exact for our purpose if we assume that as 1 is to 3.1416, so is the diameter of a circle to its circumference. If, therefore, we divide the last number I gave by 3.1416, and extract the square root, we find that 1,731

inches, or 48 yards 3 inches, is the required length of the tether " to the nearest inch."

197.—THE COMPASSES PUZZLE.

LET A B in the following diagram be the given straight line. With the centres A and B and radius A B describe the two circles. Mark off D E and E F equal to A D. With the centres A and F and radius D F describe arcs intersecting at G. With the centres A and B and distance B G describe arcs G H K and N. Make H K equal to A B and H L equal to H B. Then with centres K and L and radius A B describe arcs intersecting at I. Make B M equal to B I. Finally, with

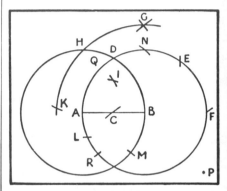

the centre M and radius M B cut the line in C, and the point C is the required middle of the line A B. For greater exactitude you can mark off R from A (as you did M from B), and from R describe another arc at C. This also solves the problem, to find a point midway between two given points without the straight line.

I will put the young geometer in the way of a rigid proof. First prove that twice the square of the line A B equals the square of the distance B G, from which it follows that H A B N are the four corners of a square. To prove that I is the centre of this square, draw a line from H to P through Q I B and continue the arc H K to P. Then, conceiving the necessary lines to be drawn, the angle H K P, being in a semicircle, is a right angle. Let fall the perpendicular K Q, and by similar triangles, and from the fact that H K I is an isosceles triangle by the construction, it can be proved that H I is half of H B. We can similarly prove that C is the centre of the square of which A I B are three corners.

I am aware that this is not the simplest possible solution.

198.—THE EIGHT STICKS.

THE first diagram is the answer that nearly every one will give to this puzzle, and at first sight it seems quite satisfactory. But consider the conditions. We have to lay " every one of the sticks on the table." Now, if a ladder be placed against a wall with only one end on the ground, it can hardly be said that it is " laid on

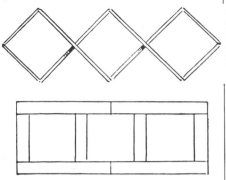

the ground." And if we place the sticks in the above manner, it is only possible to make one end of two of them touch the table : to say that every one lies on the table would not be correct. To obtain a solution it is only necessary to have our sticks of proper dimensions. Say the long sticks are each 2 ft. in length and the short ones 1 ft. Then the sticks must be 3 in. thick, when the three equal squares may be enclosed, as shown in the second diagram. If I had said " matches " instead of " sticks," the puzzle would be impossible, because an ordinary match is about twenty-one times as long as it is broad, and the enclosed rectangles would not be squares.

199.—PAPA'S PUZZLE.

I HAVE found that a large number of people imagine that the following is a correct solution of the problem. Using the letters in the diagram below, they argue that if you make the distance B A one-third of B C, and therefore the area of the rectangle A B E equal to that of the triangular remainder, the card must hang

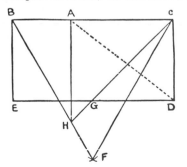

with the long side horizontal. Readers will remember the jest of Charles II., who induced the Royal Society to meet and discuss the

reason why the water in a vessel will not rise if you put a live fish in it ; but in the middle of the proceedings one of the least distinguished among them quietly slipped out and made the experiment, when he found that the water *did* rise ! If my correspondents had similarly made the experiment with a piece of cardboard, they would have found at once their error. Area is one thing, but gravitation is quite another. The fact of that triangle sticking its leg out to D has to be compensated for by additional area in the rectangle. As a matter of fact, the ratio of B A to A C is as 1 is to the square root of 3, which latter cannot be given in an exact numerical measure, but is approximately 1.732. Now let us look at the correct general solution. There are many ways of arriving at the desired result, but the one I give is, I think, the simplest for beginners.

Fix your card on a piece of paper and draw the equilateral triangle B C F, B F and C F being equal to B C. Also mark off the point G so that D G shall equal D C. Draw the line C G and produce it until it cuts the line B F in H. If we now make H A parallel to B E, then A is the point from which our cut must be made to the corner D, as indicated by the dotted line.

A curious point in connection with this problem is the fact that the position of the point A is independent of the side C D. The reason for this is more obvious in the solution I have given than in any other method that I have seen, and (although the problem may be solved with all the working on the cardboard) that is partly why I have preferred it. It will be seen at once that however much you may reduce the width of the card by bringing E nearer to B and D nearer to C, the line C G, being the diagonal of a square, will always lie in the same direction, and will cut B F in H. Finally, if you wish to get an approximate measure for the distance B A, all you have to do is to multiply the length of the card by the decimal .366. Thus, if the card were 7 inches long, we get 7×.366=2.562, or a little more than 2½ inches, for the distance from B to A.

But the real joke of the puzzle is this : We have seen that the position of the point A is independent of the width of the card, and depends entirely on the length. Now, in the illustration it will be found that both cards have the same length ; consequently all the little maid had to do was to lay the clipped card on top of the other one and mark off the point A at precisely the same distance from the top left-hand corner ! So, after all, Pappus' puzzle, as he presented it to his little maid, was quite an infantile problem, when he was able to show her how to perform the feat without first introducing her to the elements of statics and geometry.

200.—A KITE-FLYING PUZZLE.

SOLVERS of this little puzzle, I have generally found, may be roughly divided into two classes : those who get within a mile of the correct answer by means of more or less complex calculations, involving "*pi*," and those whose arithmetical kites fly hundreds and thousands of miles away from the truth. The comparatively easy method that I shall show does not involve any consideration of the ratio that the diameter of a circle bears to its circumference. I call it the "hat-box method."

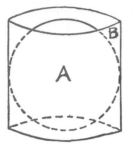

Supposing we place our ball of wire, A, in a cylindrical hat-box, B, that exactly fits it, so that it touches the side all round and exactly touches the top and bottom, as shown in the illustration. Then, by an invariable law that should be known by everybody, that box contains exactly half as much again as the ball. Therefore, as the ball is 24 in. in diameter, a hat-box of the same circumference but two-thirds of the height (that is, 16 in. high) will have exactly the same contents as the ball.

Now let us consider that this reduced hatbox is a cylinder of metal made up of an immense number of little wire cylinders close together like the hairs in a painter's brush. By the conditions of the puzzle we are allowed to consider that there are no spaces between the wires. How many of these cylinders one onehundredth of an inch thick are equal to the large cylinder, which is 24 in. thick? Circles are to one another as the squares of their diameters. The square of $\frac{1}{100}$ is $\frac{1}{10000}$, and the square of 24 is 576; therefore the large cylinder contains 5,760,000 of the little wire cylinders. But we have seen that each of these wires is 16 in. long; hence $16 \times 5,760,000 = 92,160,000$ inches as the complete length of the wire. Reduce this to miles, and we get 1,454 miles 2,880 ft. as the length of the wire attached to the professor's kite.

Whether a kite would fly at such a height, or support such a weight, are questions that do not enter into the problem.

201.—HOW TO MAKE CISTERNS.

HERE is a general formula for solving this problem. Call the two sides of the rectangle *a* and *b*. Then $\dfrac{a+b-\sqrt{a^2+b^2-ab}}{6}$ equals the side of the little square pieces to cut away. The measurements given were 8 ft. by 3 ft., and the above rule gives 8 in. as the side of the square pieces that have to be cut away. Of course,

it will not always come out exact, as in this case (on account of that square root), but you can get as near as you like with decimals.

202.—THE CONE PUZZLE.

THE simple rule is that the cone must be cut at one-third of its altitude.

203.—CONCERNING WHEELS.

IF you mark a point A on the circumference of a wheel that runs on the surface of a level road, like an ordinary cart-wheel, the curve described by that point will be a common cycloid, as in Fig. 1. But if you mark a point B on the circumference of the flange of a locomotive-wheel, the curve will be a curtate cycloid, as in Fig. 2, terminating in nodes. Now, if we consider one of these nodes or loops, we shall see that "at any given moment" certain points at the bottom of the loop must be moving in the opposite direction to the train. As there is an infinite

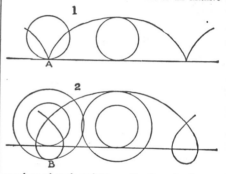

number of such points on the flange's circumference, there must be an infinite number of these loops being described while the train is in motion. In fact, at any given moment certain points on the flanges are always moving in a direction opposite to that in which the train is going.

In the case of the two wheels, the wheel that runs round the stationary one makes two revolutions round its own centre. As both wheels are of the same size, it is obvious that if at the start we mark a point on the circumference of the upper wheel, at the very top, this point will be in contact with the lower wheel at its lowest part when half the journey has been made. Therefore this point is again at the top of the moving wheel, and one revolution has been made. Consequently there are two such revolutions in the complete journey.

204.—A NEW MATCH PUZZLE.

1. THE easiest way is to arrange the eighteen matches as in Diagrams 1 and 2, making the length of the perpendicular A B equal to a match and a half. Then, if the matches are an inch in

length, Fig. 1 contains two square inches and
Fig. 2 contains six square inches—4 × 1½. The
second case (2) is a little more difficult to solve.
The solution is given in Figs. 3 and 4. For the

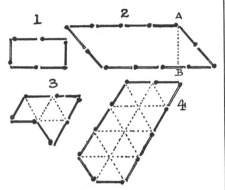

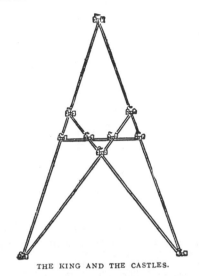

THE KING AND THE CASTLES.

purpose of construction, place matches tempo-
rarily on the dotted lines. Then it will be seen
that as 3 contains five equal equilateral triangles and 4 contains fifteen similar triangles,
one figure is three times as large as the other,
and exactly eighteen matches are used.

205.—THE SIX SHEEP-PENS.

PLACE the twelve matches in the manner shown
in the illustration, and you will have six pens of
equal size.

206.—THE KING AND THE CASTLES.

THERE are various ways of building the ten
castles so that they shall form five rows with
four castles in every row, but the arrangement
in the next column is the only one that also
provides that two castles (the greatest number
possible) shall not be approachable from the
outside. It will be seen that you must cross
the walls to reach these two.

207.—CHERRIES AND PLUMS.

THERE are several ways in which this problem
might be solved were it not for the condition
that as few cherries and plums as possible shall
be planted on the north and east sides of the
orchard. The best possible arrangement is that
shown in the diagram, where the cherries, plums,

and apples are indicated respectively by the
letters C, P, and A. The dotted lines connect
the cherries, and the other lines the plums. It
will be seen that the ten cherry trees and the
ten plum trees are so planted that each fruit
forms five lines with four trees of its kind in

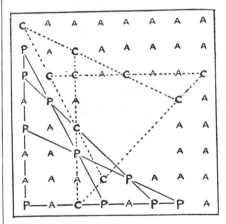

line. This is the only arrangement that allows
of so few as two cherries or plums being planted
on the north and east outside rows.

208.—A PLANTATION PUZZLE.

THE illustration shows the ten trees that must
be left to form five rows with four trees in every

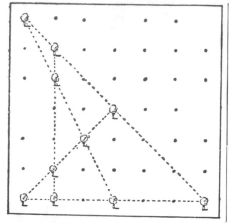

row. The dots represent the positions of the trees that have been cut down.

209.—THE TWENTY-ONE TREES.

I GIVE two pleasing arrangements of the trees. In each case there are twelve straight rows with five trees in every row.

As all the points and lines puzzles that I have given so far, excepting the last, are variations of the case of ten points arranged to form five lines of four, it will be well to consider this particular case generally. There are six fundamental solutions, and no more, as shown in the six diagrams. These, for the sake of convenience, I named some years ago the Star, the Dart,

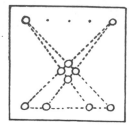

the Compasses, the Funnel, the Scissors, and the Nail. (See next page.) Readers will understand that any one of these forms may be distorted in an infinite number of different ways without destroying its real character.

In "The King and the Castles" we have the Star, and its solution gives the Compasses. In the "Cherries and Plums" solution we find that the Cherries represent the Funnel and the Plums the Dart. The solution of the "Plan-

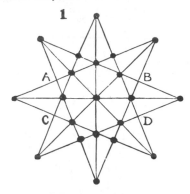

1

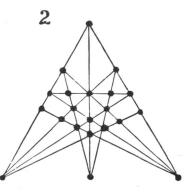

2

210.—THE TEN COINS.

THE answer is that there are just 2,400 different ways. Any three coins may be taken from one side to combine with one coin taken from the other side. I give four examples on this and the next page. We may thus select three from the top in ten ways and one from the bottom in five ways, making fifty. But we may also select three from the bottom and one from the top in fifty ways. We may thus select the four coins in one hundred ways, and the four removed may be arranged by permutation in twenty-four ways. Thus there are $24 \times 100 = 2,400$ different solutions.

tation Puzzle" is an example of the Dart distorted. Any solution to the "Ten Coins" will represent the Scissors. Thus examples of all have been given except the Nail.

On a reduced chessboard, 7 by 7, we may place the ten pawns in just three different ways, but they must all represent the Dart. The "Plantation" shows one way, the Plums a second way, and the reader may like to find the third way for himself. On an ordinary chessboard, 8 by 8, we can also get in a beautiful example of the Funnel—symmetrical in relation to the diagonal of the board. The smallest board that will take a Star is one 9 by 7. The Nail requires a board 11 by 7, the Scissors

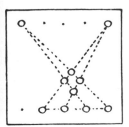
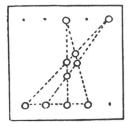

11 by 9, and the Compasses 17 by 12. At least these are the best results recorded in my note-book. They may be beaten, but I do not think so.

If you divide a chessboard into two parts by a diagonal zigzag line, so that the larger part contains 36 squares and the smaller part 28

tions, it is clearly necessary to find what is the smallest number of heads that could form sixteen lines with three heads in every line. Note that I say sixteen, and not thirty-two, because every line taken by a bullet may be also taken by another bullet fired in exactly

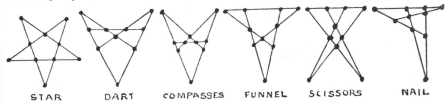

STAR   DART   COMPASSES   FUNNEL   SCISSORS   NAIL

squares, you can place three separate schemes on the larger part and one on the smaller part (all Darts) without their conflicting—that is, they occupy forty different squares. They can be placed in other ways without a division of the board. The smallest square board that will contain six different schemes (not funda-mentally different), without any line of one scheme crossing the line of another, is 14 by 14; and the smallest board that will contain one scheme entirely enclosed within the lines of a second scheme, without any of the lines of the one, when drawn from point to point, crossing a line of the other, is 14 by 12.

211.—THE TWELVE MINCE-PIES.

IF you ignore the four black pies in our illus-tration, the remaining twelve are in their origi-nal positions. Now remove the four detached pies to the places occupied by the black ones, and you will have your seven straight rows of four, as shown by the dotted lines.

212.—THE BURMESE PLANTATION.

THE arrangement on the next page is the most symmetrical answer that can probably be found for twenty-one rows, which is, I believe, the greatest number of rows possible. There are several ways of doing it.

213.—TURKS AND RUSSIANS.

THE main point is to discover the smallest pos-sible number of Russians that there could have been. As the enemy opened fire from all direc-

the opposite direction. Now, as few as eleven points, or heads, may be arranged to form the

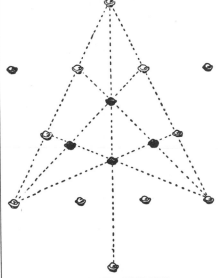

THE TWELVE MINCE PIES.

required sixteen lines of three, but the discovery of this arrangement is a hard nut. The diagram

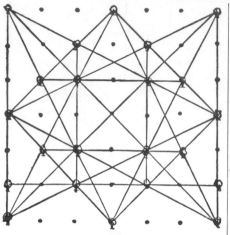

THE BURMESE PLANTATION.

at the foot of this page will show exactly how the thing is to be done.

If, therefore, eleven Russians were in the positions shown by the stars, and the thirty-two Turks in the positions indicated by the black dots, it will be seen, by the lines shown, that each Turk may fire exactly over the heads of three Russians. But as each bullet kills a man, it is essential that every Turk shall shoot one of his comrades and be shot by him in turn; otherwise we should have to provide extra Russians to be shot, which would be destructive of the correct solution of our problem. As the firing was simultaneous, this point presents no difficulties. The answer we thus see is that there were at least eleven Russians amongst whom there was no casualty, and that all the thirty-two Turks were shot by one another. It was not stated whether the Russians fired any shots, but it will be evident that even if they did their firing could not have been effective: for if one of their bullets killed a Turk, then we have immediately to provide another man for one of the Turkish bullets to kill; and as the Turks were known to be thirty-two in number, this

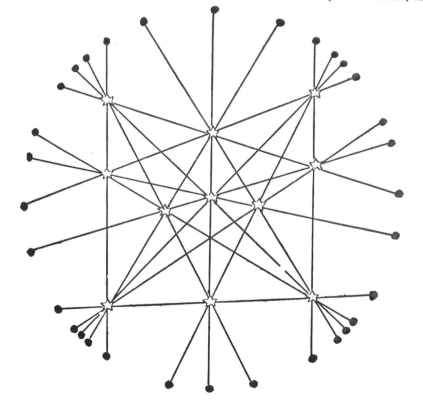

would necessitate our introducing another Russian soldier and, of course, destroying the solution. I repeat that the difficulty of the puzzle consists in finding how to arrange eleven points so that they shall form sixteen lines of three. I am told that the possibility of doing this was first discovered by the Rev. Mr. Wilkinson some twenty years ago.

214.—THE SIX FROGS.

MOVE the frogs in the following order : 2, 4, 6, 5, 3, 1 (repeat these moves in the same order twice more), 2, 4, 6. This is a solution in twenty-one moves—the fewest possible.

If n, the number of frogs, be even, we require $\frac{n^2+n}{2}$ moves, of which $\frac{n^2-n}{2}$ will be leaps and n simple moves. If n be odd, we shall need $\frac{n^2+3n}{2}-4$ moves, of which $\frac{n^2-n}{2}$ will be leaps and $2n-4$ simple moves.

In the even cases write, for the moves, all the even numbers in ascending order and the odd numbers in descending order. This series must be repeated $\frac{1}{2}n$ times and followed by the even numbers in ascending order once only. Thus the solution for 14 frogs will be (2, 4, 6, 8, 10, 12, 14, 13, 11, 9, 7, 5, 3, 1) repeated 7 times and followed by 2, 4, 6, 8, 10, 12, 14=105 moves.

In the odd cases, write the even numbers in ascending order and the odd numbers in descending order, repeat this series $\frac{1}{2}(n-1)$ times, follow with the even numbers in ascending order (omitting $n-1$), the odd numbers in descending order (omitting 1), and conclude with all the numbers (odd and even) in their natural order (omitting 1 and n). Thus for 11 frogs: (2, 4, 6, 8, 10, 11, 9, 7, 5, 3, 1) repeated 5 times, 2, 4, 6, 8, 11, 9, 7, 5, 3, and 2, 3, 4, 5, 6, 7, 8, 9, 10=73 moves.

This complete general solution is published here for the first time.

215.—THE GRASSHOPPER PUZZLE.

MOVE the counters in the following order. The moves in brackets are to be made four times in succession. 12, 1, 3, 2, 12, 11, 1, 3, 2 (5, 7, 9, 10, 8, 6, 4), 3, 2, 12, 11, 2, 1, 2. The grasshoppers will then be reversed in forty-four moves.

The general solution of this problem is very difficult. Of course it can always be solved by the method given in the solution of the last puzzle, if we have no desire to use the fewest possible moves. But to employ a full economy of moves we have two main points to consider. There are always what I call a lower movement (L) and an upper movement (U). L consists in exchanging certain of the highest numbers, such as 12, 11, 10 in our " Grasshopper Puzzle," with certain of the lower numbers, 1, 2, 3; the former moving in a clockwise direction, the latter in a non-clockwise direction. U consists in reversing the intermediate counters. In the above solution for 12, it will be seen that 12, 11, and 1, 2, 3 are engaged in the L movement, and 4, 5, 6, 7, 8, 9, 10 in the U movement. The L movement needs 16 moves and U 28, making together 44. We might also involve 10 in the L movement, which would result in L 23, U 21, making also together 44 moves. These I call the first and second methods. But any other scheme will entail an increase of moves. You always get these two methods (of equal economy) for odd or even counters, but the point is to determine just how many to involve in L and how many in U. Here is the solution in table form. But first note, in giving values to n, that 2, 3, and 4 counters are special cases, requiring respectively 3, 3, and 6 moves, and that 5 and 6 counters do not give a minimum solution by the second method—only by the first.

FIRST METHOD.

Total No. of Counters.	L MOVEMENT.		U MOVEMENT.		Total No. of Moves.
	No. of Counters.	No. of Moves.	No. of Counters.	No. of Moves.	
$4n$	$n-1$ and n	$2(n-1)^2+5n-7$	$2n+1$	$2n^2+3n+1$	$4(n^2+n-1)$
$4n-2$	$n-1$ „ n	$2(n-1)^2+5n-7$	$2n-1$	$2(n-1)^2+3n-2$	$4n^2-5$
$4n+1$	n „ $n+1$	$2n^2+5n-2$	$2n$	$2n^2+3n-4$	$2(2n^2+4n-3)$
$4n-1$	$n-1$ „ n	$2(n-1)^2+5n-7$	$2n$	$2n^2+3n-4$	$4n^2+4n-9$

SECOND METHOD.

Total No. of Counters.	L MOVEMENT.		U MOVEMENT.		Total No. of Moves.
	No. of Counters.	No. of Moves.	No. of Counters.	No. of Moves.	
$4n$	n and n	$2n^2+3n-4$	$2n$	$2(n-1)^2+5n-2$	$4(n^2+n-1)$
$4n-2$	$n-1$ „ $n-1$	$2(n-1)^2+3n-7$	$2n$	$2(n-1)^2+5n-2$	$4n^2-5$
$4n+1$	n „ n	$2n^2+3n-4$	$2n+1$	$2n^2+5n-2$	$2(2n^2+4n-3)$
$4n-1$	n „ n	$2n^2+3n-4$	$2n-1$	$2(n-1)^2+5n-7$	$4n^2+4n-9$

More generally we may say that with m counters, where m is even and greater than 4, we require $\dfrac{m^2 + 4m - 16}{4}$ moves ; and where m is odd and greater than 3, $\dfrac{m^2 + 6m - 31}{4}$ moves. I have thus shown the reader how to find the minimum number of moves for any case, and the character and direction of the moves. I will leave him to discover for himself how the actual order of moves is to be determined. This is a hard nut, and requires careful adjustment of the L and the U movements, so that they may be mutually accommodating.

216.—THE EDUCATED FROGS.

THE following leaps solve the puzzle in ten moves : 2 to 1, 5 to 2, 3 to 5, 6 to 3, 7 to 6, 4 to 7, 1 to 4, 3 to 1, 6 to 3, 7 to 6.

217.—THE TWICKENHAM PUZZLE.

PLAY the counters in the following order : K C E K W T C E H M K W T A N C E H M I K C E H M T, and there you are, at Twickenham. The position itself will always determine whether you are to make a leap or a simple move.

218.—THE VICTORIA CROSS PUZZLE.

IN solving this puzzle there were two things to be achieved : first, so to manipulate the counters that the word VICTORIA should read round the cross in the same direction, only with the V on one of the dark arms ; and secondly, to perform the feat in the fewest possible moves. Now, as a matter of fact, it would be impossible to perform the first part in any way whatever if all the letters of the word were different ; but as there are two I's, it can be done by making these letters change places—that is, the first I changes from the 2nd place to the 7th, and the second I from the 7th place to the 2nd. But the point I referred to, when introducing the puzzle, as a little remarkable is this : that a solution in twenty-two moves is obtainable by moving the letters in the order of the following words : " A VICTOR! A VICTOR! A VICTOR I ! "

There are, however, just six solutions in eighteen moves, and the following is one of them : I (1), V, A, I (2), R, O, T, I (1), I (2), A, V, I (2), I (1), C, I (2), V, A, I (1). The first and second I in the word are distinguished by the numbers 1 and 2.

It will be noticed that in the first solution given above one of the I's never moves, though the movements of the other letters cause it to change its relative position. There is another peculiarity I may point out—that there is a solution in twenty-eight moves requiring no letter to move to the central division except the I's. I may also mention that, in each of the solutions in eighteen moves, the letters C, T, O, R move once only, while the second I always moves four times, the V always being transferred to the right arm of the cross.

219.—THE LETTER BLOCK PUZZLE.

THIS puzzle can be solved in 23 moves—the fewest possible. Move the blocks in the following order : A, B, F, E, C, A, B, F, E, C, A, B, D, H, G, A, B, D, H, G, D, E, F.

220.—A LODGING-HOUSE DIFFICULTY.

THE shortest possible way is to move the articles in the following order : Piano, bookcase, wardrobe, piano, cabinet, chest of drawers, piano, wardrobe, bookcase, cabinet, wardrobe, piano, chest of drawers, wardrobe, cabinet, bookcase, piano. Thus seventeen removals are necessary. The landlady could then move chest of drawers, wardrobe, and cabinet. Mr. Dobson did not mind the wardrobe and chest of drawers changing rooms so long as he secured the piano.

221.—THE EIGHT ENGINES.

THE solution to the Eight Engines Puzzle is as follows : The engine that has had its fire drawn and therefore cannot move is No. 5. Move the other engines in the following order : 7, 6, 3, 7, 6, 1, 2, 4, 1, 3, 8, 1, 3, 2, 4, 3, 2, seventeen moves in all, leaving the eight engines in the required order.

There are two other slightly different solutions.

222.—A RAILWAY PUZZLE.

THIS little puzzle may be solved in as few as nine moves. Play the engines as follows : From 9 to 10, from 6 to 9, from 5 to 6, from 2 to 5, from 1 to 2, from 7 to 1, from 8 to 7, from 9 to 8, and from 10 to 9. You will then have engines A, B, and C on each of the three circles and on each of the three straight lines. This is the shortest solution that is possible.

223.—A RAILWAY MUDDLE.

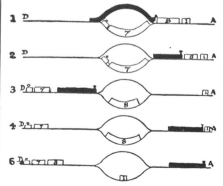

ONLY six reversals are necessary. The white train (from A to D) is divided into three sections, engine and 7 wagons, 8 wagons, and 1 wagon. The black train (D to A) never uncouples anything throughout. Fig. 1 is original position with **8** and **1** uncoupled. The black train proceeds to position in Fig. 2 (no reversal). The engine and **7** proceed towards D, and black train backs, leaves **8** on loop, and takes up position in Fig. 3 (first reversal). Black train goes to position in Fig. 4 to fetch single wagon (second reversal). Black train pushes **8** off loop and leaves single wagon there, proceeding on its journey, as in Fig. 5 (third and fourth reversals). White train now backs on to loop to pick up single car and goes right away to D (fifth and sixth reversals).

224.—THE MOTOR-GARAGE PUZZLE.

The exchange of cars can be made in forty-three moves, as follows : 6—G, 2—B, 1—E, 3—H, 4—I, 3—L, 6—K, 4—G, 1—I, 2—J, 5—H, 4—A, 7—F, 8—E, 4—D, 8—C, 7—A, 8—G, 5—C, 2—B, 1—E, 8—I, 1—G, 2—J, 7—H, 1—A, 7—G, 2—B, 6—E, 3—H, 8—L, 3—I, 7—K, 3—G, 6—I, 2—J, 5—H, 3—C, 5—G, 2—B, 6—E, 5—I, 6—J. Of course, " 6—G " means that the car numbered " 6 " moves to the point " G." There are other ways in forty-three moves.

225.—THE TEN PRISONERS.

It will be seen in the illustration how the prisoners may be arranged so as to produce as many as sixteen even rows. There are 4 such vertical rows, 4 horizontal rows, 5 diagonal rows in one direction, and 3 diagonal rows in the other direction. The arrows here show the movements of the four prisoners, and it will be seen that the infirm man in the bottom corner has not been moved.

226.—ROUND THE COAST.

In order to place words round the circle under the conditions, it is necessary to select words in which letters are repeated in certain relative positions. Thus, the word that solves our puzzle is " Swansea," in which the first and fifth letters are the same, and the third and seventh are the same. We make out jumps as follows, taking the letters of the word in their proper order : 2—5, 7—2, 4—7, 1—4, 6—1, 3—6, 8—3. Or we could place a word like " Tarapur " (in which the second and fourth

letters, and the third and seventh, are alike) with these moves : 6—1, 7—4, 2—7, 5—2, 8—5, 3—6, 8—3. But " Swansea " is the only word, apparently, that will fulfil the conditions of the puzzle.

This puzzle should be compared with Sharp's Puzzle, referred to in my solution to No. 341, " The Four Frogs." The condition " touch and jump over two " is identical with " touch and move along a line."

227.—CENTRAL SOLITAIRE.

Here is a solution in nineteen moves ; the moves enclosed in brackets count as one move only : 19—17, 16—18, (29—17, 17—19), 30—18, 27—25, (22—24, 24—26), 31—23, (4—16, 16—28), 7—9, 10—8, 12—10, 3—11, 18—6, (1—3, 3—11), (13—27, 27—25), (21—7, 7—9), (33—31, 31—23), (10—8, 8—22, 22—24, 24—26, 26—12, 12—10), 5—17. All the counters are now removed except one, which is left in the central hole. The solution needs judgment, as one is tempted to make several jumps in one move, where it would be the reverse of good play. For example, after playing the first 3—11 above, one is inclined to increase the length of the move by continuing with 11—25, 25—27, or with 11—9, 9—7.

I do not think the number of moves can be reduced.

228.—THE TEN APPLES.

Number the plates (1, 2, 3, 4), (5, 6, 7, 8), (9, 10, 11, 12), (13, 14, 15, 16) in successive rows from the top to the bottom. Then transfer the apple from 8 to 10 and play as follows, always removing the apple jumped over : 9—11, 1—9, 13—5, 16—8, 4—12, 12—10, 3—1, 1—9, 9—11.

229.—THE NINE ALMONDS.

This puzzle may be solved in as few as four moves, in the following manner : Move 5 over 8, 9, 3, 1. Move 7 over 4. Move 6 over 2 and 7. Move 5 over 6, and all the counters are removed except 5, which is left in the central square that it originally occupied.

230.—THE TWELVE PENNIES.

Here is one of several solutions. Move 12 to 3, 7 to 4, 10 to 6, 8 to 1, 9 to 5, 11 to 2.

231.—PLATES AND COINS.

Number the plates from 1 to 12 in the order that the boy is seen to be going in the illustration. Starting from 1, proceed as follows, where " 1 to 4 " means that you take the coin from plate No. 1 and transfer it to plate No. 4 : 1 to 4, 5 to 8, 9 to 12, 3 to 6, 7 to 10, 11 to 2, and complete the last revolution to 1, making three revolutions in all. Or you can proceed this way : 4 to 7, 8 to 11, 12 to 3, 2 to 5, 6 to 9, 10 to 1. It is easy to solve in four revolutions, but the solutions in three are more difficult to discover.

This is " The Riddle of the Fishpond " (No. 41, *Canterbury Puzzles*) in a different dress.

232.--CATCHING THE MICE.

In order that the cat should eat every thirteenth mouse, and the white mouse last of all, it is necessary that the count should begin at the seventh mouse (calling the white one the first)—that is, at the one nearest the tip of the cat's tail. In this case it is not at all necessary to try starting at all the mice in turn until you come to the right one, for you can just start anywhere and note how far distant the last one eaten is from the starting point. You will find it to be the eighth, and therefore must start at the eighth, counting backwards from the white mouse. This is the one I have indicated.

In the case of the second puzzle, where you have to find the smallest number with which the cat may start at the white mouse and eat this one last of all, unless you have mastered the general solution of the problem, which is very difficult, there is no better course open to you than to try every number in succession until you come to one that works correctly. The smallest number is twenty-one. If you have to proceed by trial, you will shorten your labour a great deal by only counting out the remainders when the number is divided successively by 13, 12, 11, 10, etc. Thus, in the case of 21, we have the remainders 8, 9, 10, 1, 3, 5, 7, 3, 1, 1, 3, 1, 1. Note that I do not give the remainders of 7, 3, and 1 as nought, but as 7, 3, and 1. Now, count round each of these numbers in turn, and you will find that the white mouse is killed last of all. Of course, if we wanted simply any number, not the smallest, the solution is very easy, for we merely take the least common multiple of 13, 12, 11, 10, etc. down to 2. This is 360360, and you will find that the first count kills the thirteenth mouse, the next the twelfth, the next the eleventh, and so on down to the first. But the most arithmetically inclined cat could not be expected to take such a big number when a small one like twenty-one would equally serve its purpose.

In the third case, the smallest number is 100. The number 1,000 would also do, and there are just seventy-two other numbers between these that the cat might employ with equal success.

233.—THE ECCENTRIC CHEESEMONGER.

To leave the three piles at the extreme ends of the rows, the cheeses may be moved as follows—the numbers refer to the cheeses and not to their positions in the row : 7—2, 8—7, 9—8, 10—15, 6—10, 5—6, 14—16, 13—14, 12—13, 3—1, 4—3, 11—4. This is probably the easiest solution of all to find. To get three of the piles on cheeses 13, 14, and 15, play thus : 9—4, 10—9, 11—10, 6—14, 5—6, 12—15, 8—12, 7—8, 16—5, 5—13, 2—3, 1—2. To leave the piles on cheeses 3, 5, 12, and 14, play thus : 8—3, 9—14, 16—12, 1—5, 10—9, 7—10, 11—8, 2—1, 4—16, 13—2, 6—11, 15—4.

234.—THE EXCHANGE PUZZLE.

Make the following exchanges of pairs : H—K, H—E, H—C, H—A, I—L, I—F, I—D, K—L, G—J, J—A, F—K, L—E, D—K, E—F, E—D, E—B, B—K. It will be found that, although the white counters can be moved to their proper places in 11 moves, if we omit all consideration of exchanges, yet the black cannot be so moved in fewer than 17 moves. So we have to introduce waste moves with the white counters to equal the minimum required by the black. Thus fewer than 17 moves must be impossible. Some of the moves are, of course, interchangeable.

235.—TORPEDO PRACTICE.

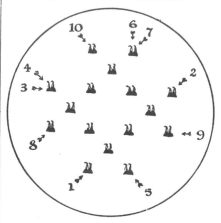

If the enemy's fleet be anchored in the formation shown in the illustration, it will be seen that as many as ten out of the sixteen ships may be blown up by discharging the torpedoes in the order indicated by the numbers and in the directions indicated by the arrows. As each torpedo in succession passes under three ships and sinks the fourth, strike out each vessel with the pencil as it is sunk.

236.—THE HAT PUZZLE.

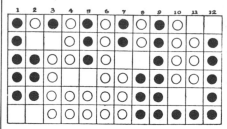

I suggested that the reader should try this puzzle with counters, so I give my solution in that form. The silk hats are represented by black counters and the felt hats by white counters. The first row shows the hats in their original positions, and then each successive row shows how they appear after one of the five manipulations. It will thus be seen that we first move hats 2 and 3, then 7 and 8, then 4 and 5, then 10 and 11, and, finally, 1 and 2, leaving the four silk hats together, the four felt hats together, and the two vacant pegs at one end of the row. The first three pairs moved are dissimilar hats, the last two pairs being similar. There are other ways of solving the puzzle.

237.—BOYS AND GIRLS.

THERE are a good many different solutions to this puzzle. Any contiguous pair, except 7–8, may be moved first, and after the first move there are variations. The following solution shows the position from the start right through each successive move to the end :—

```
. . 1 2 3 4 5 6 7 8    4 3 1 2 7 . . 5 6 8
4 3 1 2 . . 5 6 7 8    4 . . 2 7 1 3 5 6 8
4 3 1 2 7 6 5 . . 8    4 8 6 2 7 1 3 5 . .
```

238.—ARRANGING THE JAM POTS.

Two of the pots, 13 and 19, were in their proper places. As every interchange may result in a pot being put in its place, it is clear that twenty-two interchanges will get them all in order. But this number of moves is not the fewest possible, the correct answer being seventeen. Exchange the following pairs : (3—1, 2—3), (15—4, 16—15), (17—7, 20—17), (24—10, 11—24, 12—11), (8—5, 6—8, 21—6, 23—21, 22—23, 14—22, 9—14, 18—9). When you have made the interchanges within any pair of brackets, all numbers within those brackets are in their places. There are five pairs of brackets, and 5 from 22 gives the number of changes required—17.

239.—A JUVENILE PUZZLE.

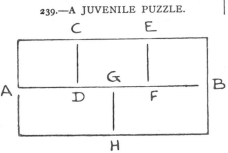

As the conditions are generally understood, this puzzle is incapable of solution. This can be demonstrated quite easily. So we have to look for some catch or quibble in the statement of what we are asked to do. Now if you fold the paper and then push the point of your pencil down between the fold, you can with one stroke make the two lines C D and E F in our diagram. Then start at A, and describe the line ending at B. Finally put in the last line G H, and the thing is done strictly within the

conditions, since folding the paper is not actually forbidden. Of course the lines are here left unjoined for the purpose of clearness.

In the rubbing out form of the puzzle, first rub out A to B with a single finger in one stroke. Then rub out the line G H with one finger. Finally, rub out the remaining two vertical lines with two fingers at once! That is the old trick.

240.—THE UNION JACK.

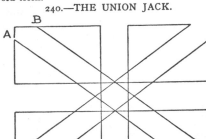

THERE are just sixteen points (all on the outside) where three roads may be said to join. These are called by mathematicians "odd nodes." There is a rule that tells us that in the case of a drawing like the present one, where there are sixteen odd nodes, it requires eight separate strokes or routes (that is, half as many as there are odd nodes) to complete it. As we have to produce as much as possible with only one of these eight strokes, it is clearly necessary to contrive that the seven strokes from odd node to odd node shall be as short as possible. Start at A and end at B, or go the reverse way.

241.—THE DISSECTED CIRCLE.

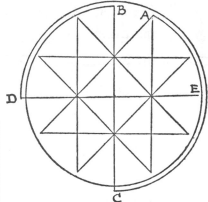

IT can be done in twelve continuous strokes, thus : Start at A in the illustration, and eight strokes, forming the star, will bring you back to A ; then one stroke round the circle to B, one stroke to C, one round the circle to D, and one final stroke to E—twelve in all. Of course, in practice the second circular stroke will be over the first one ; it is separated in the diagram, and the points of the star not joined to the circle, to make the solution clear to the eye.

242.—THE TUBE INSPECTOR'S PUZZLE.

THE inspector need only travel nineteen miles if he starts at B and takes the following route : BADGDEFIFCBEHKLIHGJK. Thus the only portions of line travelled over twice are the two sections D to G and F to I. Of course, the route may be varied, but it cannot be shortened.

243.—VISITING THE TOWNS.

NOTE that there are six towns, from which only two roads issue. Thus 1 must lie between 9 and 12 in the circular route. Mark these two roads as settled. Similarly mark 9, 5, 14, and 4, 8, 14, and 10, 6, 15, and 10, 2, 13, and 3, 7, 13. All these roads must be taken. Then you will find that he must go from 4 to 15, as 13 is closed, and that he is compelled to take 3, 11, 16, and also 16, 12. Thus, there is only one route, as follows : 1, 9, 5, 14, 8, 4, 15, 6, 10, 2, 13, 7, 3, 11, 16, 12, 1, or its reverse—reading the line the other way. Seven roads are not used.

244.—THE FIFTEEN TURNINGS.

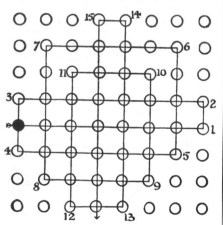

IT will be seen from the illustration (where the roads not used are omitted) that the traveller can go as far as seventy miles in fifteen turnings. The turnings are all numbered in the order in which they are taken. It will be seen that he never visits nineteen of the towns. He might visit them all in fifteen turnings, never entering any town twice, and end at the black town from which he starts (see "The Rook's Tour,"

No. 320), but such a tour would only take him sixty-four miles.

245.—THE FLY ON THE OCTAHEDRON.

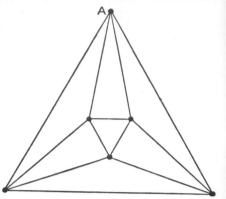

THOUGH we cannot really see all the sides of the octahedron at once, we can make a projection of it that suits our purpose just as well. In the diagram the six points represent the six angles of the octahedron, and four lines proceed from every point under exactly the same conditions as the twelve edges of the solid. Therefore if we start at the point A and go over all the lines once, we must always end our route at A. And the number of different routes is just 1,488, counting the reverse way of any route as different. It would take too much space to show how I make the count. It can be done in about five minutes, but an explanation of the method is difficult. The reader is therefore asked to accept my answer as correct.

246.—THE ICOSAHEDRON PUZZLE.

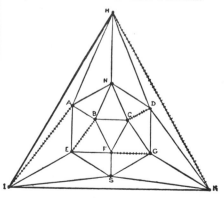

THERE are thirty edges, of which eighteen were visible in the original illustration, represented in the following diagram by the hexagon NAESGD. By this projection of the solid we get an imaginary view of the remaining twelve edges, and are able to see at once their direction and the twelve points at which all the edges meet. The difference in the length of the lines is of no importance ; all we want is to present their direction in a graphic manner. But in case the novice should be puzzled at only finding nineteen triangles instead of the required twenty, I will point out that the apparently missing triangle is the outline HIK.

In this case there are twelve odd nodes ; therefore six distinct and disconnected routes will be needful if we are not to go over any lines twice. Let us therefore find the greatest distance that we may so travel in one route.

It will be noticed that I have struck out with little cross strokes five lines or edges in the diagram. These five lines may be struck out anywhere so long as they do not join one another, and so long as none of them does not connect with N, the North Pole, from which we are to start. It will be seen that the result of striking out these five lines is that all the nodes are now even except N and S. Consequently if we begin at N and stop at S we may go over all the lines, except the five crossed out, without traversing any line twice. There are many ways of doing this. Here is one route : N to H, I, K, S, I, E, S, G, K, D, H, A, N, B, A, E, F, B, C, G, D, N, C, F, S. By thus making five of the routes as short as is possible—simply from one node to the next—we are able to get the greatest possible length for our sixth line. A greater distance in one route, without going over the same ground twice, it is not possible to get.

It is now readily seen that those five crased lines must be gone over twice, and they may be " picked up," so to speak, at any points of our route. Thus, whenever the traveller happens to be at I he can run up to A and back before proceeding on his route, or he may wait until he is at A and then run down to I and back to A. And so with the other lines that have to be traced twice. It is, therefore, clear that he can go over 25 of the lines once only (25 × 10,000 miles = 250,000 miles) and 5 of the lines twice (5 × 20,000 miles = 100,000 miles), the total, 350,000 miles, being the length of his travels and the shortest distance that is possible in visiting the whole body.

It will be noticed that I have made him end his travels at S, the South Pole, but this is not imperative. I might have made him finish at any of the other nodes, except the one from which he started. Suppose it had been required to bring him home again to N at the end of his travels. Then instead of suppressing the line AI we might leave that open and close IS. This would enable him to complete his 350,000 miles tour at A, and another 10,000 miles would take him to his own fireside. There are a great many different routes, but as the lengths of the edges are all alike, one course is as good as another. To make the complete 350,000 miles tour from N to S absolutely clear to everybody, I will give it entire : N to H, I, A, I, K, H, K,

S, I, E, S, G, F, G, K, D, C, D, H, A, N, B, E, B, A, E, F, B, C, G, D, N, C, F, S—that is, thirty-five lines of 10,000 miles each.

247.—INSPECTING A MINE.

STARTING from A, the inspector need only travel 36 furlongs if he takes the following route : A to B, G, H, C, D, I, H, M, N, I, J, O, N, S, R, M, L, G, F, K, L, Q, R, S, T, O, J, E, D, C, B, A, F, K, P, Q. He thus passes between A and B twice, between C and D twice, between F and K twice, between J and O twice, and between R and S twice—five repetitions. Therefore 31 passages plus 5 repeated equal 36 furlongs. The little pitfall in this puzzle lies in the fact that we start from an even node. Otherwise we need only travel 35 furlongs.

248.—THE CYCLIST'S TOUR.

WHEN Mr. Maggs replied, " No way, I'm sure," he was not saying that the thing was impossible, but was really giving the actual route by which the problem can be solved. Starting from the star, if you visit the towns in the order, NO WAY, I'M SURE, you will visit every town once, and only once, and end at E. So both men were correct. This was the little joke of the puzzle, which is not by any means difficult.

249.—THE SAILOR'S PUZZLE.

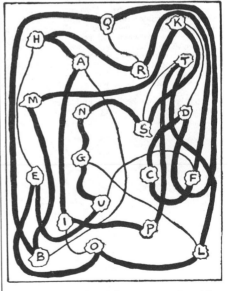

THERE are only four different routes (or eight, if we count the reverse ways) by which the sailor can start at the island marked A, visit all the islands once, and once only, and return again to A. Here they are :—

```
A I P T L O E H R Q D C F U G N S K M B A
A I P T S N G L O E U F C D K M B Q R H A
A B M K S N G L T P I O E U F C D Q R H A
A I P T L O E U G N S K M B Q D C F R H A
```

Now, if the sailor takes the first route he will make C his 12th island (counting A as 1); by the second route he will make C his 13th island; by the third route, his 16th island; and by the fourth route, his 17th island. If he goes the reverse way, C will be respectively his 10th, 9th, 6th, and 5th island. As these are the only possible routes, it is evident that if the sailor puts off his visit to C as long as possible, he must take the last route reading from left to right. This route I show by the dark lines in the diagram, and it is the correct answer to the puzzle.

The map may be greatly simplified by the "buttons and string" method, explained in the solution to No. 341, "The Four Frogs."

250.—THE GRAND TOUR.

THE first thing to do in trying to solve a puzzle like this is to attempt to simplify it. If you look at Fig. 1, you will see that it is a simplified version of the map. Imagine the circular towns to be buttons and the railways to be connecting strings. (See solution to No. 341.) Then, it will be seen, we have simply "straightened out" the previous diagram without affecting the conditions. Now we can further simplify by converting Fig. 1 into Fig. 2, which is a portion of a chessboard. Here the directions of the railways will resemble the moves of a rook in chess—that is, we may move in any direction parallel to the sides of the diagram, but not diagonally. Therefore the first town (or square) visited must be a black one; the second must be a white; the third must be a black; and so on. Every odd square visited will thus be black and every even one white. Now, we have 23 squares to visit (an

odd number), so the last square visited must be black. But Z happens to be white, so the puzzle would seem to be impossible of solution.

As we were told that the man "succeeded" in carrying out his plan, we must try to find some loophole in the conditions. He was to "enter every town once and only once," and we find no prohibition against his entering once the town A after leaving it, especially as he has never left it since he was born, and would thus be "entering" it for the first time in his life. But he must return at once from the first town he visits, and then he will have only 22 towns to visit, and as 22 is an even number, there is no reason why he should not end on the white square Z. A possible route for him is indicated by the dotted line from A to Z. This route is repeated by the dark lines in Fig. 1, and the reader will now have no difficulty in applying it to the original map. We have thus proved that the puzzle can only be solved by a return to A immediately after leaving it.

251.—WATER, GAS, AND ELECTRICITY.

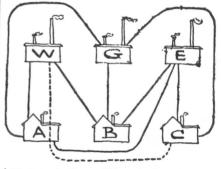

ACCORDING to the conditions, in the strict sense in which one at first understands them, there

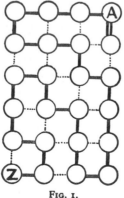

FIG. 1.

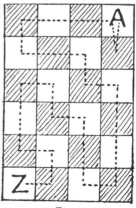

FIG. 2.

is no possible solution to this puzzle. In such a dilemma one always has to look for some verbal quibble or trick. If the owner of house A will allow the water company to run their pipe for house C through his property (and we are not bound to assume that he would object), then the difficulty is got over, as shown in our illustration. It will be seen that the dotted line from W to C passes through house A, but no pipe ever crosses another pipe.

252.—A PUZZLE FOR MOTORISTS.

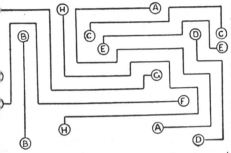

THE routes taken by the eight drivers are shown in the illustration, where the dotted line roads are omitted to make the paths clearer to the eye.

253.—A BANK HOLIDAY PUZZLE.

THE simplest way is to write in the number of routes to all the towns in this manner. Put a 1 on all the towns in the top row and in the first column. Then the number of routes to any town will be the sum of the routes to the town immediately above and to the town immediately to the left. Thus the routes in the second row will be 1, 2, 3, 4, 5, 6, etc., in the third row, 1, 3, 6, 10, 15, 21, etc.; and so on with the other rows. It will then be seen that the only town to which there are exactly 1,365 different routes is the twelfth town in the fifth row—the one immediately over the letter E. This town was therefore the cyclist's destination.

The general formula for the number of routes from one corner to the corner diagonally opposite on any such rectangular reticulated arrangement, under the conditions as to direction, is $|m+n / |m \,|n$, where m is the number of towns on one side, less one, and n the number on the other side, less one. Our solution involves the case where there are 12 towns by 5. Therefore $m=11$ and $n=4$. Then the formula gives us the answer 1,365 as above.

254.—THE MOTOR-CAR TOUR.

FIRST of all I will ask the reader to compare the original square diagram with the circular one shown in Figs. 1, 2, and 3 below. If for the moment we ignore the shading (the purpose of which I shall proceed to explain), we find that the circular diagram in each case is merely a simplification of the original square one—that is, the roads from A lead to B, E, and M in both cases, the roads from L (London) lead to I, K, and S, and so on. The form below, being circular and symmetrical, answers my purpose better in applying a mechanical solution, and I therefore adopt it without altering in any way the conditions of the puzzle. If such a question as distances from town to town came into the problem, the new diagrams might require the addition of numbers to indicate these distances, or they might conceivably not be at all practicable.

Now, I draw the three circular diagrams, as shown, on a sheet of paper and then cut out three pieces of cardboard of the forms indicated by the shaded parts of these diagrams. It can be shown that every route, if marked out with a red pencil, will form one or other of the designs indicated by the edges of the cards, or a reflection thereof. Let us direct our attention to Fig. 1. Here the card is so placed that the star is at the town T; it therefore gives us (by following the edge of the card) one of the circular routes from London: L, S, R, T, M, A, E, P, O, J, D, C, B, G, N, Q, K, H, F, I, L. If we went the other way, we should get L, I, F, H, K, Q, etc., but these reverse routes were not to be counted. When we have written out this first route we revolve the card until the

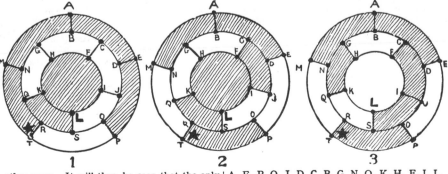

| 1 | 2 | 3 |

star is at M, when we get another different route, at A a third route, at E a fourth route, and at P a fifth route. We have thus obtained five different routes by revolving the card as it lies. But it is evident that if we now take up the card and replace it with the other side uppermost, we shall in the same manner get five other routes by revolution.

We therefore see how, by using the revolving card in Fig. 1, we may, without any difficulty, at once write out ten routes. And if we employ the cards in Figs. 2 and 3, we similarly obtain in each case ten other routes. These thirty routes are all that are possible. I do not give the actual proof that the three cards exhaust all the possible cases, but leave the reader to reason that out for himself. If he works out any route at haphazard, he will certainly find that it falls into one or other of the three categories.

255.—THE LEVEL PUZZLE.

LET us confine our attention to the L in the top left-hand corner. Suppose we go by way of the E on the right : we must then go straight on to the V, from which letter the word may be completed in four ways, for there are four E's available through which we may reach an L. There are therefore four ways of reading through the right-hand E. It is also clear that there must be the same number of ways through the E that is immediately below our starting point. That makes eight. If, however, we take the third route through the E on the diagonal, we then have the option of any one of the three V's, by means of each of which we may complete the word in four ways. We can therefore spell LEVEL in twelve ways through the diagonal E. Twelve added to eight gives twenty readings, all emanating from the L in the top left-hand corner ; and as the four corners are equal, the answer must be four times twenty, or eighty different ways.

256.—THE DIAMOND PUZZLE.

THERE are 252 different ways. The general formula is that, for words of n letters (not palindromes, as in the case of the next puzzle), when grouped in this manner, there are always $2^{n+1} - 4$ different readings. This does not allow diagonal readings, such as you would get if you used instead such a word as DIGGING, where it would be possible to pass from one G to another G by a diagonal step.

257.—THE DEIFIED PUZZLE.

THE correct answer is 1,992 different ways. Every F is either a corner F or a side F—standing next to a corner in its own square of F's. Now, FIED may be read *from* a corner F in 16 ways ; therefore DEIF may be read *into* a corner F also in 16 ways ; hence DEIFIED may be read *through* a corner F in 16 × 16 = 256 ways. Consequently, the four corner F's give 4 × 256 = 1,024 ways. Then FIED may be read from a side F in 11 ways, and DEIFIED therefore in 121 ways. But there are eight

side F's ; consequently these give together 8 × 121 = 968 ways. Add 968 to 1,024 and we get the answer, 1,992.

In this form the solution will depend on whether the number of letters in the palindrome be odd or even. For example, if you apply the word NUN in precisely the same manner, you will get 64 different readings ; but if you use the word NOON, you will only get 56, because you cannot use the same letter twice in immediate succession (since you must " always pass from one letter to another ") or diagonal readings, and every reading must involve the use of the central N.

The reader may like to find for himself the general formula in this case, which is complex and difficult. I will merely add that for such a case as MADAM, dealt with in the same way as DEIFIED, the number of readings is 400.

258.—THE VOTERS' PUZZLE.

THE number of readings here is 63,504, as in the case of "WAS IT A RAT I SAW" (No. 30, *Canterbury Puzzles*). The general formula is that for palindromic sentences containing $2n + 1$ letters there are $[4(2^n - 1)]^2$ readings.

259.—HANNAH'S PUZZLE.

STARTING from any one of the N's, there are 17 different readings of NAH, or 68 (4 times 17) for the 4 N's. Therefore there are also 68 ways of spelling HAN. If we were allowed to use the same N twice in a spelling, the answer would be 68 times 68, or 4,624 ways. But the conditions were, " always passing from one letter to another." Therefore, for every one of the 17 ways of spelling HAN with a particular N, there would be 51 ways (3 times 17) of completing the NAH, or 867 (17 times 51) ways for the complete word. Hence, as there are four N's to use in HAN, the correct solution of the puzzle is 3,468 (4 times 867) different ways.

260.—THE HONEYCOMB PUZZLE.

THE required proverb is, " There is many a slip 'twixt the cup and the lip." Start at the T on the outside at the bottom right-hand corner, pass to the H above it, and the rest is easy.

261.—THE MONK AND THE BRIDGES.

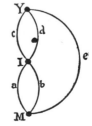

THE problem of the Bridges may be reduced to the simple diagram shown in illustration. The point **M** represents the Monk, the point **I** the Island, and the point **Y** the Monastery. Now the only direct ways from **M** to **I** are by the bridges *a* and *b* ; the only direct ways from **I** to **Y** are by the bridges *c* and *d* ; and there is a direct way from **M** to **Y** by the bridge *e*. Now, what we have to do is to count all the routes that will lead from **M** to **Y**, passing over all the bridges, *a*, *b*, *c*, *d*, and *e* once and once only. With the simple diagram under the eye it is quite easy, without any elaborate rule, to count these routes methodically. Thus, starting from *a*, *b*, we find there are only two ways of completing the route ; with *a*, *c*, there are only two routes ; with *a*, *d*, only two routes ; and so on. It will be found that there are sixteen such routes in all, as in the following list :—

a b e c d	*b c d a e*
a b e d c	*b c e a d*
a c d b e	*b d c a e*
a c e b d	*b d e a c*
a d c b e	*e c a b d*
a d e b c	*e c b a d*
b a e c d	*e d a b c*
b a e d c	*e d b a c*

If the reader will transfer the letters indicating the bridges from the diagram to the corresponding bridges in the original illustration, everything will be quite obvious.

262.—THOSE FIFTEEN SHEEP.

If we read the exact words of the writer in the cyclopædia, we find that we are not told that the pens were all necessarily empty ! In fact, if the reader will refer back to the illustration, he will see that one sheep is already in one of the pens. It was just at this point that the wily farmer said to me, " *Now* I'm going to start placing the fifteen sheep." He thereupon proceeded to drive three from his flock into the already occupied pen, and then placed four sheep in each of the other three pens. "There," says he, " you have seen me place fifteen sheep in four pens so that there shall be the same number of sheep in every pen." I was, of course, forced to admit that he was perfectly correct, according to the exact wording of the question.

263.—KING ARTHUR'S KNIGHTS.

On the second evening King Arthur arranged the knights and himself in the following order round the table : A, F, B, D, G, E, C. On the third evening they sat thus, A, E, B, G, C, F, D. He thus had B next but one to him on both occasions (the nearest possible), and G was the third from him at both sittings (the furthest position possible). No other way of sitting the knights would have been so satisfactory.

264.—THE CITY LUNCHEONS.

The men may be grouped as follows, where each line represents a day and each column a table :—

AB	CD	EF	GH	IJ	KL
AE	DL	GK	FI	CB	HJ
AG	LJ	FH	KC	DE	IB
AF	JB	KI	HD	LG	CE
AK	BE	HC	IL	JF	DG
AH	EG	ID	CJ	BK	LF
AI	GF	CL	DB	EH	JK
AC	FK	DJ	LE	GI	BH
AD	KH	LB	JG	FC	EI
AL	HI	JE	BF	KD	GC
AJ	IC	BG	EK	HL	FD

Note that in every column (except in the case of the A's) all the letters descend cyclically in the same order, B, E, G, F, up to J, which is followed by B.

265.—A PUZZLE FOR CARD-PLAYERS.

In the following solution each of the eleven lines represents a sitting, each column a table, and each pair of letters a pair of partners.

A B — I L	E J — G K	F H — C D
A C — J B	F K — H L	G I — D E
A D — K C	G L — I B	H J — E F
A E — L D	H B — J C	I K — F G
A F — B E	I C — K D	J L — G H
A G — C F	J D — L E	K B — H I
A H — D G	K E — B F	L C — I J
A I — E H	L F — C G	B D — J K
A J — F I	B G — D H	C E — K L
A K — G J	C H — E I	D F — L B
A L — H K	D I — F J	E G — B C

It will be seen that the letters B, C, D...L descend cyclically. The solution given above is absolutely perfect in all respects. It will be found that every player has every other player once as his partner and twice as his opponent.

266.—A TENNIS TOURNAMENT.

Call the men A, B, D, E, and their wives a, b, d, e. Then they may play as follows without any person ever playing twice with or against any other person :—

	First Court.		Second Court.	
1st Day	A d against B e		D a against E b	
2nd Day	A e „ D b		E a „ B d	
3rd Day	A b „ E d		B a „ D e	

It will be seen that no man ever plays with or against his own wife—an ideal arrangement. If the reader wants a hard puzzle, let him try to arrange eight married couples (in four courts on seven days) under exactly similar conditions. It can be done, but I leave the reader in this case the pleasure of seeking the answer and the general solution.

267.—THE WRONG HATS.

The number of different ways in which eight persons, with eight hats, can each take the wrong hat, is 14,833.

Here are the successive solutions for any number of persons from one to eight :—

$$1 = 0$$
$$2 = 1$$
$$3 = 2$$
$$4 = 9$$
$$5 = 44$$
$$6 = 265$$
$$7 = 1,854$$
$$8 = 14,833$$

To get these numbers, multiply successively by 2, 3, 4, 5, etc. When the multiplier is even, add 1 ; when odd, deduct 1. Thus, $3 \times 1 - 1 = 2$, $4 \times 2 + 1 = 9$; $5 \times 9 - 1 = 44$; and so on. Or you can multiply the sum of the number of ways for $n - 1$ and $n - 2$ persons by $n - 1$, and so get the solution for n persons. Thus, $4(2 + 9) = 44$; $5(9 + 44) = 265$; and so on.

268.—THE PEAL OF BELLS.

THE bells should be rung as follows :—

1	2	3	4		3	1	2	4		2	3	1	4
2	1	4	3		1	3	4	2		3	2	4	1
2	4	1	3		1	4	3	2		3	4	2	1
4	2	3	1		4	1	2	3		4	3	1	2
4	3	2	1		4	2	1	3		4	1	3	2
3	4	1	2		2	4	3	1		1	4	2	3
3	1	4	2		2	3	4	1		1	2	4	3
1	3	2	4		3	2	1	4		2	1	3	4

I have constructed peals for five and six bells respectively, and a solution is possible for any number of bells under the conditions previously stated.

269.—THREE MEN IN A BOAT.

IF there were no conditions whatever, except that the men were all to go out together, in threes, they could row in an immense number of different ways. If the reader wishes to know how many, the number is 455^7. And with the condition that no two may ever be together more than once, there are no fewer than 15,567,552,000 different solutions—that is, different ways of arranging the men. With one solution before him, the reader will realize why this must be, for although, as an example, A must go out once with B and once with C, it does not necessarily follow that he must go out with C on the same occasion that he goes with B. He might take any other letter with him on that occasion, though the fact of his taking other than B would have its effect on the arrangement of the other triplets.

Of course only a certain number of all these arrangements are available when we have that other condition of using the smallest possible number of boats. As a matter of fact we need employ only ten different boats. Here is one of the arrangements :—

	1	2	3	4	5
1st Day .	(ABC)	(DEF)	(GHI)	(JKL)	MNO)
	8	6	7	9	10
2nd Day .	(ADG)	(BKN)	(COL)	(JEI)	(MHF)
	3	5	4	1	2
3rd Day .	(AJM)	(BEH)	(CFI)	(DKO)	(GNL)
	7	6	8	9	1
4th Day .	(AEK)	(CGM)	(BOI)	(DHL)	(JNF)
	4	5	3	10	2
5th Day .	(AHN)	(CDJ)	(BFL)	(GEO)	(MKI)
	6	7	8	10	1
6th Day .	(AFO)	(BGJ)	(CKH)	(DNI)	(MEL)
	5	4	3	9	2
7th Day .	(AIL)	(BDM)	(CEN)	(GKF)	(JHO)

It will be found that no two men ever go out twice together, and that no man ever goes out twice in the same boat.

This is an extension of the well-known problem of the " Fifteen Schoolgirls," by Kirkman. The original conditions were simply that fifteen girls walked out on seven days in triplets without any girl ever walking twice in a triplet with another girl. Attempts at a general solution of this puzzle had exercised the ingenuity of mathematicians since 1850, when the question was first propounded, until recently. In 1908 and the two following years I indicated (see *Educational Times Reprints*, Vols. XIV., XV., and XVII.) that all our trouble had arisen from a failure to discover that 15 is a special case (too small to enter into the general law for all higher numbers of girls of the form $6n + 3$), and showed what that general law is and how the groups should be posed for any number of girls. I gave actual arrangements for numbers that had previously baffled all attempts to manipulate, and the problem may now be considered generally solved. Readers will find an excellent full account of the puzzle in W. W. Rouse Ball's *Mathematical Recreations*, 5th edition.

270.—THE GLASS BALLS.

THERE are, in all, sixteen balls to be broken, or sixteen places in the order of breaking. Call the four strings A, B, C, and D—order is here of no importance. The breaking of the balls on A may occupy any 4 out of these 16 places —that is, the combinations of 16 things, taken 4 together, will be $\frac{13 \times 14 \times 15 \times 16}{1 \times 2 \times 3 \times 4} = 1,820$ ways for A. In every one of these cases B may occupy any 4 out of the remaining 12 places, making $\frac{9 \times 10 \times 11 \times 12}{1 \times 2 \times 3 \times 4} = 495$ ways. Thus 1,820 $\times$ 495 = 900,000 different placings are open to A and B. But for every one of these cases C may occupy $\frac{5 \times 6 \times 7 \times 8}{1 \times 2 \times 3 \times 4} = 70$ different places ; so that 900,900 $\times$ 70 = 63,063,000 different placings are open to A, B, and C. In every one of these cases, D has no choice but to take the four places that remain. Therefore the correct answer is that the balls may be broken in 63,063,000 different ways under the conditions. Readers should compare this problem with No. 345, " The Two Pawns," which they will then know how to solve for cases where there are three, four, or more pawns on the board.

271.—FIFTEEN LETTER PUZZLE.

THE following will be found to comply with the conditions of grouping :—

ALE	MET	MOP	BLM
BAG	CAP	YOU	CLT
IRE	OIL	LUG	LNR
NAY	BIT	BUN	BPR
AIM	BEY	RUM	GMY
OAR	GIN	PLY	CGR
PEG	ICY	TRY	CMN
CUE	COB	TAU	PNT
ONE	GOT	PIU	

The fifteen letters used are A, E, I, O, U, Y, and B, C, G, L, M, N, P, R, T. The number of words is 27, and these are all shown in the first three columns. The last word, PIU, is a musical term in common use; but although it has crept into some of our dictionaries, it is Italian, meaning "a little; slightly." The remaining twenty-six are good words. Of course a TAU-cross is a T-shaped cross, also called the cross of St. Anthony, and borne on a badge in the Bishop's Palace at Exeter. It is also a name for the toad-fish.

We thus have twenty-six good words and one doubtful, obtained under the required conditions, and I do not think it will be easy to improve on this answer. Of course we are not bound by dictionaries but by common usage. If we went by the dictionary only in a case of this kind, we should find ourselves involved in prefixes, contractions, and such absurdities as I.O.U., which Nuttall actually gives as a word.

272.—THE NINE SCHOOLBOYS.

THE boys can walk out as follows :—

1st Day.	2nd Day.	3rd Day.
A B C	B F H	F A G
D E F	E I A	I D B
G H I	C G D	H C E

4th Day.	5th Day.	6th Day.
A D H	G B I	D C A
B E G	C F D	E H B
F I C	H A E	I G F

Every boy will then have walked by the side of every other boy once and once only.

Dealing with the problem generally, 12 $n+9$ boys may walk out in triplets under the conditions on 9 $n+6$ days, where n may be nought or any integer. Every possible pair will occur once. Call the number of boys m. Then every boy will pair $m-1$ times, of which $\frac{m-1}{4}$ times he will be in the middle of a triplet and $\frac{m-1}{2}$ times on the outside. Thus, if we refer to the solution above, we find that every boy is in the middle twice (making 4 pairs) and four times on the outside (making the remaining 4 pairs of his 8). The reader may now like to try his hand at solving the two next cases of 21 boys on 15

days, and 33 boys on 24 days. It is, perhaps, interesting to note that a school of 489 boys could thus walk out daily in one leap year, but it would take 731 girls (referred to in the solution to No. 269) to perform their particular feat by a daily walk in a year of 365 days.

273.—THE ROUND TABLE.

THE history of this problem will be found in *The Canterbury Puzzles* (No. 90). Since the publication of that book in 1907, so far as I know, nobody has succeeded in solving the case for that unlucky number of persons, 13, seated at a table on 66 occasions. A solution is possible for any number of persons, and I have recorded schedules for every number up to 25 persons inclusive and for 33. But as I know a good many mathematicians are still considering the case of 13, I will not at this stage rob them of the pleasure of solving it by showing the answer. But I will now display the solutions for all the cases up to 12 persons inclusive. Some of these solutions are now published for the first time, and they may afford useful clues to investigators.

The solution for the case of 3 persons seated on 1 occasion needs no remark.

A solution for the case of 4 persons on 3 occasions is as follows :—

1	2	3	4
1	3	4	2
1	4	2	3

Each line represents the order for a sitting, and the person represented by the last number in a line must, of course, be regarded as sitting next to the first person in the same line, when placed at the round table.

The case of 5 persons on 6 occasions may be solved as follows :—

1	2	3	4	5
1	2	4	5	3
1	2	5	3	4

1	3	2	5	4
1	4	2	3	5
1	5	2	4	3

The case for 6 persons on 10 occasions is solved thus :—

1	2	3	6	4	5
1	3	4	2	5	6
1	4	5	3	6	2
1	5	6	4	2	3
1	6	2	5	3	4

1	2	4	5	6	3
1	3	5	6	2	4
1	4	6	2	3	5
1	5	2	3	4	6
1	6	3	4	5	2

It will now no longer be necessary to give the solutions in full, for reasons that I will explain. It will be seen in the examples above that the 1 (and, in the case of 5 persons, also the 2) is repeated down the column. Such a number I call a "repeater." The other numbers descend in cyclical order. Thus, for 6 persons we get the cycle, 2, 3, 4, 5, 6, 2, and so on, in every column. So it is only necessary to give the two lines 1 2 3 6 4 5 and 1 2 4 5 6 3, and denote the cycle and repeaters, to enable any one to write out the full solution straight away. The reader may wonder why I do not start the last solution with the numbers in their natural order, 1 2 3 4 5 6. If I did so the numbers in the descending cycle would not be in their natural order, and it is more convenient to have a regular cycle than to consider the order in the first line.

The difficult case of 7 persons on 15 occasions is solved as follows, and was given by me in *The Canterbury Puzzles* :—

1	2	3	4	5	7	6
1	6	2	7	5	3	4
1	3	5	2	6	7	4
1	5	7	4	3	6	2
1	5	2	7	3	4	6

In this case the 1 is a repeater, and there are *two* separate cycles, 2, 3, 4, 2, and 5, 6, 7, 5. We thus get five groups of three lines each, for a fourth line in any group will merely repeat the first line.

A solution for 8 persons on 21 occasions is as follows :—

1	8	6	3	4	5	2	7
1	8	4	5	7	2	3	6
1	8	2	7	3	6	4	5

The 1 is here a repeater, and the cycle 2, 3, 4, 5, 6, 7, 8. Every one of the 3 groups will give 7 lines.

Here is my solution for 9 persons on 28 occasions :—

2	1	9	7	4	5	6	3	8
2	9	5	1	6	8	3	4	7
2	9	3	1	8	4	7	5	6
2	9	1	5	6	4	7	8	3

There are here two repeaters, 1 and 2, and the cycle is 3, 4, 5, 6, 7, 8, 9, We thus get 4 groups of 7 lines each.

The case of 10 persons on 36 occasions is solved as follows :—

1	10	8	3	6	5	4	7	2	9
1	10	6	5	2	9	7	4	3	8
1	10	2	9	3	8	6	5	7	4
1	10	7	4	8	3	2	9	5	6

The repeater is 1, and the cycle, 2, 3, 4, 5, 6, 7, 8, 9, 10. We here have 4 groups of 9 lines each.

My solution for 11 persons on 45 occasions is as follows :—

2	11	9	4	7	6	5	1	8	3	10
2	1	11	7	6	3	10	8	5	4	9
2	11	10	3	9	4	8	5	1	7	6
2	11	5	8	1	3	10	6	7	9	4
2	11	1	10	3	4	9	6	7	5	8

There are two repeaters, 1 and 2, and the cycle is, 3, 4, 5, . . . 11. We thus get 5 groups of 9 lines each.

The case of 12 persons on 55 occasions is solved thus :—

1	2	3	12	4	11	5	10	6	9	7	8
1	2	4	11	6	9	8	7	10	5	12	3
1	2	5	10	8	7	11	4	3	12	6	9
1	2	6	9	10	5	3	12	7	8	11	4
1	2	7	8	12	3	6	9	11	4	5	10

Here 1 is a repeater, and the cycle is 2, 3, 4, 5, . . . 12. We thus get 5 groups of 11 lines each.

274.—THE MOUSE-TRAP PUZZLE.

IF we interchange cards 6 and 13 and begin our count at 14, we may take up all the twenty-one cards—that is, make twenty-one "catches"—in the following order : 6, 8, 13, 2, 10, 1, 11, 4, 14, 3, 5, 7, 21, 12, 15, 20, 9, 16, 18, 17, 19. We may also exchange 10 and 14 and start at 16, or exchange 6 and 8 and start at 19.

275.—THE SIXTEEN SHEEP.

THE six diagrams on next page show solutions for the cases where we replace 2, 3, 4, 5, 6, and 7 hurdles. The dark lines indicate the hurdles that have been replaced. There are, of course, other ways of making the removals.

276.—THE EIGHT VILLAS.

THERE are several ways of solving the puzzle, but there is very little difference between them. The solver should, however, first of all bear in mind that in making his calculations he need only consider the four villas that stand at the corners, because the intermediate villas can never vary when the corners are known. One way is to place the numbers nought to 9 one at a time in the top left-hand corner, and then consider each case in turn.

Now, if we place 9 in the corner as shown in the Diagram A, two of the corners cannot be occupied, while the corner that is diagonally opposite may be filled by 0, 1, 2, 3, 4, 5, 6, 7, 8, or 9 persons. We thus see that there are 10

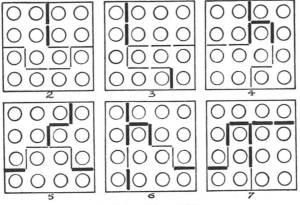

THE SIXTEEN SHEEP.

solutions with a 9 in the corner. If, however, we substitute 8, the two corners in the same row and column may contain o, o, or 1, 1, or o, 1, or 1, o. In the case of B, ten different selections may be made for the fourth corner; but in each of the cases C, D, and E, only nine selections are possible, because we cannot use the 9. Therefore with 8 in the top left-hand corner there are $10+(3\times9)=37$ different solutions. If we then try 7 in the corner, the result will be $10+27+40$, or 77 solutions. With 6 we get $10+27+40+49=126$; with 5, $10+27+40+49+54=180$; with 4, the same as with 5, $+55=235$; with 3, the same as with 4, $+52=287$; with 2, the same as with 3, $+45=332$; with 1, the same as with 2, $+34=366$, and with nought in the top left-hand corner the number

tenants may occupy some or all of the eight villas so that there shall be always nine persons living along each side of the square is 2,035. Of course, this method must obviously cover all the reversals and reflections, since each corner in turn is occupied by every number in all possible combinations with the other two corners that are in line with it.

Here is a general formula for solving the puzzle : $\dfrac{(n^2+3n+2)(n^2+3n+3)}{6}$. Whatever may be the stipulated number of residents along each of the sides (which number is represented by n), the total number of different arrangements may be thus ascertained. In our particular case the number of residents was nine. Therefore $(81+27+2)\times(81+27+3)$ and the product, divided by 6, gives 2,035. If the number of residents had been 0, 1, 2, 3, 4, 5, 6, 7, or 8, the total arrangements would be 1, 7, 26, 70, 155, 301, 532, 876, or 1,365 respectively.

277.—COUNTER CROSSES.

LET us first deal with the Greek Cross. There are just eighteen forms in which the numbers may be paired for the two arms. Here they are :—

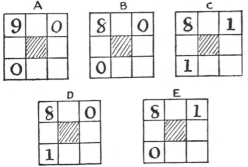

of solutions will be found to be $10+27+40+49+54+55+52+45+34+19=385$. As there is no other number to be placed in the top left-hand corner, we have now only to add these totals together thus, $10+37+77+126+180+235+287+332+366+385=2,035$. We therefore find that the total number of ways in which

1 2 9 7 8	1 3 9 6 8	1 4 9 5 8
3 4 9 5 6	2 4 9 5 7	2 3 9 6 7
2 3 9 5 8	1 3 7 6 9	1 4 7 5 9
1 4 9 6 7	2 4 7 5 8	2 3 7 6 8
1 2 5 8 9	2 3 7 5 9	1 3 5 7 9
3 4 5 6 7	1 4 7 6 8	2 4 5 6 8
1 4 5 6 9	2 3 5 6 9	1 4 3 7 9
2 3 5 7 8	1 4 5 7 8	2 5 3 6 8
1 5 3 6 9	2 4 3 6 9	2 3 1 8 9
2 4 3 7 8	1 5 3 7 8	4 5 1 6 7
2 4 1 7 9	2 5 1 6 9	3 4 1 6 9
3 5 1 6 8	3 4 1 7 8	2 5 1 7 8

Of course, the number in the middle is common to both arms. The first pair is the one I gave as an example. I will suppose that we have written out all these crosses, always placing the first row of a pair in the upright and the second row in the horizontal arm. Now, if we leave the central figure fixed, there are 24 ways in which the numbers in the upright may be varied, for the four counters may be changed in $1 \times 2 \times 3 \times 4 = 24$ ways. And as the four in the horizontal may also be changed in 24 ways for every arrangement on the other arm, we find that there are $24 \times 24 = 576$ variations for every form; therefore, as there are 18 forms, we get $18 \times 576 = 10,368$ ways. But this will include half the four reversals and half the four reflections that we barred, so we must divide this by 4 to obtain the correct answer to the Greek Cross, which is thus 2,592 different ways. The division is by 4 and not by 8, because we provided against half the reversals and reflections by always reserving one number for the upright and the other for the horizontal.

In the case of the Latin Cross, it is obvious that we have to deal with the same 18 forms of pairing. The total number of different ways in this case is the full number, 18×576. Owing to the fact that the upper and lower arms are unequal in length, permutations will repeat by reflection, but not by reversal, for we cannot reverse. Therefore this fact only entails division by 2. But in every pair we may exchange the figures in the upright with those in the horizontal (which we could not do in the case of the Greek Cross, as the arms are there all alike); consequently we must multiply by 2. This multiplication by 2 and division by 2 cancel one another. Hence 10,368 is here the correct answer.

278.—A DORMITORY PUZZLE.

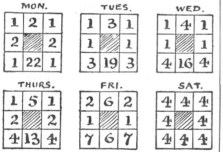

ARRANGE the nuns from day to day as shown in the six diagrams. The smallest possible number of nuns would be thirty-two, and the arrangements on the last three days admit of variation.

279.—THE BARRELS OF BALSAM.

THIS is quite easy to solve for any number of barrels—if you know how. This is the way to do it. There are five barrels in each row. Multiply the numbers 1, 2, 3, 4, 5 together; and also multiply 6, 7, 8, 9, 10 together. Divide one result by the other, and we get the number of different combinations or selections of ten things taken five at a time. This is here 252. Now, if we divide this by 6 (1 more than the number in the row) we get 42, which is the correct answer to the puzzle, for there are 42 different ways of arranging the barrels. Try this method of solution in the case of six barrels, three in each row, and you will find the answer is 5 ways. If you check this by trial, you will discover the five arrangements with 123, 124, 125, 134, 135 respectively in the top row, and you will find no others.

The general solution to the problem is, in fact, this : $\dfrac{C^n{}_{2n}}{n+1}$ where $2n$ equals the number of barrels. The symbol C, of course, implies that we have to find how many combinations, or selections, we can make of $2n$ things, taken n at a time.

280.—BUILDING THE TETRAHEDRON.

TAKE your constructed pyramid and hold it so that one stick only lies on the table. Now, four sticks must branch off from it in different directions—two at each end. Any one of the five sticks may be left out of this connection ; therefore the four may be selected in 5 different ways. But these four matches may be placed in 24 different orders. And as any match may be joined at either of its ends, they may further be varied (after their situations are settled for any particular arrangement) in 16 different ways. In every arrangement the sixth stick may be added in 2 different ways. Now multiply these results together, and we get $5 \times 24 \times 16 \times 2 = 3,840$ as the exact number of ways in which the pyramid may be constructed. This method excludes all possibility of error.

A common cause of error is this. If you calculate your combinations by working upwards from a basic triangle lying on the table, you will get half the correct number of ways, because you overlook the fact that an equal number of pyramids may be built on that triangle downwards, so to speak, through the table. They are, in fact, reflections of the others, and examples from the two sets of pyramids cannot be set up to resemble one another—except under fourth dimensional conditions !

281.—PAINTING A PYRAMID.

IT will be convenient to imagine that we are painting our pyramids on the flat cardboard, as in the diagrams, before folding up. Now, if we take any *four* colours (say red, blue, green, and yellow), they may be applied in only 2 distinctive ways, as shown in Figs. 1 and 2. Any way will only result in one of these when the pyramids are folded up. If we take any *three* colours, they may be applied in the 3 ways shown in Figs. 3, 4, and 5. If we take any *two* colours, they may be applied in the 3 ways shown in Figs. 6, 7, and 8. If we take any *single* colour, it may obviously be applied in only 1 way. But four colours may be

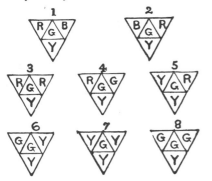

selected in 35 ways out of seven ; three in 35 ways ; two in 21 ways ; and one colour in 7 ways. Therefore 35 applied in 2 ways = 70 ; 35 in 3 ways = 105 ; 21 in 3 ways = 63 ; and 7 in 1 way = 7. Consequently the pyramid may be painted in 245 different ways (70 + 105 + 63 + 7), using the seven colours of the solar spectrum in accordance with the conditions of the puzzle.

282.—THE ANTIQUARY'S CHAIN.

THE number of ways in which nine things may be arranged in a row without any restrictions is 1 × 2 × 3 × 4 × 5 × 6 × 7 × 8 × 9 = 362,880. But we are told that the two circular rings must never be together ; therefore we must deduct the number of times that this would occur. The number is 1 × 2 × 3 × 4 × 5 × 6 × 7 × 8 = 40,320 × 2 = 80,640, because if we consider the two circular links to be inseparably joined together they become as one link, and eight links are capable of 40,320 arrangements ; but as these two links may always be put on in the orders A B or B A, we have to double this number, it being a question of arrangement and not of design. The deduction required reduces our total to 282,240. Then one of our links is of a peculiar form, like an 8. We have therefore the option of joining on either one end or the other on every occasion, so we must double the last result. This brings up our total to 564,480. We now come to the point to which I directed

the reader's attention—that every link may be put on in one of two ways. If we join the first finger and thumb of our left hand horizontally, and then link the first finger and thumb of the right hand, we see that the right thumb may be either above or below. But in the case of our chain we must remember that although that 8-shaped link has two independent *ends* it is like every other link in having only two *sides*—that is, you cannot turn over one end without turning the other at the same time. We will, for convenience, assume that each link has a black side and a side painted white. Now, if it were stipulated that (with the chain lying on the table, and every successive link falling over its predecessor in the same way, as in the diagram) only the white sides should be uppermost as in A, then the answer would be 564,480, as above—ignoring for the present all reversals of the completed chain. If, however, the first link were allowed to be placed either side up, then we could have either A or B, and the answer would be 2 × 564,480 = 1,128,960 ; if two links might be placed either way up, the answer would be 4 × 564,480 ; if three links, then 8 × 564,480, and so on. Since, therefore, every link may be placed either side up, the number will be 564,480 multiplied by 2⁹, or by 512. This raises our total to 289,013,760. But there is still one more point to be considered. We have not yet allowed for the fact that with any given arrangement three of the other arrangements may be obtained by simply turning the chain over through its entire length and by reversing the ends. Thus C is really the same as A, and if we turn this page upside down, then A and C give two other arrangements that are still really identical. Thus to get the correct answer to the puzzle we must divide our last total by 4, when we find that there are just 72,253,440 different ways in which the smith might have put those links together. In other words, if the nine links had originally formed a piece of chain, and it was known that the two circular links were separated, then it would be 72,253,439 chances to 1 that the smith would not have put the links together again precisely as they were arranged before !

283.—THE FIFTEEN DOMINOES.

THE reader may have noticed that at each end of the line I give is a four, so that, if we like, we can form a ring instead of a line. It can easily be proved that this must always be so. Every line arrangement will make a circular arrangement if we like to join the ends. Now, curious as it may at first appear, the following diagram exactly represents the conditions when we leave the doubles out of the question and devote our attention to forming circular arrangements. Each number, or half domino, is in line with every other number, so that if we start at any one of the five numbers and go over all the lines of the pentagon once and once only we shall come back to the starting place, and the order of our route will give us one of the circular arrangements for the ten dominoes. Take your pencil and follow out the following route, starting at the 4 : 41304210234. You have been over all the lines once only, and by repeating all these figures in this way, 41—13—30—04—42—21—10—02—23—34, you get an arrangement of the dominoes (without the doubles) which will be perfectly clear. Take other routes and you will get other arrangements. If, therefore, we can ascertain just how many of these circular routes are obtainable from the pentagon, then the rest is very easy.

Well, the number of different circular routes over the pentagon is 264. How I arrive at these figures I will not at present explain, because it would take a lot of space. The dominoes may, therefore, be arranged in a circle in just 264 different ways, leaving out the doubles. Now, in any one of these circles the five doubles may be inserted in $2^5=32$ different ways. Therefore when we include the doubles there are $264 \times 32 = 8,448$ different circular arrangements. But each of those circles may be broken (so as to form our straight line) in any one of 15 different places. Consequently, $8,448 \times 15$ gives 126,720 different ways as the correct answer to the puzzle.

I purposely refrained from asking the reader to discover in just how many different ways the full set of twenty-eight dominoes may be arranged in a straight line in accordance with the ordinary rules of the game, left to right

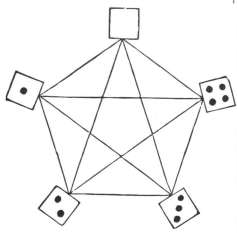

and right to left of any arrangement counting as different ways. It is an exceedingly difficult problem, but the correct answer is 7,959,229,931,520 ways. The method of solving is very complex.

284.—THE CROSS TARGET.

TWENTY-ONE different squares may be selected. Of these nine will be of the size shown by the four

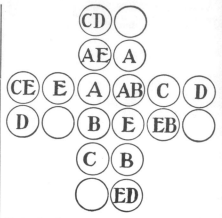

A's in the diagram, four of the size shown by the B's, four of the size shown by the C's, two of the size shown by the D's, and two of the size indicated by the upper single A, the upper single E, the lower single E, and the EB. It is an interesting fact that you cannot form any one of these twenty-one squares without using at least one of the six circles marked E.

285.—THE FOUR POSTAGE STAMPS.

REFERRING to the original diagram, the four stamps may be given in the shape 1, 2, 3, 4, in three ways ; in the shape 1, 2, 5, 6, in six ways ; in the shape 1, 2, 3, 5, or 1, 2, 3, 7, or 1, 5, 6, 7, or 3, 5, 6, 7, in twenty-eight ways ; in shape 1, 2, 3, 6, or 2, 5, 6, 7, in fourteen ways ; in shape 1, 2, 6, 7, or 2, 3, 5, 6, or 1, 5, 6, 10, or 2, 5, 6, 9, in fourteen ways. Thus there are sixty-five ways in all.

286.—PAINTING THE DIE.

THE 1 can be marked on any one of six different sides. For every side occupied by 1 we have a selection of four sides for the 2. For every situation of the 2 we have two places for the 3. (The 6, 5, and 4 need not be considered, as their positions are determined by the 1, 2, and 3.) Therefore 6, 4, and 2 multiplied together make 48 different ways—the correct answer.

287.—AN ACROSTIC PUZZLE.

THERE are twenty-six letters in the alphabet, giving 325 different pairs. Every one of these pairs may be reversed, making 650 ways. But every initial letter may be repeated as the final, producing 26 other ways. The total is therefore 676 different pairs. In other words, the answer is the square of the number of letters in the alphabet.

288.—CHEQUERED BOARD DIVISIONS.

THERE are 255 different ways of cutting the board into two pieces of exactly the same size

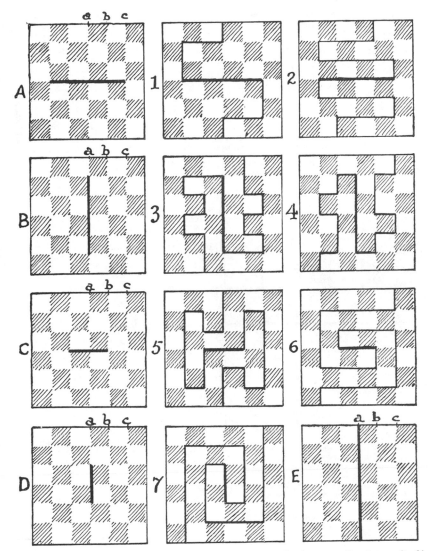

and shape. Every way must involve one of the five cuts shown in Diagrams A, B, C, D, and E. To avoid repetitions by reversal and reflection, we need only consider cuts that enter at the points *a*, *b*, and *c*. But the exit must always be at a point in a straight line from the entry through the centre. This is the most important condition to remember. In case B you cannot enter at *a*, or you will get the cut provided for in E. Similarly in C or D, you must not enter the key-line in the same direction as itself, or you will get A or B. If you are working on A or C and entering at *a*, you must consider joins at one end only of the key-line, or you will get repetitions. In other cases you must consider joins at both ends of the key; but after leaving *a* in case D, turn always either to right or left—use one direction only. Figs. 1 and 2 are examples under A; 3 and 4 are examples under B; 5 and 6 come under C;

and 7 is a pretty example of D. Of course, E is a peculiar type, and obviously admits of only one way of cutting, for you clearly cannot enter at *b* or *c*.
Here is a table of the results :—

		a		*b*		*c*		Ways.
A	=	8	+	17	+	21	=	46
B	=	0	+	17	+	21	=	38
C	=	15	+	31	+	39	=	85
D	=	17	+	29	+	39	=	85
E	=	1	+	0	+	0	=	1
		41		94		120		255

I have not attempted the task of enumerating the ways of dividing a board 8 × 8—that is, an ordinary chessboard. Whatever the method adopted, the solution would entail considerable labour.

289.—LIONS AND CROWNS.

HERE is the solution. It will be seen that each of the four pieces (after making the cuts along the thick lines) is of exactly the same size and shape, and that each piece contains a lion and a crown. Two of the pieces are shaded so as to make the solution quite clear to the eye.

290.—BOARDS WITH AN ODD NUMBER OF SQUARES.

THERE are fifteen different ways of cutting the 5 × 5 board (with the central square removed) into two pieces of the same size and shape. Limitations of space will not allow me to give diagrams of all these, but I will enable the reader to draw them all out for himself without the slightest difficulty. At whatever point on the edge your cut enters, it must always end at a point on the edge, exactly opposite in a line through the centre of the square. Thus, if you enter at point 1 (see Fig. 1) at the

top, you must leave at point 1 at the bottom. Now, 1 and 2 are the only two really different points of entry ; if we use any others they will simply produce similar solutions. The directions of the cuts in the following fifteen

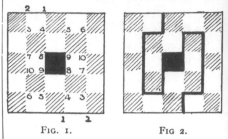

FIG. 1.   FIG 2.

solutions are indicated by the numbers on the diagram. The duplication of the numbers can lead to no confusion, since every successive number is contiguous to the previous one. But whichever direction you take from the top downwards you must repeat from the bottom upwards, one direction being an exact reflection of the other.

1, 4, 8.
1, 4, 3, 7, 8.
1, 4, 3, 7, 10, 9.
1, 4, 3, 7, 10, 6, 5, 9.
1, 4, 5, 9.
1, 4, 5, 6, 10, 9.
1, 4, 5, 6, 10, 7, 8.
2, 3, 4, 5, 9.
2, 3, 4, 5, 6, 10, 9.
2, 3, 4, 5, 6, 10, 7, 8.
2, 3, 7, 8.
2, 3, 7, 10, 9.
2, 3, 7, 10, 6, 5, 9.
2, 3, 7, 10, 6, 5, 4, 8.

It will be seen that the fourth direction (1, 4, 3, 7, 10, 6, 5, 9) produces the solution shown in Fig. 2. The thirteenth produces the solution given in propounding the puzzle, where the cut entered at the side instead of at the top. The pieces, however, will be of the same shape if turned over, which, as it was stated in the conditions, would not constitute a different solution.

291.—THE GRAND LAMA'S PROBLEM.

THE method of dividing the chessboard so that each of the four parts shall be of exactly the same size and shape, and contain one of the gems, is shown in the diagram. The method of shading the squares is adopted to make the shape of the pieces clear to the eye. Two of the pieces are shaded and two left white.
The reader may find it interesting to compare this puzzle with that of the " Weaver " (No. 14, *Canterbury Puzzles*).

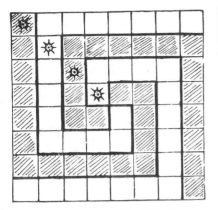

THE GRAND LAMA'S PROBLEM.

292.—THE ABBOT'S WINDOW.

THE man who was "learned in strange mysteries" pointed out to Father John that the orders of the Lord Abbot of St. Edmondsbury might be easily carried out by blocking up twelve of the lights in the window as shown by the dark squares in the following sketch :—

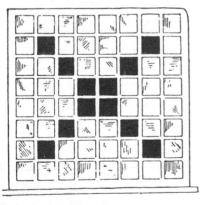

Father John held that the four corners should also be darkened, but the sage explained that it was desired to obstruct no more light than was absolutely necessary, and he said, anticipating Lord Dundreary, " A single pane can no more be in a *line* with itself than one bird can go into a corner and flock in solitude. The Abbot's condition was that no diagonal *lines* should contain an odd number of lights." Now, when the holy man saw what had been done he was well pleased, and said, " Truly, Father John, thou art a man of deep wisdom, in that thou hast done that which seemed impossible, and yet withal adorned our window

with a device of the cross of St. Andrew, whose name I received from my godfathers and godmothers." Thereafter he slept well and arose refreshed. The window might be seen intact to-day in the monastery of St. Edmondsbury, if it existed, which, alas ! the window does not.

293.—THE CHINESE CHESSBOARD.

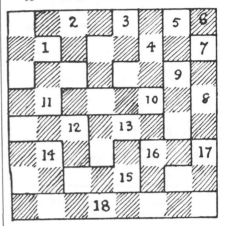

EIGHTEEN is the maximum number of pieces. I give two solutions. The numbered diagram is so cut that the eighteenth piece has the largest area—eight squares—that is possible under the conditions. The second diagram was prepared under the added condition that no piece should contain more than five squares.

No. 74 in *The Canterbury Puzzles* shows how to cut the board into twelve pieces, all different, each containing five squares, with one square piece of four squares.

294.—THE CHESSBOARD SENTENCE.

THE pieces may be fitted together, as shown in the illustration, to form a perfect chessboard.

295.—THE EIGHT ROOKS.

OBVIOUSLY there must be a rook in every row and every column. Starting with the top row, it is clear that we may put our first rook on any one of eight different squares. Wherever it is placed, we have the option of seven squares for the second rook in the second row. Then we have six squares from which to select the third row, five in the fourth, and so on. Therefore the number of our different ways must be $8 \times 7 \times 6 \times 5 \times 4 \times 3 \times 2 \times 1 = 40,320$ (that is $\lfloor 8$), which is the correct answer.

How many ways there are if mere reversals and reflections are not counted as different has not yet been determined; it is a difficult problem. But this point, on a smaller square, is considered in the next puzzle.

296.—THE FOUR LIONS.

THERE are only seven different ways under the conditions. They are as follows: 1 2 3 4, 1 2 4 3, 1 3 2 4, 1 3 4 2, 1 4 3 2, 2 1 4 3, 2 4 1 3. Taking the last example, this notation means that we place a lion in the second square of first row, fourth square of second row, first square of third row, and third square of fourth row. The first example is, of course, the one we gave when setting the puzzle.

297.—BISHOPS—UNGUARDED.

THIS cannot be done with fewer bishops than eight, and the simplest solution is to place the bishops in line along the fourth or fifth row of

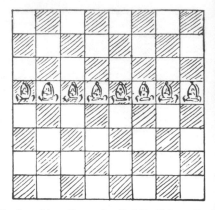

the board (see diagram). But it will be noticed that no bishop is here guarded by another, so we consider that point in the next puzzle.

298.—BISHOPS—GUARDED.

THIS puzzle is quite easy if you first of all give it a little thought. You need only consider squares of one colour, for whatever can be done in the case of the white squares can always be repeated on the black, and they are here quite independent of one another. This equality, of course, is in consequence of the fact that the number of squares on an ordinary chessboard, sixty-four, is an even number. If a square chequered board has an odd number of squares, then there will always be one more square of one colour than of the other.

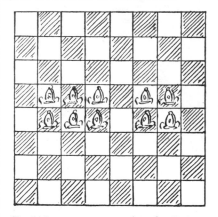

Ten bishops are necessary in order that every square shall be attacked and every bishop guarded by another bishop. I give one way of arranging them in the diagram. It will be noticed that the two central bishops in the group of six on the left-hand side of the board serve no purpose, except to protect those bishops that are on adjoining squares. Another solution would therefore be obtained by simply raising the upper one of these one square and placing the other a square lower down.

299.—BISHOPS IN CONVOCATION.

THE fourteen bishops may be placed in 256 different ways. But every bishop must always be placed on one of the sides of the board—that is, somewhere on a row or file on the extreme edge. The puzzle, therefore, consists in counting the number of different ways that we can arrange the fourteen round the edge of the board without attack. This is not a difficult

are in a straight line in any oblique direction. This is the only arrangement out of the twelve fundamentally different ways of placing eight queens without attack that fulfils the last condition.

301.—THE EIGHT STARS.

THE solution of this puzzle is shown in the first diagram. It is the only possible solution within the conditions stated. But if one of the eight stars had not already been placed as shown, there would then have been eight ways of arranging the stars according to this scheme, if we count reversals and reflections as different. If you turn this page round so that each side is in turn at the bottom, you will get the four reversals; and if you reflect each of these in a

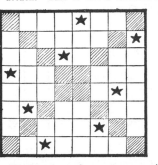

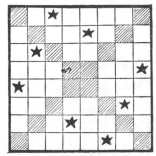

matter. On a chessboard of n^2 squares $2n-2$ bishops (the maximum number) may always be placed in 2^n ways without attacking. On an ordinary chessboard n would be 8 ; therefore 14 bishops may be placed in 256 different ways. It is rather curious that the general result should come out in so simple a form.

300.—THE EIGHT QUEENS.

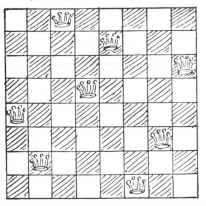

THE solution to this puzzle is shown in the diagram. It will be found that no queen attacks another, and also that no three queens

mirror, you will get the four reflections. These are, therefore, merely eight aspects of one "fundamental solution." But without that first star being so placed, there is another fundamental solution, as shown in the second diagram. But this arrangement being in a way symmetrical, only produces four different aspects by reversal and reflection.

302.—A PROBLEM IN MOSAICS.

THE diagram shows how the tiles may be rearranged. As before, one yellow and one purple tile are dispensed with. I will here point out that in the previous arrangement the yellow and purple tiles in the seventh row might have changed places, but no other arrangement was possible.

303.—UNDER THE VEIL.

SOME schemes give more diagonal readings of four letters than others, and we are at first tempted to favour these; but this is a false scent, because what you appear to gain in this direction you lose in others. Of course it immediately occurs to the solver that every LIVE or EVIL is worth twice as much as any other word, since it reads both ways and always counts as 2. This is an important consideration, though sometimes those arrangements that contain most readings of these two words are fruitless in other words, and we lose in the general count.

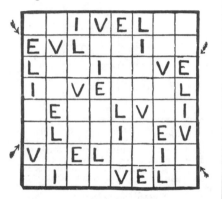

The above diagram is in accordance with the conditions requiring no letter to be in line with another similar letter, and it gives twenty readings of the five words—six horizontally, six vertically, four in the diagonals indicated by the arrows on the left, and four in the diagonals indicated by the arrows on the right. This is the maximum.

Four sets of eight letters may be placed on the board of sixty-four squares in as many as 604 different ways, without any letter ever being in line with a similar one. This does not count reversals and reflections as different, and it does not take into consideration the actual permutations of the letters among themselves; that is, for example, making the L's change places with the E's. Now it is a singular fact that not only do the twenty word-readings that I have given prove to be the real maximum, but there is actually only that one arrangement from which this maximum may be obtained. But if you make the V's change places with the I's, and the L's with the E's, in the solution given, you still get twenty readings—the same number as before in every direction. Therefore there are two ways of getting the maximum

from the same arrangement. The minimum number of readings is zero—that is, the letters can be so arranged that no word can be read in any of the directions.

304.—BACHET'S SQUARE.

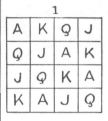

LET us use the letters A, K, Q, J, to denote ace, king, queen, jack; and D, S, H, C, to denote diamonds, spades, hearts, clubs. In Diagrams 1 and 2 we have the two available ways of arranging either group of letters so that no two similar letters shall be in line—though a quarter-turn of 1 will give us the arrangement in 2. If we superimpose or combine these two squares, we get the arrangement of Diagram 3, which is one solution. But in each square we may put the letters in the top line in twenty-four different ways without altering the scheme of arrangement. Thus, in Diagram 4 the S's are similarly placed to the D's in 2, the H's to the S's, the C's to the H's, and the D's to the C's. It clearly follows that there must be $24 \times 24 = 576$ ways of combining the two primitive arrangements. But the error that Labosne fell into was that of assuming that the A, K, Q, J must be arranged in the form 1, and the D, S, H, C in the form 2. He thus included reflections and half-turns, but not quarter-turns. They may obviously be interchanged. So that the correct answer is $2 \times 576 = 1,152$, counting reflections and reversals as different. Put in another manner, the pairs in the top row may be written in $16 \times 9 \times 4 \times 1 = 576$ different ways, and the square then completed in 2 ways, making 1,152 ways in all.

305.—THE THIRTY-SIX LETTER BLOCKS.

I POINTED out that it was impossible to get all the letters into the box under the conditions, but the puzzle was to place as many as possible.

This requires a little judgment and careful investigation, or we are liable to jump to the hasty conclusion that the proper way to solve the puzzle must be first to place all six of one letter, then all six of another letter, and so on. As there is only one scheme (with its reversals) for placing six similar letters so that no two shall be in a line in any direction, the reader will find that after he has placed four different kinds of letters, six times each, every place is occupied except those twelve that form the two long diagonals. He is, therefore, unable to place more than two each of his last two letters, and there are eight blanks left. I give such an arrangement in Diagram 1.

1

A	B	C	D	E	F
D	F	E	B	A	C
E	C			D	B
B	D			C	E
C		B	E		D
	E	D	C	B	

2

A	B	C	D	E	F
D	E	A	F	B	C
F	C			D	A
B	D			C	E
C	A	E	B	F	D
E	F	D	C	A	B

The secret, however, consists in not trying thus to place all six of each letter. It will be found that if we content ourselves with placing only five of each letter, this number (thirty in all) may be got into the box, and there will be only six blanks. But the correct solution is to place six of each of two letters and five of each of the remaining four. An examination of Diagram 2 will show that there are six each of C and D, and five each of A, B, E, and F. There are, therefore, only four blanks left, and no letter is in line with a similar letter in any direction.

306.—THE CROWDED CHESSBOARD.

Here is the solution Only 8 queens or 8 rooks can be placed on the board without attack, while the greatest number of bishops is 14, and of knights 32. But as all these knights must be placed on squares of the same colour, while the queens occupy four of each colour and the bishops 7 of each colour, it follows that only 21 knights can be placed on the same colour in this puzzle. More than 21 knights can be placed alone on the board if we use both colours, but I have not succeeded in placing more than 21 on the "crowded chessboard." I believe the above solution contains the maximum number of pieces, but possibly some ingenious reader may succeed in getting in another knight.

307.—THE COLOURED COUNTERS.

The counters may be arranged in this order :—

R1, B2, Y3, O4, G5.
Y4, O5, G1, R2, B3.
G2, R3, B4, Y5, O1.
B5, Y1, O2, G3, R4.
O3, G4, R5, B1, Y2.

308.—THE GENTLE ART OF STAMP-LICKING.

The following arrangement shows how sixteen stamps may be stuck on the card, under the conditions, of a total value of fifty pence, or 4s. 2d. :—

4	3	5	2
5	2	1	4
1	4	3	5
3	5	2	1

If, after placing the four 5d. stamps, the reader is tempted to place four 4d. stamps also, he can afterwards only place two of each of the three other denominations, thus losing two spaces and counting no more than forty-eight pence, or 4s. This is the pitfall that was hinted at. (Compare with No. 43, *Canterbury Puzzles*.)

309.—THE FORTY-NINE COUNTERS.

The counters may be arranged in this order :—

A1, B2, C3, D4, E5, F6, G7.
F4, G5, A6, B7, C1, D2, E3.
D7, E1, F2, G3, A4, B5, C6.
B3, C4, D5, E6, F7, G1, A2.
G6, A7, B1, C2, D3, E4, F5.
E2, F3, G4, A5, B6, C7, D1.
C5, D6, E7, F1, G2, A3, B4.

310.—THE THREE SHEEP.

The number of different ways in which the three sheep may be placed so that every pen shall always be either occupied or in line with at least one sheep is forty-seven.

The following table, if used with the key in Diagram 1, will enable the reader to place them in all these ways :—

Two Sheep.	Third Sheep.	No. of Ways.
A and B	C, E, G, K, L, N, or P	7
A and C	I, J, K, or O	4
A and D	M, N, or J	3
A and F	J, K, L, or P	4
A and G	H, J, K, N, O, or P	6
A and H	K, L, N, or O	4
A and O	K or L	2
B and C	N	1
B and E	F, H, K, or L	4
B and F	G, J, N, or O	4
B and G	K, L, or N	3
B and H	J or N	2
B and J	K or L	2
F and G	J	1
		47

This, of course, means that if you place sheep in the pens marked A and B, then there are seven different pens in which you may place the third sheep, giving seven different solutions. It was understood that reversals and reflections do not count as different.

If one pen at least is to be *not* in line with a sheep, there would be thirty solutions to that problem. If we counted all the reversals and reflections of these 47 and 30 cases respectively as different, their total would be 560, which is the number of different ways in which the sheep may be placed in three pens without any conditions. I will remark that there are three ways in which two sheep may be placed so that every pen is occupied or in line, as in Diagrams 2, 3, and 4, but in every case each sheep is in line with its companion. There are only two ways in which three sheep may be so placed that every pen shall be occupied or in line, so no sheep in line with another. These I show in Diagrams 5 and 6. Finally, there is only one way in which three sheep may be placed so that at least one pen shall not be in line with a sheep and yet no sheep in line with another. Place the sheep in C, E, L. This is practically all there is to be said on this pleasant pastoral subject.

311.—THE FIVE DOGS PUZZLE.

THE diagrams show four fundamentally differ. ent solutions. In the case of A we can reverse

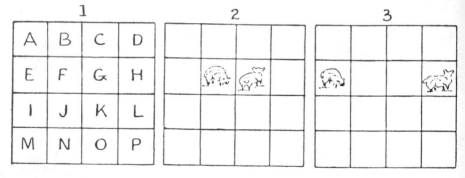

1			
A	B	C	D
E	F	G	H
I	J	K	L
M	N	O	P

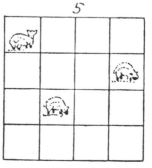

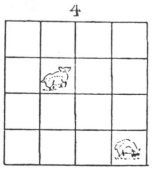

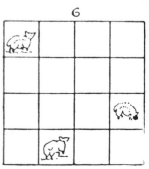

the order, so that the single dog is in the bottom row and the other four shifted up two squares. Also we may use the next column to the right and both of the two central horizontal rows. Thus A gives 8 solutions. Then B may be

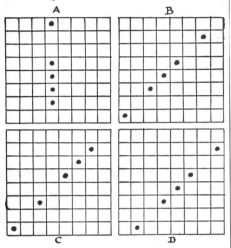

A   B

C   D

reversed and placed in either diagonal, giving 4 solutions. Similarly C will give 4 solutions. The line in D being symmetrical, its reversal will not be different, but it may be disposed in 4 different directions. We thus have in all 20 different solutions.

312.—THE FIVE CRESCENTS OF BYZANTIUM.

IF that ancient architect had arranged his five crescent tiles in the manner shown in the following diagram, every tile would have been watched

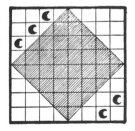

over by, or in a line with, at least one crescent, and space would have been reserved for a perfectly square carpet equal in area to exactly half of the pavement. It is a very curious fact that, although there are two or three solutions allowing a carpet to be laid down within the conditions so as to cover an area of nearly twenty-nine of the tiles, this is the only possible

solution giving exactly half the area of the pavement, which is the largest space obtainable.

313.—QUEENS AND BISHOP PUZZLE.

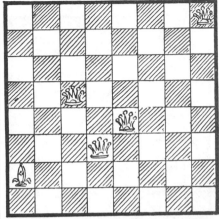

FIG. 1.

THE bishop is on the square originally occupied by the rook, and the four queens are so placed that every square is either occupied or attacked by a piece. (Fig. 1.)

I pointed out in 1899 that if four queens are placed as shown in the diagram (Fig. 2), then

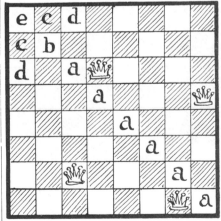

FIG. 2.

the fifth queen may be placed on any one of the twelve squares marked a, b, c, d, and e; or a rook on the two squares, c; or a bishop on the eight squares, a, b, and e; or a pawn on the square **b**; or a king on the four squares, **b**, **c**, and **e**. The only known arrangement for four queens and a knight is that given by Mr. J. Wallis in *The Strand Magazine* for August 1908, here reproduced. (Fig. 3.)

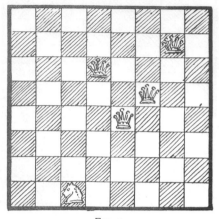

FIG. 3.

I have recorded a large number of solutions with four queens and a rook, or bishop, but the only arrangement, I believe, with three queens and two rooks in which all the pieces are guarded is that of which I give an illustration (Fig. 4),

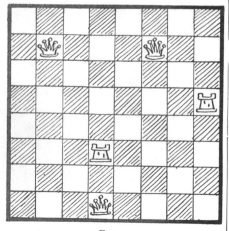

FIG. 4.

first published by Dr. C. Planck. But I have since found the accompanying solution with three queens, a rook, and a bishop, though the pieces do not protect one another. (Fig. 5.)

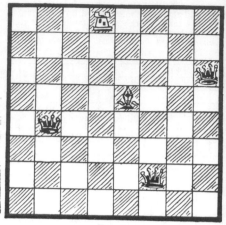

FIG. 5.

314.—THE SOUTHERN CROSS.

My readers have been so familiarized with the fact that it requires at least five planets to attack every one of a square arrangement of sixty-four stars that many of them have, perhaps, got to believe that a larger square arrangement of stars must need an increase of planets. It was to correct this possible error of reasoning, and so warn readers against another of those numerous little pitfalls in the world of puzzledom, that I devised this new stellar problem. Let me then state at once that, in the case of a square arrangement of eighty one stars, there are several ways of placing five planets so that every star shall be in line with at least one planet vertically, horizontally, or diagonally. Here is the solution to the "Southern Cross":—

It will be remembered that I said that the five planets in their new positions "will, of course, obscure five other stars in place of those at present covered." This was to exclude an easier solution in which only four planets need be moved.

315.—THE HAT-PEG PUZZLE.

THE moves will be made quite clear by a reference to the diagrams, which show the position on the board after each of the four moves. The

queen attacks any other. In the case of the last move the queen in the top row might also have been moved one square farther to the left. This is, I believe, the only solution to the puzzle.

316.—THE AMAZONS.

IT will be seen that only three queens have been removed from their positions on the edge of the board, and that, as a consequence, eleven squares (indicated by the black dots) are left unattacked by any queen. I will hazard the

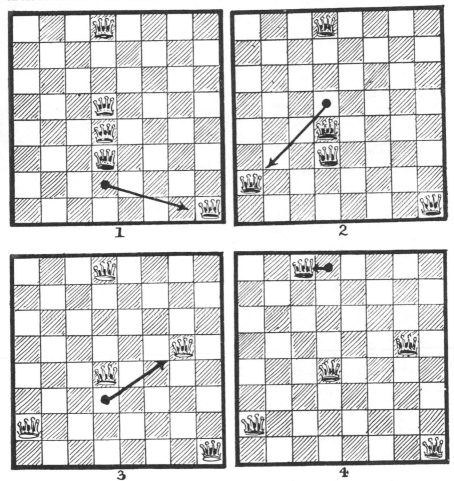

1

2

3

4

darts indicate the successive removals that have been made. It will be seen that at every stage all the squares are either attacked or occupied, and that after the fourth move no

statement that eight queens cannot be placed on the chessboard so as to leave more than eleven squares unattacked. It is true that we have no rigid proof of this yet, but I have

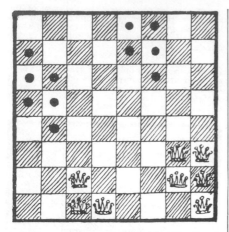

entirely convinced myself of the truth of the statement. There are at least five different ways of arranging the queens so as to leave eleven squares unattacked.

317.—A PUZZLE WITH PAWNS.

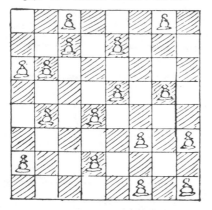

SIXTEEN pawns may be placed so that no three shall be in a straight line in any possible direction, as in the diagram. We regard, as the conditions required, the pawns as mere points on a plane.

318.—LION-HUNTING.

THERE are 6,480 ways of placing the man and the lion, if there are no restrictions whatever except that they must be on different spots. This is obvious, because the man may be placed on any one of the 81 spots, and in every case there are 80 spots remaining for the lion; therefore $81 \times 80 = 6,480$. Now, if we deduct the number of ways in which the lion and the

man may be placed on the same path, the result must be the number of ways in which they will not be on the same path. The number of ways in which they may be in line is found without much difficulty to be 816. Consequently, $6,480 - 816 = 5,664$, the required answer.

The general solution is this: $\frac{1}{3}n(n-1)$ $(3n^2 - n + 2)$. This is, of course, equivalent to saying that if we call the number of squares on the side of a " chessboard " n, then the formula shows the number of ways in which two bishops may be placed without attacking one another. Only in this case we must divide by two, because the two bishops have no distinct individuality, and cannot produce a different solution by mere exchange of places.

319.—THE KNIGHT-GUARDS.

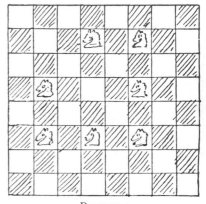

DIAGRAM 1.

THE smallest possible number of knights with which this puzzle can be solved is fourteen.

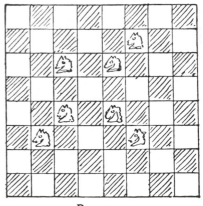

DIAGRAM 2.

It has sometimes been assumed that there are a great many different solutions. As a matter of fact, there are only three arrangements—not counting mere reversals and reflections as

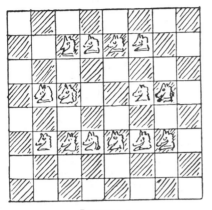

DIAGRAM 3.

different. Curiously enough, nobody seems ever to have hit on the following simple proof, or to have thought of dealing with the black and the white squares separately.

Seven knights can be placed on the board on white squares so as to attack every black square in two ways only. These are shown in Diagrams 1 and 2. Note that three knights occupy the same position in both arrangements. It is therefore clear that if we turn the board so that a black square shall be in the top left-hand

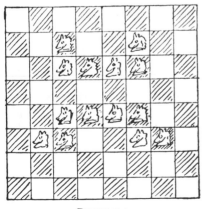

DIAGRAM 4.

corner instead of a white, and place the knights in exactly the same positions, we shall have

two similar ways of attacking all the white squares. I will assume the reader has made the two last described diagrams on transparent paper, and marked them 1a and 2a. Now, by placing the transparent Diagram 1a over 1 you will be able to obtain the solution in Diagram 3, by placing 2a over 2 you will get Diagram 4, and by placing 2a over 1 you will get Diagram 5.

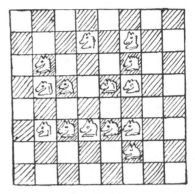

DIAGRAM 5.

You may now try all possible combinations of those two pairs of diagrams, but you will only get the three arrangements I have given, or their reversals and reflections. Therefore these three solutions are all that exist.

320.—THE ROOK'S TOUR.

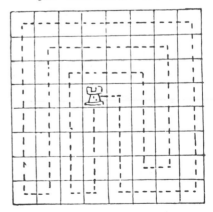

THE only possible minimum solutions are shown in the two diagrams, where it will be seen that only sixteen moves are required to perform the feat. Most people find it difficult to reduce the number of moves below seventeen.

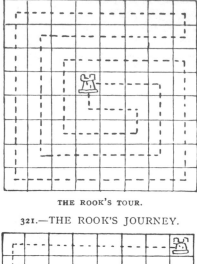

THE ROOK'S TOUR.

321.—THE ROOK'S JOURNEY.

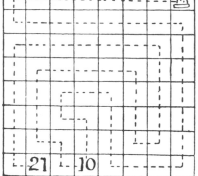

I show the route in the diagram. It will be seen that the tenth move lands us at the square marked "10," and that the last move, the twenty-first, brings us to a halt on square "21."

322.—THE LANGUISHING MAIDEN.

THE dotted line shows the route in twenty-two straight paths by which the knight may rescue the maiden. It is necessary, after entering the first cell, immediately to return before entering another. Otherwise a solution would not be possible. (See "The Grand Tour," p. 200.)

323.—A DUNGEON PUZZLE.

IF the prisoner takes the route shown in the diagram—where for clearness the doorways are omitted—he will succeed in visiting every cell once, and only once, in as many as fifty-seven straight lines. No rook's path over the chess-board can exceed this number of moves.

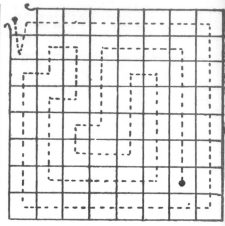

THE LANGUISHING MAIDEN.

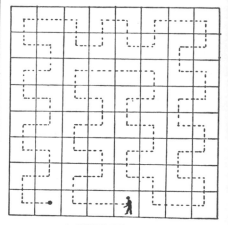

A DUNGEON PUZZLE.

324.—THE LION AND THE MAN.

FIRST of all, the fewest possible straight lines in each case are twenty-two, and in order that no cell may be visited twice it is absolutely necessary that each should pass into one cell and then immediately "visit" the one from which he started, afterwards proceeding by way of the second available cell. In the following diagram the man's route is indicated by the unbroken lines, and the lion's by the dotted lines. It will be found, if the two routes are followed cell by cell with two pencil points, that the lion and the man never meet. But there was one little point that ought not to be overlooked —"they occasionally got glimpses of one another." Now, if we take one route for the man and merely reverse it for the lion, we invariably find that, going at the same speed, they never get a glimpse of one another. But in our diagram it will be found that the man and the lion are in the cells marked A at the same moment, and may see one another through the

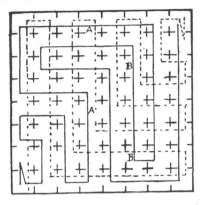

open doorways; while the same happens when they are in the two cells marked B, the upper letters indicating the man and the lower the lion. In the first case the lion goes straight for the man, while the man appears to attempt to get in the rear of the lion; in the second case it looks suspiciously like running away from one another!

325.—AN EPISCOPAL VISITATION.

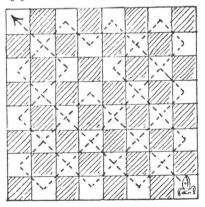

IN the diagram I show how the bishop may be made to visit every one of his white parishes in seventeen moves. It is obvious that we must start from one corner square and end at the one that is diagonally opposite to it. The puzzle cannot be solved in fewer than seventeen moves.

326.—A NEW COUNTER PUZZLE.

PLAY as follows : 2—3, 9—4, 10—7, 3—8, 4—2, 7—5, 8—6, 5—10, 6—9, 2—5, 1—6, 6—4, 5—3, 10—8, 4—7, 3—2, 8—1, 7—10. The white counters have now changed places with the red ones, in eighteen moves, without breaking the conditions.

327.—A NEW BISHOP'S PUZZLE.

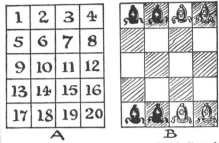

PLAY as follows, using the notation indicated by the numbered squares in Diagram A :—

White.	Black.	White.	Black.
1. 18—15	1. 3— 6	10. 20—10	10. 1—11
2. 17— 8	2. 4—13	11. 3— 9	11. 18—12
3. 19—14	3. 2— 7	12. 10—13	12. 11— 8
4. 15— 5	4. 6—16	13. 19—16	13. 2— 5
5. 8— 3	5. 13—18	14. 16— 1	14. 5—20
6. 14— 9	6. 7—12	15. 9— 6	15. 12—15
7. 5—10	7. 16—11	16. 13— 7	16. 8—14
8. 9—19	8. 12— 2	17. 6— 3	17. 15—18
9. 10— 4	9. 11—17	18. 7— 2	18. 14—19

Diagram B shows the position after the ninth move. Bishops at 1 and 20 have not yet moved, but 2 and 19 have sallied forth and returned. In the end, 1 and 19, 2 and 20, 3 and 17, and 4 and 18 will have exchanged places. Note the position after the thirteenth move.

328.—THE QUEEN'S TOUR.

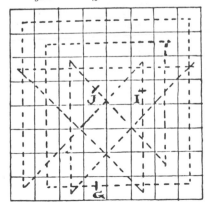

THE annexed diagram shows a second way of performing the Queen's Tour. If you break the line at the point J and erase the shorter portion of that line, you will have the required path solution for any J square. If you break the line at I, you will have a non-re-entrant solution starting from any I square. And if you break the line at G, you will have a solution for any G square. The Queen's Tour previously given may be similarly broken at three different places, but I seized the opportunity of exhibiting a second tour.

329.—THE STAR PUZZLE.

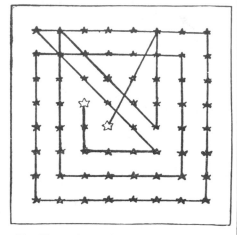

THE illustration explains itself. The stars are all struck out in fourteen straight strokes, starting and ending at a white star.

330.—THE YACHT RACE.

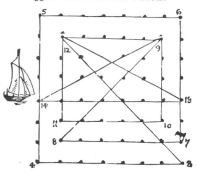

THE diagram explains itself. The numbers will show the direction of the lines in their proper order, and it will be seen that the seventh course ends at the flag-buoy, as stipulated.

331.—THE SCIENTIFIC SKATER.

IN this case we go beyond the boundary of the square. Apart from that, the moves are all queen moves. There are three or four ways in which it can be done.

Here is one way of performing the feat :—

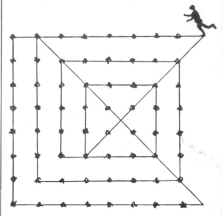

It will be seen that the skater strikes out all the stars in one continuous journey of fourteen straight lines, returning to the point from which he started. To follow the skater's course in the diagram it is necessary always to go as far as we can in a straight line before turning.

332.—THE FORTY-NINE STARS.

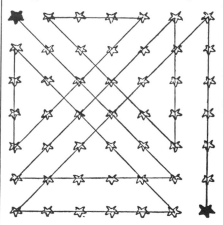

THE illustration shows how all the stars may be struck out in twelve straight strokes, beginning and ending at a black star.

333.—THE QUEEN'S JOURNEY.

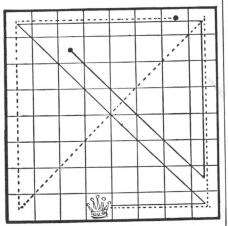

THE correct solution to this puzzle is shown in the diagram by the dark line. The five moves indicated will take the queen the greatest distance that it is possible for her to go in five moves, within the conditions. The dotted line shows the route that most people suggest, but it is not quite so long as the other. Let us assume that the distance from the centre of any square to the centre of the next in the same horizontal or vertical line is 2 inches, and that the queen travels from the centre of her original square to the centre of the one at which she rests. Then the first route will be found to exceed 67.9 inches, while the dotted route is less than 67.8 inches. The difference is small, but it is sufficient to settle the point as to the longer route. All other routes are shorter still than these two.

334.—ST. GEORGE AND THE DRAGON.

WE select for the solution of this puzzle one of the prettiest designs that can be formed by representing the moves of the knight by lines from square to square. The chequering of the squares is omitted to give greater clearness. St. George thus slays the Dragon in strict accordance with the conditions and in the elegant manner we should expect of him.

335.—FARMER LAWRENCE'S CORN-FIELDS.

THERE are numerous solutions to this little agricultural problem. The version I give in the next column is rather curious on account of the long parallel straight lines formed by some of the moves.

336.—THE GREYHOUND PUZZLE.

THERE are several interesting points involved in this question. In the first place, if we had made

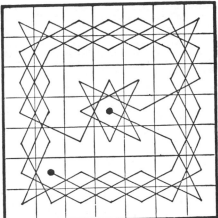

ST. GEORGE AND THE DRAGON.

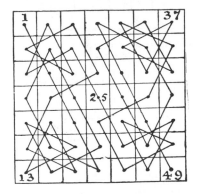

FARMER LAWRENCE'S CORNFIELDS.

no stipulation as to the positions of the two ends of the string, it is quite impossible to form any such string unless we begin and end in the top and bottom row of kennels. We may begin in the top row and end in the bottom (or, of course, the reverse), or we may begin in one of these rows and end in the same. But we can never begin or end in one of the two central rows. Our places of starting and ending, however, were fixed for us. Yet the first half of our route must be confined entirely to those squares that are distinguished in the following diagram by circles, and the second half will therefore be confined to the squares that are not circled. The squares reserved for the two half-strings will be seen to be symmetrical and similar.

The next point is that the first half-string must end in one of the central rows, and the second half-string must begin in one of these rows. This is now obvious, because they have to link together to form the complete string, and every square on an outside row is connected by a knight's move with similar squares only—that is, circled or non-circled as the case may be. The half-strings can, therefore, only be linked in the two central rows.

Now, there are just eight different first half-strings, and consequently also eight second half-strings. We shall see that these combine to form twelve complete strings, which is the total number that exist and the correct solution of our puzzle. I do not propose to give all the routes at length, but I will so far indicate them that if the reader has dropped any he will be able to discover which they are and work them out for himself without any difficulty. The following numbers apply to those in the above diagram.

The eight first half-strings are: 1 to 6 (2 routes) ; 1 to 8 (1 route) ; 1 to 10 (3 routes) ; 1 to 12 (1 route) ; and 1 to 14 (1 route). The eight second half-strings are : 7 to 20 (1 route) ; 9 to 20 (1 route) ; 11 to 20 (3 routes) ; 13 to 20 (1 route) ; and 15 to 20 (2 routes). Every different way in which you can link one half-string to another gives a different solution. These linkings will be found to be as follows : 6 to 13 (2 cases) ; 10 to 13 (3 cases) ; 8 to 11 (3 cases) ; 8 to 15 (2 cases) ; 12 to 9 (1 case) ; and 14 to 7 (1 case). There are, therefore, twelve different linkings and twelve different answers to the puzzle. The route given in the illustration with the greyhound will be found to consist of one of the three half-strings 1 to 10, linked to the half-string 13 to 20. It should be noted that ten of the solutions are produced by five distinctive routes and their reversals—that is, if you indicate these five routes by lines and then turn the diagrams upside down you will get the five other routes. The remaining two solutions are symmetrical (these are the cases where 12 to 9 and 14 to 7 are the links), and consequently they do not produce new solutions by reversal.

337.—THE FOUR KANGAROOS.

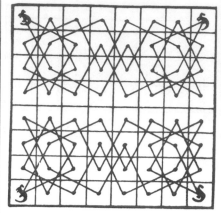

A PRETTY symmetrical solution to this puzzle is shown in the diagram. Each of the four kangaroos makes his little excursion and returns to his corner, without ever entering a square that has been visited by another kangaroo and without crossing the central line. It will at once occur to the reader, as a possible improvement of the puzzle, to divide the board by a central vertical line and make the condition that this also shall not be crossed. This would mean that each kangaroo had to confine himself to a square 4 by 4, but it would be quite impossible, as I shall explain in the next two puzzles.

338.—THE BOARD IN COMPARTMENTS.

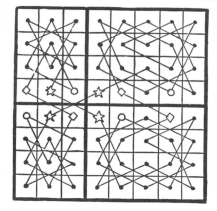

IN attempting to solve this problem it is first necessary to take the two distinctive compartments of twenty and twelve squares respectively and analyse them with a view to determining where the necessary points of entry and exit lie. In the case of the larger compartment it will be found that to complete a tour of it we must begin and end on two of the outside squares on the long sides. But though you may start at any one of these ten squares, you are restricted as to those at which you can end, or (which is the same thing) you may end at whichever of these you like, provided you begin your tour at certain particular squares. In the case of the smaller compartment you are compelled to begin and end at one of the six squares lying at the two narrow ends of the compartments, but similar restrictions apply as in the other instance. A very little thought will show that in the case of the two small compartments you must begin and finish at the ends that lie together, and it then follows that the tours in the larger compartments must also start and end on the contiguous sides.

In the diagram given of one of the possible solutions it will be seen that there are eight places at which we may start this particular tour ; but there is only one route in each case, because we must complete the compartment in which we find ourself before passing into another. In any solution we shall find that the squares distinguished by stars must be entering or exit points, but the law of reversals leaves us the option of making the other connections either at the diamonds or at the circles. In the solution worked out the diamonds are used, but other variations occur in which the circle squares are employed instead. I think these remarks explain all the essential points in the puzzle, which is distinctly instructive and interesting.

339.—THE FOUR KNIGHTS' TOURS.

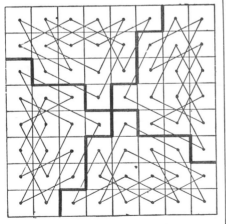

It will be seen in the illustration how a chessboard may be divided into four parts, each of the same size and shape, so that a complete

re-entrant knight's tour may be made on each portion. There is only one possible route for each knight and its reversal.

340.—THE CUBIC KNIGHT'S TOUR.

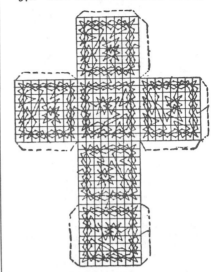

If the reader should cut out the above diagram, fold it in the form of a cube, and stick it together by the strips left for that purpose at the edges, he would have an interesting little curiosity. Or he can make one on a larger scale for himself. It will be found that if we imagine the cube to have a complete chessboard on each of its sides, we may start with the knight on any one of the 384 squares, and make a complete tour of the cube, always returning to the starting-point. The method of passing from one side of the cube to another is easily understood, but, of course, the difficulty consisted in finding the proper points of entry and exit on each board, the order in which the different boards should be taken, and in getting arrangements that would comply with the required conditions.

341.—THE FOUR FROGS.

THE fewest possible moves, counting every move separately, are sixteen. But the puzzle may be solved in seven plays, as follows, if any number of successive moves by one frog count as a single play. All the moves contained within a bracket are a single play; the numbers refer to the toadstools : (1—5), (3—7, 7—1), (8—4, 4—3, 3—7), (6—2, 2—8, 8—4, 4—3), (5—6, 6—2, 2—8), (1—5, 5—6), (7—1). This is the familiar old puzzle by Guarini, propounded in 1512, and I give it here in order to explain my " buttons and string " method of solving this class of moving-counter problem-

A

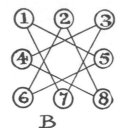

B

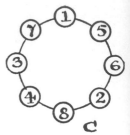

C

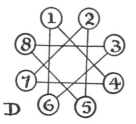

D

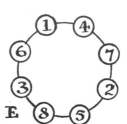

E

Diagram A shows the old way of presenting Guarini's puzzle, the point being to make the white knights change places with the black ones. In "The Four Frogs" presentation of the idea the possible directions of the moves are indicated by lines, to obviate the necessity of the reader's understanding the nature of the knight's move in chess. But it will at once be seen that the two problems are identical. The central square can, of course, be ignored, since no knight can ever enter it. Now, regard the toadstools as buttons and the connecting lines as strings, as in Diagram B. Then by disentangling these strings we can clearly present the diagram in the form shown in Diagram C, where the relationship between the buttons is precisely the same as in B. Any solution on C will be applicable to B, and to A. Place your white knights on 1 and 3 and your black knights on 6 and 8 in the C diagram, and the simplicity of the solution will be very evident. You have simply to move the knights round the circle in one direction or the other. Play over the moves given above, and you will find that every little difficulty has disappeared.

In Diagram D I give another familiar puzzle that first appeared in a book published in Brussels in 1789, *Les Petites Aventures de Jerome Sharp*. Place seven counters on seven of the eight points in the following manner. You must always touch a point that is vacant with a counter, and then move it along a straight line leading from that point to the next vacant point (in either direction), where you deposit the counter. You proceed in the same way until all the counters are placed. Remember you always touch a vacant place and slide the counter from it to the next place, which must be also vacant. Now, by the

"buttons and string" method of simplification we can transform the diagram into E. Then the solution becomes obvious. "Always move *to* the point that you last moved *from*." This is not, of course, the only way of placing the counters, but it is the simplest solution to carry in the mind.

There are several puzzles in this book that the reader will find lend themselves readily to this method.

342.—THE MANDARIN'S PUZZLE.

The rather perplexing point that the solver has to decide for himself in attacking this puzzle is whether the shaded numbers (those that are shown in their right places) are mere dummies or not. Ninety-nine persons out of a hundred might form the opinion that there can be no advantage in moving any of them, but if so they would be wrong.

The shortest solution without moving any shaded number is in thirty-two moves. But the puzzle can be solved in thirty moves. The trick lies in moving the 6, or the 15, on the second move and replacing it on the nineteenth move. Here is the solution: 2—6—13—4—1—21—4—1—10—2—21—10—2—5—22—16—1—13—6—19—11—2—5—22—16—5—13—4—10—21. Thirty moves.

343.—EXERCISE FOR PRISONERS.

There are eighty different arrangements of the numbers in the form of a perfect knight's path, but only forty of these can be reached without two men ever being in a cell at the same time. Two is the greatest number of men that can be given a complete rest, and though the knight's path can be arranged so as to leave either 7 and 13, 8 and 13, 5 and 7, or 5 and 13 in their original positions, the following four arrangements, in which 7 and 13 are unmoved, are the only ones that can be reached under the moving conditions. It therefore resolves itself into finding the fewest possible moves that will lead up to one of these positions. This is certainly no easy matter, and no rigid rules can be laid down for arriving at the correct answer. It is largely a matter for individual judgment, patient experiment, and a sharp eye for revolutions and position.

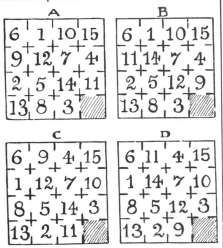

following diagram, in which it will be seen that each successive number is a knight's move from the preceding one, and that five of the dogs (1, 5, 10, 15, and 20) never leave their original kennels.

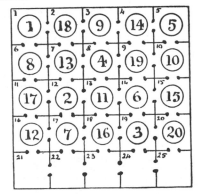

This position may be arrived at in as few as forty-six moves, as follows: 16—21, 16—22, 16—23, 17—16, 12—17, 12—22, 12—21, 7—12, 7—17, 7—22, 11—12, 11—17, 2—7, 2—12, 6—11, 8—7, 8—6, 13—8, 18—13, 11—18, 2—17, 18—12, 18—7, 18—2, 13—7, 3—8, 3—13, 4—3, 4—8, 9—4, 9—3, 14—9, 14—4, 19—14, 19—9, 3—14, 3—19, 6—12, 6—13, 6—14, 17—11, 12—16, 2—12, 7—17, 11—13, 16—18=46 moves. I am, of course, not able to say positively that a solution cannot be discovered in fewer moves, but I believe it will be found a very hard task to reduce the number.

345.—THE TWO PAWNS.

CALL one pawn A and the other B. Now, owing to that optional first move, either pawn may make either 5 or 6 moves in reaching the eighth square. There are, therefore, four cases to be considered : (1) A 6 moves and B 6 moves ; (2) A 6 moves and B 5 moves ; (3) A 5 moves and B 6 moves ; (4) A 5 moves and B 5 moves. In case (1) there are 12 moves, and we may select any 6 of these for A. Therefore 11×12 divided by $1 \times 2 \times 3 \times 4 \times 5 \times 6$ gives us the number of variations for this case—that is, 924. Similarly for case (2), 6 selections out of 11 will be 462 ; in case (3), 5 selections out of 11 will also be 462 ; and in case (4), 5 selections out of 10 will be 252. Add these four numbers together and we get 2,100, which is the correct number of different ways in which the pawns may advance under the conditions. (See No. 270, on p. 204.)

346.—SETTING THE BOARD.

THE White pawns may be arranged in 40,320 ways, the White rooks in 2 ways, the bishops in 2 ways, and the knights in 2 ways. Multiply these numbers together, and we find that the White pieces may be placed in 322,560 different

As a matter of fact, the position C can be reached in as few as sixty-six moves in the following manner: 12, 11, 15, 12, 11, 8, 4, 3, 2, 6, 5, 1, 6, 5, 10, 15, 8, 4, 3, 2, 5, 10, 15, 8, 4, 3, 2, 5, 10, 15, 8, 4, 12, 11, 3, 2, 5, 10, 15, 6, 1, 8, 4, 9, 8, 1, 6, 4, 9, 12, 2, 5, 10, 15, 4, 9, 12, 2, 5, 3, 11, 14, 2, 5, 14, 11=66 moves. Though this is the shortest that I know of, and I do not think it can be beaten, I cannot state positively that there is not a shorter way yet to be discovered. The most tempting arrangement is certainly A ; but things are not what they seem, and C is really the easiest to reach.

If the bottom left-hand corner cell might be left vacant, the following is a solution in forty-five moves by Mr. R. Elrick : 15, 11, 10, 9, 13, 14, 11, 10, 7, 8, 4, 3, 8, 6, 9, 7, 12, 4, 6, 9, 5, 13, 7, 5, 13, 1, 2, 13, 5, 7, 1, 2, 13, 8, 3, 6, 9, 12, 7, 11, 14, 1, 11, 14, 1. But every man has moved.

344.—THE KENNEL PUZZLE.

THE first point is to make a choice of the most promising knight's string and then consider the question of reaching the arrangement in the fewest moves. I am strongly of opinion that the best string is the one represented in the ways. The Black pieces may, of course, be placed in the same number of ways. Therefore the men may be set up in 322,560 × 322,560 = 104,044,953,600 ways. But the point that nearly everybody overlooks is that the board may be placed in two different ways for every arrangement. Therefore the answer is doubled, and is 208,089,907,200 different ways.

347.—COUNTING THE RECTANGLES.

THERE are 1,296 different rectangles in all, 204 of which are squares, counting the square board itself as one, and 1,092 rectangles that are not squares. The general formula is that a board of n^2 squares contains $\dfrac{(n^2+n)^2}{4}$ rectangles, of which

$$\frac{2n^3 + 3n^2 + n}{6}$$

are squares and

$$\frac{3n^4 + 2n^3 - 3n^2 - 2n}{12}$$

are rectangles that are not squares. It is curious and interesting that the total number of rectangles is always the square of the triangular number whose side is n.

348.—THE ROOKERY.

THE answer involves the little point that in the final position the numbered rooks must be in numerical order in the direction contrary to that in which they appear in the original diagram, otherwise it cannot be solved. Play the rooks in the following order of their numbers. As there is never more than one square to which a rook can move (except on the final move), the notation is obvious—5, 6, 7, 5, 6, 4, 3, 6, 4, 7, 5, 4, 7, 3, 6, 7, 3, 5, 4, 3, 1, 8, 3, 4, 5, 6, 7, 1, 8, 2, 1, and rook takes bishop, checkmate. These are the fewest possible moves—thirty-two. The Black king's moves are all forced, and need not be given.

349.—STALEMATE.

WORKING independently, the same position was arrived at by Messrs. S. Loyd, E. N. Franken-

stein, W. H. Thompson, and myself. So the following may be accepted as the best solution possible to this curious problem :—

White.	Black.
1. P—Q 4	1. P—K 4
2. Q—Q 3	2. Q—R 5
3. Q—KKt 3	3. B—Kt 5 ch
4. Kt—Q 2	4. P—QR 4
5. P—R 4	5. P—Q 3
6. P—R 3	6. B—K 3
7. R—R 3	7. P—KB 4
8. Q—R 2	8. P—B 4
9. R—KKt 3	9. B—Kt 6
10. P—QB 4	10. P—B 5
11. P—B 3	11. P—K 5
12. P—Q 5	12. P—K 6

And White is stalemated.

We give a diagram of the curious position arrived at. It will be seen that not one of White's pieces may be moved.

350.—THE FORSAKEN KING.

PLAY as follows :—

White.	Black.
1. P to K 4th	1. Any move
2. Q to Kt 4th	2. Any move except on KB file (a)
3. Q to Kt 7th	3. K moves to royal row
4. B to Kt 5th	4. Any move
5. Mate in two moves	

If 3, K other than to royal
4, Any move [row
5. Mate in two moves

(a) If 2, Any move on KB file
3, K moves to royal row
4, Any move

3. Q to Q 7th	
4. P to Q Kt 3rd	
5. Mate in two moves	

If 3, K other than to royal
4, Any move [row
5. Mate in two moves

| 4. P to Q 4th | |
| 5. Mate in two moves | |

Of course, by "royal row" is meant the row on which the king originally stands at the beginning of a game. Though, if Black plays badly, he may, in certain positions, be mated in fewer moves, the above provides for every variation he can possibly bring about.

351.—THE CRUSADER.

White.	Black.
1. Kt to QB 3rd	1. P to Q 4th
2. Kt takes QP	2. Kt to QB 3rd
3. Kt takes KP	3. P to KKt 4th
4. Kt takes B	4. Kt to KB 3rd
5. Kt takes P	5. Kt to K 5th
6. Kt takes Kt	6. Kt to B 6th
7. Kt takes Q	7. R to KKt sq
8. Kt takes BP	8. R to KKt 3rd
9. Kt takes P	9. R to K 3rd
10. Kt takes P	10. Kt to Kt 8th
11. Kt takes B	11. R to R 6th
12. Kt takes R	12. P to Kt 4th
13. Kt takes P (ch)	13. K to B 2nd

White.	Black.
14. Kt takes P	14. K to Kt 3rd
15. Kt takes R	15. K to R 4th
16. Kt takes Kt	16. K to R 5th

White now mates in three moves.

17. P to Q 4th	17. K to R 4th
18. Q to Q 3rd	18. K moves
19. Q to KR 3rd (mate)	If 17, K to Kt 5th
18. P to K 4th (dis. ch)	18, K moves
19. P to KKt 3rd (mate)	

The position after the sixteenth move, with the mate in three moves, was first given by S. Loyd in *Chess Nuts*.

352.—IMMOVABLE PAWNS.

1. Kt to KB 3	9. Kt to R 4
2. Kt to KR 4	10. Kt to Kt 6
3. Kt to Kt 6	11. Kt takes R
4. Kt takes R	12. Kt to Kt 6
5. Kt to Kt 6	13. Kt takes B
6. Kt takes B	14. Kt to Q 6
7. K takes Kt	15. Q to K sq
8. Kt to QB 3	16. Kt takes Q

17. K takes Kt, and the position is reached.

Black plays precisely the same moves as White, and therefore we give one set of moves only. The above seventeen moves are the fewest possible.

353.—THIRTY-SIX MATES.

PLACE the remaining eight White pieces thus : K at KB 4th, Q at QKt 6th, R at Q 6th, R at KKt 7th, B at Q 5th, B at KR 8th, Kt at QR 5th, and Kt at QB 5th. The following mates can then be given :—

By discovery from Q	.	.	.	.	8
By discovery from R at Q 6th	.				13
By discovery from B at R 8th	.				11
Given by Kt at R 5th	.	.	.	.	2
Given by pawns	.	.	.	.	2
Total	.	.	.	.	36

Is it possible to construct a position in which more than thirty-six different mates on the move can be given ? So far as I know, nobody has yet beaten my arrangement.

354.—AN AMAZING DILEMMA.

MR. BLACK left his king on his queen's knight's 7th, and no matter what piece White chooses for his pawn, Black cannot be checkmated. As we said, the Black king takes no notice of checks and never moves. White may queen his pawn, capture the Black rook, and bring his three pieces up to the attack, but mate is quite impossible. The Black king cannot be left on any other square without a checkmate being possible.

The late Sam Loyd first pointed out the peculiarity on which this puzzle is based.

355.—CHECKMATE !

REMOVE the White pawn from B 6th to K 4th and place a Black pawn on Black's KB 2nd. Now, White plays P to K 5th, check, and Black must play P to B 4th. Then White plays P, takes P *en passant*, checkmate. This was therefore White's last move, and leaves the position given. It is the only possible solution.

356.—QUEER CHESS.

IF you place the pieces as follows (where only a portion of the board is given, to save space), the Black king is in check, with no possible move open to him. The reader will now see why I avoided the term "checkmate," apart from the fact that there is no White king. The position is impossible in the game of chess, because Black could not be given check by both rooks at the same time, nor could he have moved into check on his last move.

I believe the position was first published by the late S. Loyd.

357.—ANCIENT CHINESE PUZZLE.

PLAY as follows :—

1. R—Q 6
2. K—R 7
3. R (R 6)—B 6 (mate).

Black's moves are forced, so need not be given.

358.—THE SIX PAWNS.

THE general formula for six pawns on all squares greater than 2^2 is this : Six times the square of the number of combinations of n things taken three at a time, where n represents the number of squares on the side of the board. Of course, where n is even the unoccupied squares in the rows and columns will be even, and where n is odd the number of squares will be odd. Here n is 8, so the answer is 18,816 different ways. This is "The Dyer's Puzzle" (*Canterbury Puzzles*, No. 27) in another form. I repeat it here in order to explain a method of solving that will be readily grasped by the novice. First of all, it is evident that if we put a pawn on any line, we must put a second one in that line in order that the remainder may be even in number. We cannot put four or six in any row without making it impossible to get an even number in all the columns interfered with. We have, therefore, to put two pawns in each of three rows and in each of three columns. Now, there are just six schemes or arrangements that fulfil these conditions, and these are shown in Diagrams A to F, inclusive, on next page.

I will just remark in passing that A and B are the only distinctive arrangements, because, if you give A a quarter-turn, you get F; and if you give B three quarter-turns in the direction that a clock hand moves, you will get successively C, D, and E. No matter how you may place

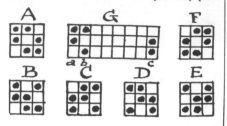

your six pawns, if you have complied with the conditions of the puzzle they will fall under one of these arrangements. Of course it will be understood that mere expansions do not destroy the essential character of the arrangements. Thus G is only an expansion of form A. The solution therefore consists in finding the number of these expansions. Supposing we confine our operations to the first three rows, as in G, then with the pairs *a* and *b* placed in the first and second columns the pair *c* may be disposed in any one of the remaining six columns, and so give six solutions. Now slide pair *b* into the third column, and there are five possible positions for *c*. Slide *b* into the fourth column, and *c* may produce four new solutions. And so on, until (still leaving *a* in the first column) you have *b* in the seventh column, and there is only one place for *c*—in the eighth column. Then you may put *a* in the second column, *b* in the third, and *c* in the fourth, and start sliding *c* and *b* as before for another series of solutions.

We find thus that, by using form A alone and confining our operations to the three top rows, we get as many answers as there are combinations of 8 things taken 3 at a time. This is $\frac{8 \times 7 \times 6}{1 \times 2 \times 3} = 56$. And it will at once strike the reader that if there are 56 different ways of selecting the columns, there must be for each of these ways just 56 ways of selecting the rows, for we may simultaneously work that " sliding " process downwards to the very bottom in exactly the same way as we have worked from left to right. Therefore the total number of ways in which form A may be applied is $56 \times 56 = 3,136$. But there are, as we have seen, six arrangements, and we have only dealt with one of these, A. We must, therefore, multiply this result by 6, which gives us $3,136 \times 6 = 18,816$, which is the total number of ways, as we have already stated.

359.—COUNTER SOLITAIRE.

PLAY as follows : 3—11, 9—10, 1—2, 7—15, 8—16, 8—7, 5—13, 1—4, 8—5, 6—14, 3—8,

6—3, 6—12, 1—6, 1—9, and all the counters will have been removed, with the exception of No. 1, as required by the conditions.

360.—CHESSBOARD SOLITAIRE.

PLAY as follows : 7—15, 8—16, 8—7, 2—10, 1—9, 1—2, 5—13, 3—4, 6—3, 11—1, 14—8, 6—12, 5—6, 5—11, 31—23, 32—24, 32—31, 26—18, 25—17, 25—26, 22—32, 14—22, 29—21, 14—29, 27—28, 30—27, 25—14, 30—20, 25—30, 25—5. The two counters left on the board are 25 and 19—both belonging to the same group, as stipulated—and 19 has never been moved from its original place.

I do not think any solution is possible in which only one counter is left on the board.

361.—THE MONSTROSITY.

White.	Black.
1. P to KB 4	P to QB 3
2. K to B 2	Q to R 4
3. K to K 3	K to Q sq
4. P to B 5	K to B 2
5. Q to K sq	K to Kt 3
6. Q to Kt 3	Kt to QR 3
7. Q to Kt 8	P to KR 4
8. Kt to KB 3	R to R 3
9. Kt to K 5	R to Kt 3
10. Q takes B	R to Kt 6, ch
11. P takes R	K to Kt 4
12. R to R 4	P to B 3
13. R to Q 4	P takes Kt
14. P to QKt 4	P takes R, ch
15. K to B 4	P to R 5
16. Q to K 8	P to R 6
17. Kt to B 3, ch	P takes Kt
18. B to R 3	P to R 7
19. R to Kt sq	P to R 8 (Q)
20. R to Kt 2	P takes R
21. K to Kt 5	Q to KKt 8
22. Q to R 5	K to R 5
23. P to Kt 5	R to B sq
24. P to Kt 6	R to B 2
25. P takes R	P to Kt 8 (B)
26. P to B 8 (R)	Q to B 2
27. B to Q 6	Kt to Kt 5
28. K to Kt 6	K to R 6
29. R to R 8	K to Kt 7
30. P to R 4	Q (Kt 8) to Kt 3
31. P to R 5	K to B 8
32. P takes Q	K to Q 8
33. P takes Q	K to K 8
34. K to B 7	Kt to KR 3, ch
35. K to K 8	B to R 7
36. P to B 6	B to Kt sq
37. P to B 7	K takes B
38. P to B 8 (B)	Kt to Kt 3
39. B to Kt 8	Kt to B 3, ch
40. K to Q 8	Kt to K sq
41. P takes Kt (R)	Kt to B 2, ch
42. K to B 7	Kt to Q sq
43. Q to B 7, ch	K to Kt 8

And the position is reached.

The order of the moves is immaterial, and this order may be greatly varied. But, although many attempts have been made, nobody has succeeded in reducing the number of my moves.

362.—THE WASSAIL BOWL.

THE division of the twelve pints of ale can be made in eleven manipulations, as below. The six columns show at a glance the quantity of ale in the barrel, the five-pint jug, the three-pint jug, and the tramps X, Y, and Z respectively after each manipulation.

Barrel.	5-pint.	3-pint.	X.	Y.	Z.
7	5	0	0	0	0
7	2	3	0	0	0
7	0	3	2	0	0
7	3	0	2	0	0
4	3	3	2	0	0
0	3	3	2	4	0
0	5	1	2	4	0
0	5	0	2	4	1
0	2	3	2	4	1
0	0	3	4	4	1
0	0	0	4	4	4

And each man has received his four pints of ale.

363.—THE DOCTOR'S QUERY.

THE mixture of spirits of wine and water is in the proportion of 40 to 1, just as in the other bottle it was in the proportion of 1 to 40.

364.—THE BARREL PUZZLE.

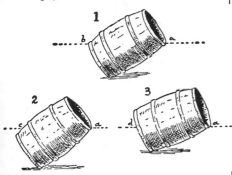

ALL that is necessary is to tilt the barrel as in Fig. 1, and if the edge of the surface of the water exactly touches the lip a at the same time that it touches the edge of the bottom b, it will be just half full. To be more exact, if the bottom is an inch or so from the ground, then we can allow for that, and the thickness of the bottom, at the top. If when the surface of the water reached the lip a it had risen to the point c in Fig. 2, then it would be more than half full. If, as in Fig. 3, some portion of the bottom were visible and the level of the water fell to the point d, then it would be less than half full.

This method applies to all symmetrically constructed vessels.

365.—NEW MEASURING PUZZLE.

THE following solution in eleven manipulations shows the contents of every vessel at the start and after every manipulation :—

10-quart.	10-quart.	5-quart.	4-quart.
10	10	0	0
5	10	5	0
5	10	1	4
9	10	1	0
9	6	1	4
9	7	0	4
9	7	4	0
9	3	4	4
9	3	5	3
9	8	0	3
4	8	5	3
4	10	3	3

366.—THE HONEST DAIRYMAN.

WHATEVER the respective quantities of milk and water, the relative proportion sent to London would always be three parts of water to one of milk. But there are one or two points to be observed. There must originally be more water than milk, or there will be no water in A to double in the second transaction. And the water must not be more than three times the quantity of milk, or there will not be enough liquid in B to effect the second transaction. The third transaction has no effect on A, as the relative proportions in it must be the same as after the second transaction. It was introduced to prevent a quibble if the quantity of milk and water were originally the same; for though double "nothing" would be "nothing," yet the third transaction in such a case could not take place.

367.—WINE AND WATER.

THE wine in small glass was one-sixth of the total liquid, and the wine in large glass two-ninths of total. Add these together, and we find that the wine was seven-eighteenths of total fluid, and therefore the water eleven-eighteenths.

368.—THE KEG OF WINE.

THE capacity of the jug must have been a little less than three gallons. To be more exact, it was 2.93 gallons.

369.—MIXING THE TEA.

THERE are three ways of mixing the teas. Taking them in the order of quality, 2s. 6d., 2s. 3d., 1s. 9d., mix 16 lbs., 1 lb., 3 lbs.; or 14 lbs., 4 lbs., 2 lbs.; or 12 lbs., 7 lbs., 1 lb. In every case the twenty pounds mixture should be worth 2s. 4½d. per pound; but the last case requires the smallest quantity of the best tea, therefore it is the correct answer.

370.—A PACKING PUZZLE.

On the side of the box, 14 by 22$\frac{2}{5}$, we can arrange 13 rows containing alternately 7 and 6 balls, or 85 in all. Above this we can place another layer consisting of 12 rows of 7 and 6 alternately, or a total of 78. In the length of 24$\frac{9}{10}$ inches 15 such layers may be packed, the alternate layers containing 85 and 78 balls. Thus 8 times 85 added to 7 times 78 gives us 1,226 for the full contents of the box.

371.—GOLD PACKING IN RUSSIA.

The box should be 100 inches by 100 inches by 11 inches deep, internal dimensions. We can lay flat at the bottom a row of eight slabs, lengthways, end to end, which will just fill one side, and nine of these rows will dispose of seventy-two slabs (all on the bottom), with a space left over on the bottom measuring 100 inches by 1 inch by 1 inch. Now make eleven depths of such seventy-two slabs, and we have packed 792, and have a space 100 inches by 1 inch by 11 inches deep. In this we may exactly pack the remaining eight slabs on edge, end to end.

372.—THE BARRELS OF HONEY.

The only way in which the barrels could be equally divided among the three brothers, so that each should receive his 3$\frac{1}{2}$ barrels of honey and his 7 barrels, is as follows :—

	Full.	Half-full.	Empty.
A . . .	3	1	3
B . . .	2	3	2
C . . .	2	3	2

There is one other way in which the division could be made, were it not for the objection that all the brothers made to taking more than four barrels of the same description. Except for this difficulty, they might have given B his quantity in exactly the same way as A above, and then have left C one full barrel, five half-full barrels, and one empty barrel. It will thus be seen that in any case two brothers would have to receive their allowance in the same way.

373.—CROSSING THE STREAM.

First, the two sons cross, and one returns. Then the man crosses and the other son returns. Then both sons cross and one returns. Then the lady crosses and the other son returns. Then the two sons cross and one of them returns for the dog. Eleven crossings in all.

It would appear that no general rule can be given for solving these river-crossing puzzles. A formula can be found for a particular case (say on No. 375 or 376) that would apply to any number of individuals under the restricted conditions; but it is not of much use, for some little added stipulation will entirely upset it. As in the case of the measuring puzzles, we generally have to rely on individual ingenuity.

374.—CROSSING THE RIVER AXE.

Here is the solution :—

		···{ J 5)···	···G	T 8	3
5	···	···(J }	···G	T 8	3
5	···	···{ G 3)	···	J T 8
5 3	··	···(G }	···	J T 8
5 3	··	···{ J T)	···G	8
J	5	···	···(T 3 }	···G	8
J	5	···	···{ G 8)	···	T	3
G	8	···	···(J 5 }	···	T	3
G	8	···	···{ J T)	···	5 3
J T 8	···	···(G }	···	5 3
J T 8	···	···{ G 3)	···	5
G T 8	3	··	···(J }	···	5
G T 8	3	··	···{ J 5)···	

G, J, and T stand for Giles, Jasper, and Timothy ; and 8, 5, 3, for £800, £500, and £300 respectively. The two side columns represent the left bank and the right bank, and the middle column the river. Thirteen crossings are necessary, and each line shows the position when the boat is in mid-stream during a crossing, the point of the bracket indicating the direction.

It will be found that not only is no person left alone on the land or in the boat with more than his share of the spoil, but that also no two persons are left with more than their joint shares, though this last point was not insisted upon in the conditions.

375.—FIVE JEALOUS HUSBANDS.

It is obvious that there must be an odd number of crossings, and that if the five husbands had not been jealous of one another the party might have all got over in nine crossings. But no wife was to be in the company of a man or men unless her husband was present. This entails two more crossings, eleven in all.

The following shows how it might have been done. The capital letters stand for the husbands, and the small letters for their respective wives. The position of affairs is shown at the start, and after each crossing between the left bank and the right, and the boat is represented by the asterisk. So you can see at a glance that a, b, and c went over at the first crossing, that b and c returned at the second crossing, and so on.

ABCDE	abcde *	..		
1. ABCDE	de	.. *		abc
2. ABCDE	bcde *	.. *		a
3. ABCDE	e	.. *		abcd
4. ABCDE	de *	..		abc
5. DE	de	..	* ABC	abc
6. CDE	cde *	..	AB	ab
7.	cde	..	* ABCDE	ab
8.	bcde *	..	ABCDE	a
9.	e	..	* ABCDE	abcd
10.	bc e *	..	ABCDE	a d
11.		..	* ABCDE	abcde

There is a little subtlety concealed in the words " show the *quickest* way."

Everybody correctly assumes that, as we are told nothing of the rowing capabilities of the party, we must take it that they all row equally well. But it is obvious that two such persons should row more quickly than one.

Therefore in the second and third crossings two of the ladies should take back the boat to fetch d, not one of them only. This does not affect the number of landings, so no time is lost on that account. A similar opportunity occurs in crossings 10 and 11, where the party again had the option of sending over two ladies or one only.

To those who think they have solved the puzzle in nine crossings I would say that in every case they will find that they are wrong. No such jealous husband would, in the circumstances, send his wife over to the other bank to a man or men, even if she assured him that she was coming back next time in the boat. If readers will have this fact in mind, they will at once discover their errors.

376.—THE FOUR ELOPEMENTS.

IF there had been only three couples, the island might have been dispensed with, but with four or more couples it is absolutely necessary in order to cross under the conditions laid down. It can be done in seventeen passages from land to land (though French mathematicians have declared in their books that in such circumstances twenty-four are needed), and it cannot be done in fewer. I will give one way. A, B, C, and D are the young men, a, b, c, and d are the girls to whom they are respectively engaged. The three columns show the positions of the different individuals on the lawn, the island, and the opposite shore before starting and after each passage, while the asterisk indicates the position of the boat on every occasion.

Having found the fewest possible passages, we should consider two other points in deciding on the " quickest method ": Which persons were the most expert in handling the oars, and which method entails the fewest possible delays in getting in and out of the boat ? We have no data upon which to decide the first point, though it is probable that, as the boat belonged to the girls' household, they would be capable oarswomen. The other point, however, is im-

Lawn.	Island.	Shore.
ABCD abcd *		
ABCD cd		ab *
ABCD bcd *		a
ABCD d	bc *	a
ABCD cd *	b	a
CD cd	b	AB a *
BCD cd *	b	A a
BCD	bcd *	A a
BCD d *	bc	A a
D d	bc	ABC a *
D d	abc *	ABC
D d	b	ABC a c *
B D d *	b	AC a c
d	b	ABCD a c *
d	bc *	ABCD a
d		ABCD abc *
cd *		ABCD ab
		ABCD abcd *

portant, and in the solution I have given (where the girls do 8-13ths of the rowing and A and D need not row at all) there are only sixteen gettings-in and sixteen gettings-out. A man and a girl are never in the boat together, and no man ever lands on the island. There are other methods that require several more exchanges of places.

377.—STEALING THE CASTLE TREASURE.

HERE is the best answer, in eleven manipulations :—

Treasure down.
Boy down—treasure up.
Youth down—boy up.
Treasure down.
Man down—youth and treasure up.
Treasure down.
Boy down—treasure up.
Treasure down.
Youth down—boy up.
Boy down—treasure up.
Treasure down.

378.—DOMINOES IN PROGRESSION.

THERE are twenty-three different ways. You may start with any domino, except the 4—4 and those that bear a 5 or 6, though only certain initial dominoes may be played either way round. If you are given the common difference and the first domino is played, you have no option as to the other dominoes. Therefore all I need do is to give the initial domino for all the twenty-three ways, and state the common difference. This I will do as follows :—

With a common difference of 1, the first domino may be either of these : 0—0, 0—1, 1—0, 0—2, 1—1, 2—0, 0—3, 1—2, 2—1, 3—0, 0—4, 1—3, 2—2, 3—1, 1—4, 2—3, 3—3, 3—4. With a difference of 2, the first domino may be 0—0, 2, or 0—1. Take the last case of all as an example. Having played the 0—1, and the difference being 2, we are compelled to continue with 1—2, 2—3, 3—4, 4—5, 5—6. There are three dominoes that can never be used at all. These are 0—5, 0—6, and 1—6. If we used a box of dominoes extending to 9—9, there would be forty different ways.

379.—THE FIVE DOMINOES.

THERE are just ten different ways of arranging the dominoes. Here is one of them :—

(2—0) (0—0) (0—1) (1—4) (4—0).

I will leave my readers to find the remaining nine for themselves.

380.—THE DOMINO FRAME PUZZLE.

THE illustration is a solution. It will be found that all four sides of the frame add up 44. The sum of the pips on all the dominoes is 168, and if we wish to make the sides sum to 44, we must take care that the four corners sum to 8, because these corners are counted twice, and 168 added to 8 will equal 4 times 44, which is necessary. There are many different solutions. Even in the example given certain interchanges are possible to produce different arrangements. For example, on the left-hand side the string of dominoes from 2—2 down to 3—2 may be reversed, or from 2—6 to 3—2, or from 3—0 to 5—3. Also, on the right-hand side we may reverse from 4—3 to 1—4. These changes will not affect the correctness of the solution.

381.—THE CARD FRAME PUZZLE.

THE sum of all the pips on the ten cards is 55. Suppose we are trying to get 14 pips on every side. Then 4 times 14 is 56. But each of the four corner cards is added in twice, so that 55 deducted from 56, or 1, must represent the sum of the four corner cards. This is clearly impossible ; therefore 14 is also impossible. But suppose we came to trying 18. Then 4 times 18 is 72, and if we deduct 55 we get 17 as the sum of the corners. We need then only try

different arrangements with the four corners always summing to 17, and we soon discover the following solution :—

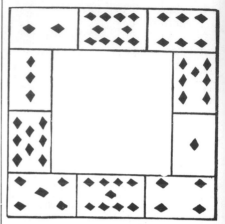

The final trials are very limited in number, and must with a little judgment either bring us to a correct solution or satisfy us that a solution is impossible under the conditions we are attempting. The two centre cards on the upright sides can, of course, always be interchanged, but I do not call these different solutions. If you reflect in a mirror you get another arrangement, which also is not considered different. In the answer given, however, we may exchange the 5 with the 8 and the 4 with the 1. This is a different solution. There are two solutions with 18, four with 19, two with 20, and two with 22—ten arrangements in all. Readers may like to find all these for themselves.

382.—THE CROSS OF CARDS.

THERE are eighteen fundamental arrangements, as follows, where I only give the numbers in the horizontal bar, since the remainder must naturally fall into their places.

5	6	1	7	4		2	4	5	6	8
3	5	1	6	8		3	4	5	6	7
3	4	1	7	8		1	4	7	6	8
2	5	1	7	8		2	3	7	6	8
2	5	3	6	8		2	4	7	5	8
1	5	3	7	8		3	4	9	5	6
2	4	3	7	8		2	4	9	5	7
1	4	5	7	8		1	4	9	6	7
2	3	5	7	8		2	3	9	6	7

It will be noticed that there must always be an odd number in the centre, that there are four ways each of adding up 23, 25, and 27, but only three ways each of summing to 24 and 26.

383.—THE " T " CARD PUZZLE.

If we remove the ace, the remaining cards may be divided into two groups (each adding up alike) in four ways; if we remove 3, there are three ways; if 5, there are four ways; if 7, there are three ways; and if we remove 9, there are four ways of making two equal groups. There are thus eighteen different ways of grouping, and if we take any one of these and keep the odd card (that I have called "removed") at the head of the column, then one set of numbers can be varied in order in twenty-four ways in the column and the other four twenty-four ways in the horizontal, or together they may be varied in 24 × 24 = 576 ways. And as there are eighteen such cases, we multiply this number by 18 and get 10,368, the correct number of ways of placing the cards. As this number includes the reflections, we must divide by 2, but we have also to remember that every horizontal row can change places with a vertical row, necessitating our multiplying by 2; so one operation cancels the other.

384.—CARD TRIANGLES.

The following arrangements of the cards show (1) the smallest possible sum, 17; and (2) the largest possible, 23.

```
      1              7
    9 6            4 2
   4   8          3   6
   3 7 5 2        9 5 1 8
```

It will be seen that the two cards in the middle of any side may always be interchanged without affecting the conditions. Thus there are eight ways of presenting every fundamental arrangement. The number of fundamentals is eighteen, as follows: two summing to 17, four summing to 19, six summing to 20, four summing to 21, and two summing to 23. These eighteen fundamentals, multiplied by eight (for the reason stated above), give 144 as the total number of different ways of placing the cards.

385.—"STRAND" PATIENCE.

The reader may find a solution quite easy in a little over 200 moves, but, surprising as it may at first appear, not more than 62 moves are required. Here is the play: By "4 C up" I mean a transfer of the 4 of clubs with all the cards that rest on it. 1 D on space, 2 S on space, 3 D on space, 2 S on 3 D, 1 H on 2 S, 2 C on space, 1 D on 2 C, 4 S on space, 3 H on 4 S (9 moves so far), 2 S up on 3 H (3 moves), 5 H and 5 D exchanged, and 4 C on 5 D (6 moves), 3 D on 4 C (1), 6 S (with 5 H) on space (3), 4 C up on 5 H (3), 2 C up on 3 D (3), 7 D on space (1), 6 C up on 7 D (3), 8 S on space (1), 7 H on 8 S (1), 8 C on 9 D (1), 7 H on 8 C (1), 8 S on 9 H (1), 7 H on 8 S (1), 7 D up on 8 C (5), 4 C up on 5 D (9), 6 S up on 7 H (3), 4 S up on 5 H (7) = 62 moves in all. This is my record; perhaps the reader can beat it.

386.—A TRICK WITH DICE.

All you have to do is to deduct 250 from the result given, and the three figures in the answer will be the three points thrown with the dice. Thus, in the throw we gave, the number given would be 386; and when we deduct 250 we get 136, from which we know that the throws were 1, 3, and 6.

The process merely consists in giving $100a + 10b + c + 250$, where a, b, and c represent the three throws. The result is obvious.

387.—THE VILLAGE CRICKET MATCH.

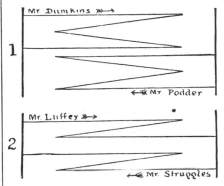

The diagram No. 1 will show that as neither Mr. Podder nor Mr. Dumkins can ever have been within the crease opposite to that from which he started, Mr. Dumkins would score nothing by his performance. Diagram No. 2 will, however, make it clear that since Mr. Luffey and Mr. Struggles have, notwithstanding their energetic but careless movements, contrived to change places, the manœuvre must increase Mr. Struggles's total by one run.

388.—SLOW CRICKET.

The captain must have been "not out" and scored 21. Thus :—

2 men (each lbw)	19
4 men (each caught)	17
1 man (run out)	0
3 men (each bowled)	9
1 man (captain—not out). . .	21
11	66

The captain thus scored exactly 15 more than the average of the team. The "others" who were bowled could only refer to three men, as the eleventh man would be "not out." The reader can discover for himself why the captain must have been that eleventh man. It would not necessarily follow with any figures.

389.—THE FOOTBALL PLAYERS.

THE smallest possible number of men is seven. They could be accounted for in three different ways : 1. Two with both arms sound, one with broken right arm, and four with both arms broken. 2. One with both arms sound, one with broken left arm, two with broken right arm, and three with both arms broken. 3. Two with left arm broken, three with right arm broken, and two with both arms broken. But if every man was injured, the last case is the only one that would apply.

390.—THE HORSE-RACE PUZZLE.

THE answer is : £12 on Acorn, £15 on Blue-bottle, £20 on Capsule.

391.—THE MOTOR-CAR RACE.

THE first point is to appreciate the fact that, in a race round a circular track, there are the same number of cars behind one as there are before. All the others are both behind and before. There were thirteen cars in the race, including Gogglesmith's car. Then one-third of twelve added to three-quarters of twelve will give us thirteen—the correct answer.

392.—THE PEBBLE GAME.

IN the case of fifteen pebbles, the first player wins if he first takes two. Then when he holds an odd number and leaves 1, 8, or 9 he wins, and when he holds an even number and leaves 4, 5, or 12 he also wins. He can always do one or other of these things until the end of the game, and so defeat his opponent. In the case of thirteen pebbles the first player must lose if his opponent plays correctly. In fact, the only numbers with which the first player ought to lose are 5 and multiples of 8 added to 5, such as 13, 21, 29, etc.

393.—THE TWO ROOKS.

THE second player can always win, but to ensure his doing so he must always place his rook, at the start and on every subsequent move, on the same diagonal as his opponent's rook. He can then force his opponent into a corner and win. Supposing the diagram to represent the positions of the rooks at the start, then, if Black played first, White might have placed his rook at A and won next move. Any square on that diagonal from A to H will win, but the best play is always to restrict the moves of the opposing rook as much as possible. If White played first, then Black should have placed his rook at B (F would not be so good, as it gives White more scope) ; then if White goes to C, Black moves to D ; White to E, Black to F ; White to G, Black to C ; White to H, Black to I ; and Black must win next move. If at any time Black had failed to move on to the same diagonal as White, then White could take Black's diagonal and win.

THE TWO ROOKS.

394.—PUSS IN THE CORNER.

No matter whether he plays first or second, the player A, who starts the game at 55, must win. Assuming that B adopts the very best lines of play in order to prolong as much as possible his existence, A, if he has first move, can always on his 12th move capture B ; and if he has the second move, A can always on his 14th move make the capture. His point is always to get diagonally in line with his opponent, and by going to 33, if he has first move, he prevents B getting diagonally in line with himself. Here are two good games. The number in front of the hyphen is always A's move ; that after the hyphen is B's :—

33-8, 32-15, 31-22, 30-21, 29-14, 22-7, 15-6, 14-2, 7-3, 6-4, 11-, and A must capture on his next (12th) move, -13, 54-20, 53-27, 52-34, 51-41, 50-34, 42-27, 35-20, 28-13, 21-6, 14-2, 7-3, 6-4, 11-, and A must capture on his next (14th) move.

395.—A WAR PUZZLE GAME.

THE Britisher can always catch the enemy, no matter how clever and elusive that astute individual may be ; but curious though it may seem, the British general can only do so after he has paid a somewhat mysterious visit to the particular town marked " 1 " in the map, going in by 3 and leaving by 2, or entering by 2 and leaving by 3. The three towns that are shaded and have no numbers do not really come into the question, as some may suppose, for the simple reason that the Britisher never needs to enter any one of them, while the enemy cannot be forced to go into them, and would be clearly ill-advised to do so voluntarily. We may therefore leave these out of consideration altogether. No matter what the enemy may do, the Britisher should make the following first nine moves : He should visit towns 24, 20, 19, 15, 11, 7, 3, 1, 2. If the enemy takes it into his head also to go to town 1, it will be found that he will have to beat a precipitate retreat *the same way that he went in*, or the

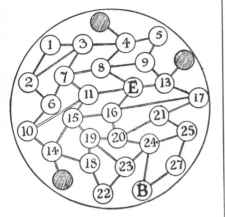

Britisher will infallibly catch him in towns 2 or 3, as the case may be. So the enemy will be wise to avoid that north-west corner of the map altogether.

Now, when the British general has made the nine moves that I have given, the enemy will be, after his own ninth move, in one of the towns marked 5, 8, 11, 13, 14, 16, 19, 21, 24, or 27. Of course, if he imprudently goes to 3 or 6 at this point he will be caught at once. Wherever he may happen to be, the Britisher " goes for him," and has no longer any difficulty in catching him in eight more moves at most (seventeen in all) in one of the following ways. The Britisher will get to 8 when the enemy is at 5, and win next move ; or he will get to 19 when the enemy is at 22, and win next move ; or he will get to 24 when the enemy is at 27, and so win next move. It will be found that he *can* be forced into one or other of these fatal positions.

In short, the strategy really amounts to this : the Britisher plays the first nine moves that I have given, and although the enemy does his very best to escape, our general goes after his antagonist and always driving him away from that north-west corner ultimately closes in with him, and wins. As I have said, the Britisher never need make more than seventeen moves in all, and may win in fewer moves if the enemy plays badly. But after playing those first nine moves it does not matter even if the Britisher makes a few bad ones. He may lose time, but cannot lose his advantage so long as he now keeps the enemy from town 1, and must eventually catch him.

This is a complete explanation of the puzzle. It may seem a little complex in print, but in practice the winning play will now be quite

easy to the reader. Make those nine moves, and there ought to be no difficulty whatever in finding the concluding line of play. Indeed, it might almost be said that then it is difficult for the British general *not* to catch the enemy. It is a question of what in chess we call the " opposition," and the visit by the Britisher to town 1 " gives him the jump " on the enemy, as the man in the street would say.

Here is an illustrative example in which the enemy avoids capture as long as it is possible for him to do so. The Britisher's moves are above the line and the enemy's below it. Play them alternately.

24	20	19	15	11		7		3		1		2		6	10	14	18	19	20	24
13		9	13	17	21	20	24	23	19	15	19	23	24	25	27

The enemy must now go to 25 or B, in either of which towns he is immediately captured.

396.—A MATCH MYSTERY.

IF you form the three heaps (and are therefore the second to draw), any one of the following thirteen groupings will give you a win if you play correctly : 15, 14, 1 ; 15, 13, 2 ; 15, 12, 3 ; 15, 11, 4 ; 15, 10, 5 ; 15, 9, 6 ; 15, 8, 7 ; 14, 13, 3 ; 14, 11, 5 ; 14, 9, 7 ; 13, 11, 6 ; 13, 10, 7 ; 12, 11, 7.

The beautiful general solution of this problem is as follows. Express the number in every heap in powers of 2, avoiding repetitions and remembering that $2° = 1$. Then if you so leave the matches to your opponent that there is an even number of every power, you can win. And if at the start you leave the powers even, you can always continue to do so throughout the game. Take, as example, the last grouping given above—12, 11, 7. Expressed in powers of 2 we have—

12	=	8	4	—	—
11	=	8	—	2	1
7	=	—	4	2	1
		2	2	2	2

As there are thus two of every power, you must win. Say your opponent takes 7 from the 12 heap. He then leaves—

5	=	—	4	—	1
11	=	8	—	2	1
7	=	—	4	2	1
		1	2	2	3

Here the powers are not all even in number, but by taking 9 from the 11 heap you immediately restore your winning position, thus—

5	=	—	4	—	1
2	=	—	—	2	—
7	=	—	4	2	1
		—	2	2	2

And so on to the end. This solution is quite

general, and applies to any number of matches and any number of heaps. A correspondent informs me that this puzzle game was first propounded by Mr. W. M. F. Mellor, but when or where it was published I have not been able to ascertain.

397.—THE MONTENEGRIN DICE GAME.

THE players should select the pairs 5 and 9, and 13 and 15, if the chances of winning are to be quite equal. There are 216 different ways in which the three dice may fall. They may add up 5 in 6 different ways and 9 in 25 different ways, making 31 chances out of 216 for the player who selects these numbers. Also the dice may add up 13 in 21 different ways, and 15 in 10 different ways, thus giving the other player also 31 chances in 216.

398.—THE CIGAR PUZZLE.

NOT a single member of the club mastered this puzzle, and yet I shall show that it is so simple that the merest child can understand its solution—when it is pointed out to him! The large majority of my friends expressed their entire bewilderment. Many considered that "the theoretical result, in any case, is determined by the relationship between the table and the cigars;" others, regarding it as a problem in the theory of Probabilities, arrived at the conclusion that the chances are slightly in favour of the first or second player, as the case may be. One man took a table and a cigar of particular dimensions, divided the table into equal sections, and proceeded to make the two players fill up these sections so that the second player should win. But why should the first player be so accommodating? At any stage he has only to throw down a cigar obliquely across several of these sections entirely to upset Mr. 2's calculations! We have to assume that each player plays the best possible; not that one accommodates the other.

The theories of some other friends would be quite sound if the shape of the cigar were that of a torpedo—perfectly symmetrical and pointed at both ends.

I will show that the first player should infallibly win, if he always plays in the best possible manner. Examine carefully the following diagram, No. 1, and all will be clear.

by the little circle. Now, whatever the second player may do throughout, the first player must always repeat it in an exactly diametrically opposite position. Thus, if the second player places a cigar at A, I put one at AA; he places one at B, I put one at BB; he places one at C, I put one at CC; he places one at D, I put one at DD; he places one at E, I put one at EE; and so on until no more cigars can be placed without touching. As the cigars are supposed to be exactly alike in every respect, it is perfectly clear that for every move that the second player may choose to make, it is possible exactly to repeat it on a line drawn through the centre of the table. The second player can always duplicate the first player's move, no matter where he may place a cigar, or whether he places it in an end or on its side. As the cigars are all alike in every respect, one will obviously balance over the edge of the table at precisely the same point as another. Of course, as each player is supposed to play in the best possible manner, it becomes a matter of theory. It is no valid objection to say that in actual practice one would not be sufficiently exact to be sure of winning. If as the first player you did not win, it would be in consequence of your *not* having played the best possible.

The second diagram will serve to show why the first cigar must be placed on end. (And here I will say that the first cigar that I selected from a box I was able so to stand on end, and I am allowed to assume that all the other cigars would do the same.) If the first cigar were placed on its side, as at F, then the second player could place a cigar as at G—as near as possible, but not actually touching F. Now, in this position you cannot repeat his play on the opposite side, because the two ends of the cigar are not alike. It will be seen that GG, when placed on the opposite side in the same relation to the centre, intersects, or lies on top of, F, whereas the cigars are not allowed to touch. You must therefore put the cigar farther away from the centre, which would result in your having insufficient room between the centre and the bottom left-hand corner to repeat everything that the other player would do between G and the top right-hand corner. Therefore the result would not be a certain win for the first player.

399.—THE TROUBLESOME EIGHT.

$4\frac{1}{2}$	8	$2\frac{1}{2}$
3	5	7
$7\frac{1}{2}$	2	$5\frac{1}{2}$

THE conditions were to place a different number in each of the nine cells so that the three rows,

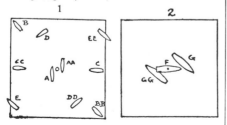

1   2

The first player must place his first cigar *on end in* the exact centre of the table, as indicated

three columns, and two diagonals should each add up 15. Probably the reader at first set himself an impossible task through reading into these conditions something which is not there—a common error in puzzle-solving. If I had said "a different figure," instead of "a different number," it would have been quite impossible with the 8 placed anywhere but in a corner. And it would have been equally impossible if I had said "a different whole number." But a number may, of course, be fractional, and therein lies the secret of the puzzle. The arrangement shown in the figure will be found to comply exactly with the conditions : all the numbers are different, and the square adds up 15 in all the required eight ways.

400.—THE MAGIC STRIPS.

THERE are of course six different places between the seven figures in which a cut may be made, and the secret lies in keeping one strip intact and cutting each of the other six in a different place. After the cuts have been made there are a large number of ways in which the thirteen pieces may be placed together so as to form a magic square. Here is one of them :—

The arrangement has some rather interesting features. It will be seen that the uncut strip is at the top, but it will be found that if the bottom row of figures be placed at the top the numbers will still form a magic square, and that every successive removal from the bottom to the top (carrying the uncut strip stage by stage to the bottom) will produce the same result. If we imagine the numbers to be on seven complete *perpendicular* strips, it will be found that these columns could also be moved in succession from left to right or from right to left, each time producing a magic square.

401.—EIGHT JOLLY GAOL-BIRDS.

THERE are eight ways of forming the magic square—all merely different aspects of one fundamental arrangement. Thus, if you give our first square a quarter turn you will get the second square ; and as the four sides may be in turn brought to the top, there are four aspects. These four in turn reflected in a mirror produce the remaining four aspects. Now, of these eight arrangements only four can possibly be reached under the conditions, and only two of these four can be reached in the fewest possible moves, which is nineteen. These two arrangements

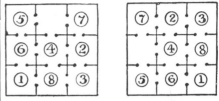

are shown. Move the men in the following order : 5, 3, 2, 5, 7, 6, 4, 1, 5, 7, 6, 4, 1, 6, 4, 8, 3, 2, 7, and you get the first square. Move them thus : 4, 1, 2, 4, 1, 6, 7, 1, 5, 8, 1, 5, 6, 7, 5, 6, 4, 2, 7, and you have the arrangement in the second square. In the first case every man has moved, but in the second case the man numbered 3 has never left his cell. Therefore No. 3 must be the obstinate prisoner, and the second square must be the required arrangement.

402.—NINE JOLLY GAOL BIRDS.

THERE is a pitfall set for the unwary in this little puzzle. At the start one man is allowed to be placed on the shoulders of another, so as to give always one empty cell to enable the prisoners to move about without any two ever being in a cell together. The two united prisoners are allowed to add their numbers together, and are, of course, permitted to remain together at the completion of the magic square. But they are obviously not compelled so to remain together, provided that one of the pair on his final move does not break the condition of entering a cell already occupied. After the acute solver has noticed this point, it is for him

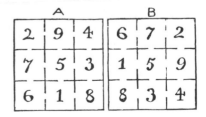

to determine which method is the better one—for the two to be together at the count or to separate. As a matter of fact, the puzzle can be solved in seventeen moves if the men are to remain together ; but if they separate at the end, they may actually save a move and perform the feat in sixteen! The trick consists in placing the man in the centre on the back of one of the corner men, and then working the pair into the centre before their final separation.

Here are the moves for getting the men into one or other of the above two positions. The numbers are those of the men in the order in which they move into the cell that is for the time being vacant. The pair is shown in brackets :—

Place 5 on 1. Then, 6, 9, 8, 6, 4, (6), 2, 4, 9, 3, 4, 9, (6), 7, 6, 1.
Place 5 on 9. Then, 4, 1, 2, 4, 6, (14), 8, 6, 1, 7, 6, 1, (14), 3, 4, 9.
Place 5 on 3. Then, 6, (8), 2, 6, 4, 7, 8, 4, 7, 1, 6, 7, (8), 9, 4, 3.
Place 5 on 7. Then, 4, (12), 8, 4, 6, 3, 2, 6, 3, 9, 4, 3, (12), 1, 6, 7.

The first and second solutions produce Diagram A ; the second and third produce Diagram B. There are only sixteen moves in every case. Having found the fewest moves, we had to consider how we were to make the burdened man do as little work as possible. It will at once be seen that as the pair have to go into the centre before separating they must take at fewest two moves. The labour of the burdened man can only be reduced by adopting the other method of solution, which, however, forces us to take another move.

403.—THE SPANISH DUNGEON.

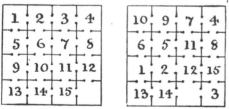

THIS can best be solved by working backwards— that is to say, you must first catch your square, and then work back to the original position. We must first construct those squares which are found to require the least amount of readjustment of the numbers. Many of these we know cannot possibly be reached. When we have before us the most favourable possible arrangements, it then becomes a question of careful analysis to discover which position can be reached in the fewest moves. I am afraid, however, it is only after considerable study and experience that the solver is able to get such a grasp of the various " areas of disturbance " and methods of circulation that his judgment is of much value to him.

The second diagram is a most favourable magic square position. It will be seen that prisoners 4, 8, 13, and 14 are left in their original cells. This position may be reached in as few as thirty-seven moves. Here are the moves : 15, 14, 10, 6, 7, 3, 2, 7, 6, 11, 3, 2, 7, 6, 11, 10, 14, 3, 2, 11, 10, 9, 5, 1, 6, 10, 9, 5, 1, 6, 10, 9, 5, 2, 12, 15, 3. This short solution will probably surprise many readers who may not find a way under from sixty to a hundred moves. The clever prisoner was No. 6, who in the original illustration will be seen with his arms extended calling out the moves. He and No. 10 did most of the work, each changing his cell five times. No. 12, the man with the crooked leg, was lame, and therefore fortunately had only to pass from his cell into the next one when his time came round.

404.—THE SIBERIAN DUNGEONS.

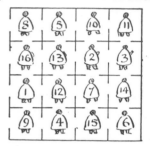

IN attempting to solve this puzzle it is clearly necessary to seek such magic squares as seem the most favourable for our purpose, and then carefully examine and try them for " fewest moves." Of course it at once occurs to us that if we can adopt a square in which a certain number of men need not leave their original cells, we may save moves on the one hand, but we may obstruct our movements on the other. For example, a magic square may be formed with the 6, 7, 13, and 16 unmoved ; but in such case it is obvious that a solution is impossible, since cells 14 and 15 can neither be left nor entered without breaking the condition of no two men ever being in the same cell together.

The following solution in fourteen moves was found by Mr. G. Wotherspoon : 8—17, 16—21, 6—16, 14—8, 5—18, 4—14, 3—24, 11—20, 10—19, 2—23, 13—22, 12—6, 1—5, 9—13. As this solution is in what I consider the theoretical minimum number of moves, I am confident that it cannot be improved upon, and on this point Mr. Wotherspoon is of the same opinion.

405.—CARD MAGIC SQUARES.

ARRANGE the cards as follows for the three new squares :—

3 2 4	6 5 7	9 8 10
4 3 2	7 6 5	10 9 8
2 4 3	5 7 6	8 10 9

Three aces and one ten are not used. The summations of the four squares are thus : 9, 15, 18, and 27—all different, as required.

406.—THE EIGHTEEN DOMINOES.

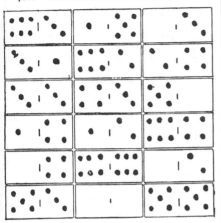

THE illustration explains itself. It will be found that the pips in every column, row, and long diagonal add up 18, as required.

407.—TWO NEW MAGIC SQUARES.

HERE are two solutions that fulfil the conditions :—

SUBTRACTING

11	4	14	13
16	7	1	2
6	5	3	12
9	10	8	15

DIVIDING

36	8	54	27
216	12	1	2
6	3	4	72
9	18	24	108

The first, by subtracting, has a constant 8, and the associated pairs all have a difference of 4. The second square, by dividing, has a constant 9, and all the associated pairs produce 3 by division. These are two remarkable and instructive squares.

408.—MAGIC SQUARES OF TWO DEGREES.

THE following is the square that I constructed. As it stands the constant is 260. If for every number you substitute, in its allotted place, its square, then the constant will be 11,180. Readers can write out for themselves the second degree square.

The main key to the solution is the pretty law that if eight numbers sum to 260 and their squares to 11,180, then the same will happen in the case of the eight numbers that are complementary to 65. Thus 1+18+23+26+31+ 48+56+57=260, and the sum of their squares is 11,180. Therefore 64+47+42+39+34+ 17+9+8 (obtained by subtracting each of the above numbers from 65) will sum to 260 and their squares to 11,180. Note that in every

7	53	41	27	2	52	48	30
12	58	38	24	13	63	35	17
51	1	29	47	54	8	28	42
64	14	18	36	57	11	23	37
25	43	55	5	32	46	50	4
22	40	60	10	19	33	61	15
45	31	3	49	44	26	6	56
34	20	16	62	39	21	9	59

one of the sixteen smaller squares the two diagonals sum to 65. There are four columns and four rows with their complementary columns and rows. Let us pick out the numbers found in the 2nd, 1st, 4th, and 3rd rows and arrange them thus :—

1	8	28	29	42	47	51	54
2	7	27	30	41	48	52	53
3	6	26	31	44	45	49	56
4	5	25	32	43	46	50	55

Here each column contains four consecutive numbers cyclically arranged, four running in one direction and four in the other. The numbers in the 2nd, 5th, 3rd, and 8th columns of the square may be similarly grouped. The great difficulty lies in discovering the conditions governing these groups of numbers, the pairing of the complementaries in the squares of four and the formation of the diagonals. But when a correct solution is shown, as above, it discloses all the more important keys to the mystery. I am inclined to think this square of two degrees the most elegant thing that exists in magics. I believe such a magic square cannot be constructed in the case of any order lower than 8.

409.—THE BASKETS OF PLUMS.

As the merchant told his man to distribute the contents of one of the baskets of plums " among some children," it would not be permissible to give the complete basketful to one child ; and as it was also directed that the man was to give " plums to every child, so that each should receive an equal number," it would also not be allowed to select just as many children as there were plums in a basket and give each child a single plum. Consequently, if the number of plums in every basket was a prime number, then the man would be correct in saying that the proposed distribution was quite impossible. Our puzzle, therefore, resolves itself into forming a magic square with nine different prime numbers.

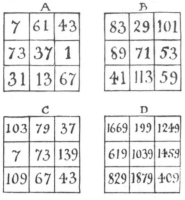

A

7	61	43
73	37	1
31	13	67

B

83	29	101
89	71	53
41	113	59

C

103	79	37
7	73	139
109	67	43

D

1669	199	1249
619	1039	1459
829	1879	409

In Diagram A we have a magic square in prime numbers, and it is the one giving the smallest constant sum that is possible. As to the little trap I mentioned, it is clear that Diagram A is barred out by the words " every basket contained plums," for one plum is not plums. And as we were referred to the baskets, " as shown in the illustration," it is perfectly evident, without actually attempting to count the plums, that there are at any rate more than 7 plums in every basket. Therefore C is also, strictly speaking, barred. Numbers over 20 and under, say, 250 would certainly come well within the range of possibility, and a large number of arrangements would come within these limits. Diagram B is one of them. Of course we can allow for the false bottoms that are so frequently used in the baskets of fruitsellers to make the basket appear to contain more fruit than it really does.

Several correspondents assumed (on what grounds I cannot think) that in the case of this problem the numbers cannot be in consecutive arithmetical progression, so I give Diagram D to show that they were mistaken. The numbers 199, 409, 619, 829, 1,039, 1,249, 1,459, 1,669, and 1,879—all primes with a common difference of 210.

410.—THE MANDARIN'S " T " PUZZLE.

THERE are many different ways of arranging the numbers, and either the 2 or the 3 may be omitted from the " T " enclosure. The arrangement that I give is a " nasik " square. Out of the total of 28,800 nasik squares of the fifth order this is the only one (with its one reflection) that fulfils the " T " condition. This puzzle was suggested to me by Dr. C. Planck.

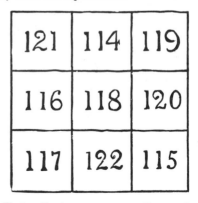

19	23	11	5	7
1	10	17	24	13
22	14	3	6	20
8	16	25	12	4
15	2	9	18	21

THE MANDARIN'S " T " PUZZLE.

411.—A MAGIC SQUARE OF COMPOSITES.

THE problem really amounts to finding the smallest prime such that the next higher prime shall exceed it by 10 at least. If we write out a little list of primes, we shall not need to exceed 150 to discover what we require, for after 113 the next prime is 127. We can then form the square in the diagram, where every number is composite. This is the solution in the smallest numbers. We thus see that the answer is arrived at quite easily, in a square of the third order, by trial. But I propose to show how we may get an answer (not, it is true, the one in smallest numbers) without any tables or trials, but in a very direct and rapid manner.

121	114	119
116	118	120
117	122	115

First write down any consecutive numbers, the smallest being greater than 1—say, 2, 3, 4, 5, 6, 7, 8, 9, 10. The only factors in these numbers are 2, 3, 5, and 7. We therefore multiply these four numbers together and add the product, 210, to each of the nine numbers. The result is the nine consecutive composite numbers, 212 to 220 inclusive, with which we can form the required square. Every number will necessarily be divisible by its difference from 210. It will be very obvious that by this method we may find as many consecutive composites as ever we please. Suppose, for example, we wish to form a magic square of sixteen such numbers; then the numbers 2 to 17 contain the factors 2, 3, 5, 7, 11, 13, and 17, which, multiplied together, make 510510 to be added to produce the sixteen numbers 510512 to 510527 inclusive, all of which are composite as before.

But, as I have said, these are not the answers in the smallest numbers : for if we add 523 to the numbers 1 to 16, we get sixteen consecutive composites ; and if we add 1,327 to the numbers 1 to 25, we get twenty-five consecutive composites, in each case the smallest numbers possible. Yet if we required to form a magic square of a hundred such numbers, we should find it a big task by means of tables, though by the process I have shown it is quite a simple matter. Even to find thirty-six such numbers you will search the tables up to 10,000 without success, and the difficulty increases in an accelerating ratio with each square of a larger order.

412.—THE MAGIC KNIGHT'S TOUR.

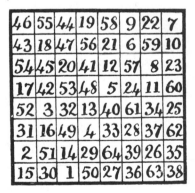

HERE each successive number (in numerical order) is a knight's move from the preceding number, and as 64 is a knight's move from 1, the tour is " re-entrant." All the columns and rows add up 260. Unfortunately, it is not a perfect magic square, because the diagonals are incorrect, one adding up 264 and the other 256 — requiring only the transfer of 4 from one diagonal to the other. I think this is the best result that has ever been obtained (either re-entrant or not), and nobody can yet say whether a perfect solution is possible or impossible.

413.—A CHESSBOARD FALLACY.

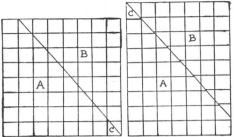

THE explanation of this little fallacy is as follows. The error lies in assuming that the little triangular piece, marked C, is exactly the same height as one of the little squares of the board. As a matter of fact, its height (if we make the sixty-four squares each a square inch) will be 1⅛ in. Consequently the rectangle is really 9⅛ by 7 in., so that the area is sixty-four square inches in either case. Now, although the pieces do fit together exactly to form the perfect rectangle, yet the directions of the horizontal lines in the pieces will not coincide. The new diagram above will make everything quite clear to the reader.

414.—WHO WAS FIRST ?

BIGGS, who saw the smoke, would be first ; Carpenter, who saw the bullet strike the water, would be second ; and Anderson, who heard the report, would be last of all.

415.—A WONDERFUL VILLAGE.

WHEN the sun is in the horizon of any place (whether in Japan or elsewhere), he is the length of half the earth's diameter more distant from that place than in his meridian at noon. As the earth's semi-diameter is nearly 4,000 miles, the sun must be considerably more than 3,000 miles nearer at noon than at his rising, there being no valley even the hundredth part of 1,000 miles deep.

416.—A CALENDAR PUZZLE.

THE first day of a century can never fall on a Sunday ; nor on a Wednesday or a Friday.

417.—THE TIRING-IRONS.

I WILL give my complete working of the solution, so that readers may see how easy it is when you know how to proceed. And first of all, as there is an even number of rings, I will say that they may all be taken off in one-third of $(2^{n+1} - 2)$ moves ; and since n in our case is 14, all the rings may be taken off in 10,922 moves. Then I say 10,922 − 9,999 = 923, and proceed to find the position when only 923 out of the 10,922 moves remain to be made. Here is the curious method of doing this. It is based on the binary scale method used by Monsieur L. Gros, for an account of which see W. W. Rouse Ball's *Mathematical Recreations.*

Divide 923 by 2, and we get 461 and the remainder 1 ; divide 461 by 2, and we get 230 and the remainder 1 ; divide 230 by 2, and we get 115 and the remainder nought. Keep on dividing by 2 in this way as long as possible, and all the remainders will be found to be 1, 1, 1, 0, 0, 1, 1, 0, 1, 1, the last remainder being to the left and the first remainder to the right. As there are fourteen rings and only ten figures, we place the difference, in the form of four noughts, in brackets to the left, and bracket all those figures that repeat a figure on their left. Then we get the following arrangement : (0 0 0 0) 1 (1 1) 0 (0) 1 (1) 0 1 (1). This is the correct answer to the puzzle, for if we now place rings below the line to represent the figures in brackets and rings on the line for the other figures, we get the solution in the required form, as below :—

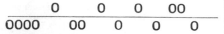

This is the exact position of the rings after the 9,999th move has been made, and the reader will find that the method shown will solve any similar question, no matter how many rings are on the tiring-irons. But in working the inverse process, where you are required to ascertain the number of moves necessary in order to reach a given position of the rings, the rule will require a little modification, because it does not necessarily follow that the position is one that is actually reached in course of taking off all the rings on the irons, as the reader will presently see. I will here state that where the total number of rings is odd the number of moves required to take them all off is one-third of $(2^{n+1}-1)$.

With n rings (where n is *odd*) there are 2^n positions counting all on and all off. In $\frac{1}{3}$ $(2^{n+1}+2)$ positions they are all removed. The number of positions not used is $\frac{1}{3}(2^n-2)$.

With n rings (where n is *even*) there are 2^n positions counting all on and all off. In $\frac{1}{3}$ $(2^{n+1}+1)$ positions they are all removed. The number of positions not used is here $\frac{1}{3}(2^n-1)$.

It will be convenient to tabulate a few cases.

No. of Rings.	Total Positions.	Positions used.	Positions not used.
1	2	2	0
3	8	6	2
5	32	22	10
7	128	86	42
9	512	342	170
2	4	3	1
4	16	11	5
6	64	43	21
8	256	171	85
10	1024	683	341

Note first that the number of *positions used* is one more than the number of *moves* required to take all the rings off, because we are including " all on " which is a position but not a move. Then note that the number of *positions not used* is the same as the number of *moves used* to take off a set that has one ring fewer. For example, it takes 85 moves to remove 7 rings, and the 42 positions not used are exactly the number of moves required to take off a set of 6 rings. The fact is that if there are 7 rings and you take off the first 6, and then wish to remove the 7th ring, there is no course open to you but to reverse all those 42 moves that never ought to have been made. In other words, you must replace all the 7 rings on the loop and start afresh ! You ought first to have taken off 5 rings, to do which you should have taken off 3 rings, and previously to that 1 ring. To take off 6 you first remove 2 and then 4 rings.

418.—SUCH A GETTING UPSTAIRS.

NUMBER the treads in regular order upwards, 1 to 8. Then proceed as follows : 1 (step back to floor), 1, 2, 3 (2), 3, 4, 5 (4), 5, 6, 7 (6), 7, 8, landing (8), landing. The steps in brackets are taken in a backward direction. It will thus be seen that by returning to the floor after the first step, and then always going three steps forward for one step backward, we perform the required feat in nineteen steps.

419.—THE FIVE PENNIES.

FIRST lay three of the pennies in the way shown in Fig. 1. Now hold the remaining two pennies in the position shown in Fig. 2, so that they touch one another at the top, and at the base are in contact with the three horizontally placed coins. Then the five pennies will be equidistant, for every penny will touch every other penny.

420.—THE INDUSTRIOUS BOOKWORM.

THE hasty reader will assume that the bookworm, in boring from the first to the last page of a book in three volumes, standing in their proper order on the shelves, has to go through all three volumes and four covers. This, in our case, would mean a distance of 9½ in., which is a long way from the correct answer. You will find, on examining any three consecutive volumes on your shelves, that the first page of Vol. I. and the last page of Vol. III. are actually the pages that are nearest to Vol. II., so that the worm would only have to penetrate four covers (together, ¼ in.) and the leaves in the second volume (3 in.), or a distance of 3¼ inches, in order to tunnel from the first page to the last.

421.—A CHAIN PUZZLE.

To open and rejoin a link costs threepence. Therefore to join the nine pieces into an endless chain would cost 2s. 3d., whereas a new chain would cost 2s. 2d. But if we break up the piece of eight links, these eight will join together the remaining eight pieces at a cost of 2s. But there is a subtle way of even improving on this. Break up the two pieces containing three and four links respectively, and these seven will join together the remaining seven pieces at a cost of only 1s. 9d.

422.—THE SABBATH PUZZLE.

THE way the author of the old poser proposed to solve the difficulty was as follows : From the Jew's abode let the Christian and the Turk set out on a tour round the globe, the Christian going due east and the Turk due west. Readers of Edgar Allan Poe's story, *Three Sundays in a Week*, or of Jules Verne's *Round the World in Eighty Days*, will know that such a proceeding will result in the Christian's gaining a day and in the Turk's losing a day, so that when they meet again at the house of the Jew their reckoning will agree with his, and all three may keep their Sabbath on the same day. The correctness of this answer, of course, depends on the popular notion as to the definition of a day —the average duration between successive sunrises. It is an old quibble, and quite sound enough for puzzle purposes. Strictly speaking, the two travellers ought to change their reckonings on passing the 180th meridian ; otherwise we have to admit that at the North or South Pole there would only be one Sabbath in seven years.

423.—THE RUBY BROOCH.

IN this case we were shown a sketch of the brooch exactly as it appeared after the four rubies had been stolen from it. The reader was asked to show the positions from which the stones " may have been taken ; " for it is not possible to show precisely how the gems were originally placed, because there are many such ways. But an important point was the statement by Lady Littlewood's brother : " I know the brooch well. It originally contained forty-five stones, and there are now only forty-one. Somebody has stolen four rubies, and then reset as small a number as possible in such a

way that there shall always be eight stones in any of the directions you have mentioned."

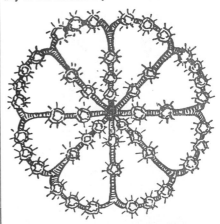

The diagram shows the arrangement before the robbery. It will be seen that it was only necessary to reset one ruby—the one in the centre. Any solution involving the resetting of more than one stone is not in accordance with the brother's statement, and must therefore be wrong. The original arrangement, of course, a little unsymmetrical, and for this reason the brooch was described as " rather eccentric."

424.—THE DOVETAILED BLOCK.

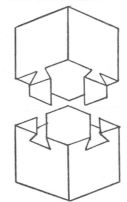

THE mystery is made clear by the illustration. It will be seen at once how the two pieces slide together in a diagonal direction.

425.—JACK AND THE BEANSTALK.

THE serious blunder that the artist made in this drawing was in depicting the tendrils of

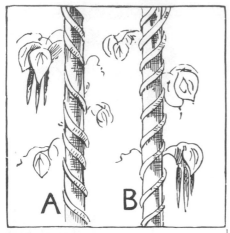

the bean climbing spirally as at A above, whereas the French bean, or scarlet runner, the variety clearly selected by the artist in the absence of any authoritative information on the point, always climbs as shown at B. Very few seem to be aware of this curious little fact. Though the bean always insists on a sinistrorsal growth, as B, the hop prefers to climb in a dextrorsal manner, as A. Why, is one of the mysteries that Nature has not yet unfolded.

426.—THE HYMN-BOARD POSER.

THIS puzzle is not nearly so easy as it looks at first sight. It was required to find the smallest possible number of plates that would be necessary to form a set for three hymn-boards, each of which would show the five hymns sung at any particular service, and then to discover the lowest possible cost for the same. The hymn-book contains 700 hymns, and therefore no higher number than 700 could possibly be needed.

Now, as we are required to use every legitimate and practical method of economy, it should at once occur to us that the plates must be painted on both sides; indeed, this is such a common practice in cases of this kind that it would readily occur to most solvers. We should also remember that some of the figures may possibly be reversed to form other figures; but as we were given a sketch of the actual shapes of these figures when painted on the plates, it would be seen that though the 6's may be turned upside down to make 9's, none of the other figures can be so treated.

It will be found that in the case of the figures 1, 2, 3, 4, and 5, thirty-three of each will be required in order to provide for every possible emergency; in the case of 7, 8, and 0, we can only need thirty of each; while in the case of the figure 6 (which may be reversed for the

figure 9) it is necessary to provide exactly forty-two.

It is therefore clear that the total number of figures necessary is 297; but as the figures are painted on both sides of the plates, only 149 such plates are required. At first it would appear as if one of the plates need only have a number on one side, the other side being left blank. But here we come to a rather subtle point in the problem.

Readers may have remarked that in real life it is sometimes cheaper when making a purchase to buy more articles than we require, on the principle of a reduction on taking a quantity: we get more articles and we pay less. Thus, if we want to buy ten apples, and the price asked is a penny each if bought singly, or ninepence a dozen, we should both save a penny and get two apples more than we wanted by buying the full twelve. In the same way, since there is a regular scale of reduction for plates painted alike, we actually save by having two figures painted on that odd plate. Supposing, for example, that we have thirty plates painted alike with 5 on one side and 6 on the other. The rate would be 4¾d., and the cost 11s. 10½d. But if the odd plate with, say, only a 5 on one side of it have a 6 painted on the other side, we get thirty-one plates at the reduced rate of 4¼d., thus saving a farthing on each of the previous thirty, and reducing the cost of the last one from 1s. to 4½d.

But even after these points are all seen there comes in a new difficulty: for although it will be found that all the 8's may be on the backs of the 7's, we cannot have all the 2's on the backs of the 1's, nor all the 4's on the backs of the 3's, etc. There is a great danger, in our attempts to get as many as possible painted alike, of our so adjusting the figures that some particular combination of hymns cannot be represented.

Here is the solution of the difficulty that was sent to the vicar of Chumpley St. Winifred. Where the sign × is placed between two figures, it implies that one of these figures is on one side of the plate and the other on the other side.

							d.	£	s.	d.
31 plates painted	5	×	9	@	4½	=		0	11	7½
30	,,	7	×	8	@	4¾	=	0	11	10½
21	,,	1	×	2	@	7	=	0	12	3
21	,,	3	×	0	@	7	=	0	12	3
12	,,	1	×	3	@	9¼	=	0	9	3
12	,,	2	×	4	@	9¼	=	0	9	3
12	,,	9	×	4	@	9¼	=	0	9	3
8	,,	4	×	0	@	10¼	=	0	6	10
1	,,	5	×	4	@	12	=	0	1	0
1	,,	5	×	0	@	12	=	0	1	0
149 plates @ 6d. each						=	3	14	6
							£7	19	1

Of course, if we could increase the number of plates, we might get the painting done for nothing, but such a contingency is prevented by the condition that the fewest possible plates must be provided.

This puzzle appeared in *Tit-Bits*, and the following remarks, made by me in the issue for 11th December 1897, may be of interest. The "Hymn-Board Poser" seems to have created extraordinary interest. The immense number of attempts at its solution sent to me from all parts of the United Kingdom and from several Continental countries show a very kind disposition amongst our readers to help the worthy vicar of Chumpley St. Winifred over his parochial difficulty. Every conceivable estimate, from a few shillings up to as high a sum as £1,347, 10s., seems to have come to hand. But the astonishing part of it is that, after going carefully through the tremendous pile of correspondence, I find that only one competitor has succeeded in maintaining the reputation of the *Tit-Bits* solvers for their capacity to solve anything, and his solution is substantially the same as the one given above, the cost being identical. Some of his figures are differently combined, but his grouping of the plates, as shown in the first column, is exactly the same. Though a large majority of competitors clearly hit upon all the essential points of the puzzle, they completely collapsed in the actual arrangement of the figures. According to their methods, some possible selection of hymns, such as 111, 112, 121, 122, 211, cannot be set up. A few correspondents suggested that it might be possible so to paint the 7's that upside down they would appear as 2's or 4's; but this would, of course, be barred out by the fact that a representation of the actual figures to be used was given.

427.—PHEASANT-SHOOTING.

THE arithmetic of this puzzle is very easy indeed. There were clearly 24 pheasants at the start. Of these 16 were shot dead, 1 was wounded in the wing, and 7 got away. The reader may have concluded that the answer is, therefore, that "seven remained." But as they flew away it is clearly absurd to say that they "remained." Had they done so they would certainly have been killed. Must we then conclude that the 17 that were shot remained, because the others flew away? No; because the question was not "how many remained?" but "how many still remained?" Now the poor bird that was wounded in the wing, though unable to fly, was very active in its painful struggles to run away. The answer is, therefore, that the 16 birds that were shot dead "still remained," or "remained still."

428.—THE GARDENER AND THE COOK.

NOBODY succeeded in solving the puzzle, so I had to let the cat out of the bag—an operation that was dimly foreshadowed by the puss in the original illustration. But I first reminded the reader that this puzzle appeared on April 1, a day on which none of us ever resents being made an "April Fool;" though, as I practically ' gave the thing away" by specially drawing attention to the fact that it was All Fools' Day,

it was quite remarkable that my correspondents, without a single exception, fell into the trap.

One large body of correspondents held that what the cook loses in stride is exactly made up in greater speed; consequently both advance at the same rate, and the result must be a tie. But another considerable section saw that, though this might be so in a race 200 ft. straight away, it could not really be, because they each go a stated distance at "every bound," and as 100 is not an exact multiple of 3, the gardener at his thirty-fourth bound will go 2 ft. beyond the mark. The gardener will, therefore, run to a point 102 ft. straight away and return (204 ft. in all), and so lose by 4 ft. This point certainly comes into the puzzle. But the most important fact of all is this, that it so happens that the gardener was a pupil from the Horticultural College for Lady Gardeners at, if I remember aright, Swanley; while the cook was a very accomplished French chef of the hemale persuasion! Therefore "she (the gardener) made three bounds to his (the cook's) two." It will now be found that while the gardener is running her 204 ft. in 68 bounds of 3 ft., the somewhat infirm old cook can only make 45½ of his 2 ft. bounds, which equals 90 ft. 8 in. The result is that the lady gardener wins the race by 109 ft. 4 in. at a moment when the cook is in the air, one-third through his 46th bound.

The moral of this puzzle is twofold : (1) Never take things for granted in attempting to solve puzzles ; (2) always remember All Fools' Day when it comes round. I was not writing of *any* gardener and cook, but of a *particular* couple, in "a race that I witnessed." The statement of the eye-witness must therefore be accepted : as the reader was not there, he cannot contradict it. Of course the information supplied was insufficient, but the correct reply was : "Assuming the gardener to be the ' he,' the cook wins by 4 ft. ; but if the gardener is the ' she,' then the gardener wins by 109 ft. 4 in." This would have won the prize. Curiously enough, one solitary competitor got on to the right track, but failed to follow it up. He said : "Is this a regular April 1 catch, meaning that they only ran 6 ft. each, and consequently the race was unfinished ? If not, I think the following must be the solution, supposing the gardener to be the ' he ' and the cook the ' she.' " Though his solution was wrong even in the case he supposed, yet he was the only person who suspected the question of sex.

429.—PLACING HALFPENNIES.

THIRTEEN coins may be placed as shown on page 252.

430.—FIND THE MAN'S WIFE.

THERE is no guessing required in this puzzle. It is all a question of elimination. If we can pair off any five of the ladies with their respective husbands, other than husband No. 10, then the remaining lady must be No. 10's **wife**.

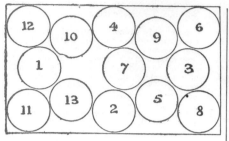

PLACING HALFPENNIES.

I will show how this may be done. No. 8 is seen carrying a lady's parasol in the same hand with his walking-stick. But every lady is provided with a parasol, except No. 3; therefore No. 3 may be safely said to be the wife of No. 8. Then No. 12 is holding a bicycle, and the dressguard and make disclose the fact that it is a lady's bicycle. The only lady in a cycling skirt is No. 5; therefore we conclude that No. 5 is No. 12's wife. Next, the man No. 6 has a dog, and lady No. 11 is seen carrying a dog chain. So we may safely pair No. 6 with No. 11. Then we see that man No. 2 is paying a newsboy for a paper. But we do not pay for newspapers in this way before receiving them, and the gentleman has apparently not taken one from the boy. But lady No. 9 is seen reading a paper. The inference is obvious—that she has sent the boy to her husband for a penny. We therefore pair No. 2 with No. 9. We have now disposed of all the ladies except Nos. 1 and 7, and of all the men except Nos. 4 and 10. On looking at No. 4 we find that he is carrying a coat over his arm, and that the buttons are on the left side—not on the right, as a man wears them. So it is a lady's coat. But the coat clearly does not belong to No. 1, as she is seen to be wearing a coat already, while No. 7 lady is very lightly clad. We therefore pair No. 7 lady with man No. 4, and we are consequently forced to the conclusion that she is the wife of No. 10. This is therefore the correct answer.

INDEX.

THE END.

A CATALOG OF SELECTED
DOVER BOOKS
IN ALL FIELDS OF INTEREST

A CATALOG OF SELECTED DOVER BOOKS IN ALL FIELDS OF INTEREST

DRAWINGS OF REMBRANDT, edited by Seymour Slive. Updated Lippmann, Hofstede de Groot edition, with definitive scholarly apparatus. All portraits, biblical sketches, landscapes, nudes. Oriental figures, classical studies, together with selection of work by followers. 550 illustrations. Total of 630pp. 9⅛ × 12¼.
21485-0, 21486-9 Pa., Two-vol. set $25.00

GHOST AND HORROR STORIES OF AMBROSE BIERCE, Ambrose Bierce. 24 tales vividly imagined, strangely prophetic, and decades ahead of their time in technical skill: "The Damned Thing," "An Inhabitant of Carcosa," "The Eyes of the Panther," "Moxon's Master," and 20 more. 199pp. 5⅜ × 8½. 20767-6 Pa. $3.95

ETHICAL WRITINGS OF MAIMONIDES, Maimonides. Most significant ethical works of great medieval sage, newly translated for utmost precision, readability. Laws Concerning Character Traits, Eight Chapters, more. 192pp. 5⅜ × 8½.
24522-5 Pa. $4.50

THE EXPLORATION OF THE COLORADO RIVER AND ITS CANYONS, J. W. Powell. Full text of Powell's 1,000-mile expedition down the fabled Colorado in 1869. Superb account of terrain, geology, vegetation, Indians, famine, mutiny, treacherous rapids, mighty canyons, during exploration of last unknown part of continental U.S. 400pp. 5⅜ × 8½. 20094-9 Pa. $6.95

HISTORY OF PHILOSOPHY, Julián Marías. Clearest one-volume history on the market. Every major philosopher and dozens of others, to Existentialism and later. 505pp. 5⅜ × 8½. 21739-6 Pa. $8.50

ALL ABOUT LIGHTNING, Martin A. Uman. Highly readable non-technical survey of nature and causes of lightning, thunderstorms, ball lightning, St. Elmo's Fire, much more. Illustrated. 192pp. 5⅜ × 8½. 25237-X Pa. $5.95

SAILING ALONE AROUND THE WORLD, Captain Joshua Slocum. First man to sail around the world, alone, in small boat. One of great feats of seamanship told in delightful manner. 67 illustrations. 294pp. 5⅜ × 8½. 20326-3 Pa. $4.50

LETTERS AND NOTES ON THE MANNERS, CUSTOMS AND CONDITIONS OF THE NORTH AMERICAN INDIANS, George Catlin. Classic account of life among Plains Indians: ceremonies, hunt, warfare, etc. 312 plates. 572pp. of text. 6⅛ × 9¼. 22118-0, 22119-9 Pa. Two-vol. set $15.90

ALASKA: The Harriman Expedition, 1899, John Burroughs, John Muir, et al. Informative, engrossing accounts of two-month, 9,000-mile expedition. Native peoples, wildlife, forests, geography, salmon industry, glaciers, more. Profusely illustrated. 240 black-and-white line drawings. 124 black-and-white photographs. 3 maps. Index. 576pp. 5⅜ × 8½. 25109-8 Pa. $11.95

HOW TO WRITE, Gertrude Stein. Gertrude Stein claimed anyone could understand her unconventional writing—here are clues to help. Fascinating improvisations, language experiments, explanations illuminate Stein's craft and the art of writing. Total of 414pp. 4⅜ × 6⅜. 23144-5 Pa. $5.95

ADVENTURES AT SEA IN THE GREAT AGE OF SAIL: Five Firsthand Narratives, edited by Elliot Snow. Rare true accounts of exploration, whaling, shipwreck, fierce natives, trade, shipboard life, more. 33 illustrations. Introduction. 353pp. 5⅜ × 8½. 25177-2 Pa. $7.95

THE HERBAL OR GENERAL HISTORY OF PLANTS, John Gerard. Classic descriptions of about 2,850 plants—with over 2,700 illustrations—includes Latin and English names, physical descriptions, varieties, time and place of growth, more. 2,706 illustrations. xlv + 1,678pp. 8½ × 12¼. 23147-X Cloth. $75.00

DOROTHY AND THE WIZARD IN OZ, L. Frank Baum. Dorothy and the Wizard visit the center of the Earth, where people are vegetables, glass houses grow and Oz characters reappear. Classic sequel to *Wizard of Oz*. 256pp. 5⅜ × 8. 24714-7 Pa. $4.95

SONGS OF EXPERIENCE: Facsimile Reproduction with 26 Plates in Full Color, William Blake. This facsimile of Blake's original "Illuminated Book" reproduces 26 full-color plates from a rare 1826 edition. Includes "The Tyger," "London," "Holy Thursday," and other immortal poems. 26 color plates. Printed text of poems. 48pp. 5¼ × 7. 24636-1 Pa. $3.50

SONGS OF INNOCENCE, William Blake. The first and most popular of Blake's famous "Illuminated Books," in a facsimile edition reproducing all 31 brightly colored plates. Additional printed text of each poem. 64pp. 5¼ × 7. 22764-2 Pa. $3.50

PRECIOUS STONES, Max Bauer. Classic, thorough study of diamonds, rubies, emeralds, garnets, etc.: physical character, occurrence, properties, use, similar topics. 20 plates, 8 in color. 94 figures. 659pp. 6⅛ × 9¼. 21910-0, 21911-9 Pa., Two-vol. set $15.90

ENCYCLOPEDIA OF VICTORIAN NEEDLEWORK, S. F. A. Caulfeild and Blanche Saward. Full, precise descriptions of stitches, techniques for dozens of needlecrafts—most exhaustive reference of its kind. Over 800 figures. Total of 679pp. 8⅛ × 11. Two volumes. Vol. 1 22800-2 Pa. $11.95
Vol. 2 22801-0 Pa. $11.95

THE MARVELOUS LAND OF OZ, L. Frank Baum. Second Oz book, the Scarecrow and Tin Woodman are back with hero named Tip, Oz magic. 136 illustrations. 287pp. 5⅜ × 8½. 20692-0 Pa. $5.95

WILD FOWL DECOYS, Joel Barber. Basic book on the subject, by foremost authority and collector. Reveals history of decoy making and rigging, place in American culture, different kinds of decoys, how to make them, and how to use them. 140 plates. 156pp. 7⅞ × 10¾. 20011-6 Pa. $8.95

HISTORY OF LACE, Mrs. Bury Palliser. Definitive, profusely illustrated chronicle of lace from earliest times to late 19th century. Laces of Italy, Greece, England, France, Belgium, etc. Landmark of needlework scholarship. 266 illustrations. 672pp. 6⅛ × 9¼. 24742-2 Pa. $14.95

ILLUSTRATED GUIDE TO SHAKER FURNITURE, Robert Meader. All furniture and appurtenances, with much on unknown local styles. 235 photos. 146pp. 9 × 12. 22819-3 Pa. $7.95

WHALE SHIPS AND WHALING: A Pictorial Survey, George Francis Dow. Over 200 vintage engravings, drawings, photographs of barks, brigs, cutters, other vessels. Also harpoons, lances, whaling guns, many other artifacts. Comprehensive text by foremost authority. 207 black-and-white illustrations. 288pp. 6 × 9. 24808-9 Pa. $8.95

THE BERTRAMS, Anthony Trollope. Powerful portrayal of blind self-will and thwarted ambition includes one of Trollope's most heartrending love stories. 497pp. 5⅜ × 8½. 25119-5 Pa. $8.95

ADVENTURES WITH A HAND LENS, Richard Headstrom. Clearly written guide to observing and studying flowers and grasses, fish scales, moth and insect wings, egg cases, buds, feathers, seeds, leaf scars, moss, molds, ferns, common crystals, etc.—all with an ordinary, inexpensive magnifying glass. 209 exact line drawings aid in your discoveries. 220pp. 5⅜ × 8½. 23330-8 Pa. $3.95

RODIN ON ART AND ARTISTS, Auguste Rodin. Great sculptor's candid, wide-ranging comments on meaning of art; great artists; relation of sculpture to poetry, painting, music; philosophy of life, more. 76 superb black-and-white illustrations of Rodin's sculpture, drawings and prints. 119pp. 8⅜ × 11¼. 24487-3 Pa. $6.95

FIFTY CLASSIC FRENCH FILMS, 1912–1982: A Pictorial Record, Anthony Slide. Memorable stills from Grand Illusion, Beauty and the Beast, Hiroshima, Mon Amour, many more. Credits, plot synopses, reviews, etc. 160pp. 8¼ × 11. 25256-6 Pa. $11.95

THE PRINCIPLES OF PSYCHOLOGY, William James. Famous long course complete, unabridged. Stream of thought, time perception, memory, experimental methods; great work decades ahead of its time. 94 figures. 1,391pp. 5⅜ × 8½. 20381-6, 20382-4 Pa., Two-vol. set $19.90

BODIES IN A BOOKSHOP, R. T. Campbell. Challenging mystery of blackmail and murder with ingenious plot and superbly drawn characters. In the best tradition of British suspense fiction. 192pp. 5⅜ × 8½. 24720-1 Pa. $3.95

CALLAS: PORTRAIT OF A PRIMA DONNA, George Jellinek. Renowned commentator on the musical scene chronicles incredible career and life of the most controversial, fascinating, influential operatic personality of our time. 64 black-and-white photographs. 416pp. 5⅜ × 8¼. 25047-4 Pa. $7.95

GEOMETRY, RELATIVITY AND THE FOURTH DIMENSION, Rudolph Rucker. Exposition of fourth dimension, concepts of relativity as Flatland characters continue adventures. Popular, easily followed yet accurate, profound. 141 illustrations. 133pp. 5⅜ × 8½. 23400-2 Pa. $3.50

HOUSEHOLD STORIES BY THE BROTHERS GRIMM, with pictures by Walter Crane. 53 classic stories—Rumpelstiltskin, Rapunzel, Hansel and Gretel, the Fisherman and his Wife, Snow White, Tom Thumb, Sleeping Beauty, Cinderella, and so much more—lavishly illustrated with original 19th century drawings. 114 illustrations. x + 269pp. 5⅜ × 8½. 21080-4 Pa. $4.50

CATALOG OF DOVER BOOKS

SUNDIALS, Albert Waugh. Far and away the best, most thorough coverage of ideas, mathematics concerned, types, construction, adjusting anywhere. Over 100 illustrations. 230pp. 5⅜ × 8½. 22947-5 Pa. $4.00

PICTURE HISTORY OF THE NORMANDIE: With 190 Illustrations, Frank O. Braynard. Full story of legendary French ocean liner: Art Deco interiors, design innovations, furnishings, celebrities, maiden voyage, tragic fire, much more. Extensive text. 144pp. 8⅞ × 11¾. 25257-4 Pa. $9.95

THE FIRST AMERICAN COOKBOOK: A Facsimile of "American Cookery," 1796, Amelia Simmons. Facsimile of the first American-written cookbook published in the United States contains authentic recipes for colonial favorites—pumpkin pudding, winter squash pudding, spruce beer, Indian slapjacks, and more. Introductory Essay and Glossary of colonial cooking terms. 80pp. 5⅜ × 8½. 24710-4 Pa. $3.50

101 PUZZLES IN THOUGHT AND LOGIC, C. R. Wylie, Jr. Solve murders and robberies, find out which fishermen are liars, how a blind man could possibly identify a color—purely by your own reasoning! 107pp. 5⅜ × 8½. 20367-0 Pa. $2.00

THE BOOK OF WORLD-FAMOUS MUSIC—CLASSICAL, POPULAR AND FOLK, James J. Fuld. Revised and enlarged republication of landmark work in musico-bibliography. Full information about nearly 1,000 songs and compositions including first lines of music and lyrics. New supplement. Index. 800pp. 5⅜ × 8¼. 24857-7 Pa. $14.95

ANTHROPOLOGY AND MODERN LIFE, Franz Boas. Great anthropologist's classic treatise on race and culture. Introduction by Ruth Bunzel. Only inexpensive paperback edition. 255pp. 5⅜ × 8½. 25245-0 Pa. $5.95

THE TALE OF PETER RABBIT, Beatrix Potter. The inimitable Peter's terrifying adventure in Mr. McGregor's garden, with all 27 wonderful, full-color Potter illustrations. 55pp. 4¼ × 5½. (Available in U.S. only) 22827-4 Pa. $1.75

THREE PROPHETIC SCIENCE FICTION NOVELS, H. G. Wells. *When the Sleeper Wakes, A Story of the Days to Come* and *The Time Machine* (full version). 335pp. 5⅜ × 8½. (Available in U.S. only) 20605-X Pa. $5.95

APICIUS COOKERY AND DINING IN IMPERIAL ROME, edited and translated by Joseph Dommers Vehling. Oldest known cookbook in existence offers readers a clear picture of what foods Romans ate, how they prepared them, etc. 49 illustrations. 301pp. 6⅛ × 9¼. 23563-7 Pa. $6.00

SHAKESPEARE LEXICON AND QUOTATION DICTIONARY, Alexander Schmidt. Full definitions, locations, shades of meaning of every word in plays and poems. More than 50,000 exact quotations. 1,485pp. 6½ × 9¼. 22726-X, 22727-8 Pa., Two-vol. set $27.90

THE WORLD'S GREAT SPEECHES, edited by Lewis Copeland and Lawrence W. Lamm. Vast collection of 278 speeches from Greeks to 1970. Powerful and effective models; unique look at history. 842pp. 5⅜ × 8½. 20468-5 Pa. $10.95

THE BLUE FAIRY BOOK, Andrew Lang. The first, most famous collection, with many familiar tales: Little Red Riding Hood, Aladdin and the Wonderful Lamp, Puss in Boots, Sleeping Beauty, Hansel and Gretel, Rumpelstiltskin; 37 in all. 138 illustrations. 390pp. 5⅜ × 8½. 21437-0 Pa. $5.95

THE STORY OF THE CHAMPIONS OF THE ROUND TABLE, Howard Pyle. Sir Launcelot, Sir Tristram and Sir Percival in spirited adventures of love and triumph retold in Pyle's inimitable style. 50 drawings, 31 full-page. xviii + 329pp. 6½ × 9¼. 21883-X Pa. $6.95

AUDUBON AND HIS JOURNALS, Maria Audubon. Unmatched two-volume portrait of the great artist, naturalist and author contains his journals, an excellent biography by his granddaughter, expert annotations by the noted ornithologist, Dr. Elliott Coues, and 37 superb illustrations. Total of 1,200pp. 5⅜ × 8.
Vol. I 25143-8 Pa. $8.95
Vol. II 25144-6 Pa. $8.95

GREAT DINOSAUR HUNTERS AND THEIR DISCOVERIES, Edwin H. Colbert. Fascinating, lavishly illustrated chronicle of dinosaur research, 1820's to 1960. Achievements of Cope, Marsh, Brown, Buckland, Mantell, Huxley, many others. 384pp. 5¼ × 8¼. 24701-5 Pa. $6.95

THE TASTEMAKERS, Russell Lynes. Informal, illustrated social history of American taste 1850's–1950's. First popularized categories Highbrow, Lowbrow, Middlebrow. 129 illustrations. New (1979) afterword. 384pp. 6 × 9.
23993-4 Pa. $6.95

DOUBLE CROSS PURPOSES, Ronald A. Knox. A treasure hunt in the Scottish Highlands, an old map, unidentified corpse, surprise discoveries keep reader guessing in this cleverly intricate tale of financial skullduggery. 2 black-and-white maps. 320pp. 5⅜ × 8½. (Available in U.S. only) 25032-6 Pa. $5.95

AUTHENTIC VICTORIAN DECORATION AND ORNAMENTATION IN FULL COLOR: 46 Plates from "Studies in Design," Christopher Dresser. Superb full-color lithographs reproduced from rare original portfolio of a major Victorian designer. 48pp. 9¼ × 12¼. 25083-0 Pa. $7.95

PRIMITIVE ART, Franz Boas. Remains the best text ever prepared on subject, thoroughly discussing Indian, African, Asian, Australian, and, especially, Northern American primitive art. Over 950 illustrations show ceramics, masks, totem poles, weapons, textiles, paintings, much more. 376pp. 5⅜ × 8. 20025-6 Pa. $6.95

SIDELIGHTS ON RELATIVITY, Albert Einstein. Unabridged republication of two lectures delivered by the great physicist in 1920–21. Ether and Relativity and Geometry and Experience. Elegant ideas in non-mathematical form, accessible to intelligent layman. vi + 56pp. 5⅜ × 8½. 24511-X Pa. $2.95

THE WIT AND HUMOR OF OSCAR WILDE, edited by Alvin Redman. More than 1,000 ripostes, paradoxes, wisecracks: Work is the curse of the drinking classes, I can resist everything except temptation, etc. 258pp. 5⅜ × 8½. 20602-5 Pa. $3.95

ADVENTURES WITH A MICROSCOPE, Richard Headstrom. 59 adventures with clothing fibers, protozoa, ferns and lichens, roots and leaves, much more. 142 illustrations. 232pp. 5⅜ × 8½. 23471-1 Pa. $3.95

PLANTS OF THE BIBLE, Harold N. Moldenke and Alma L. Moldenke. Standard reference to all 230 plants mentioned in Scriptures. Latin name, biblical reference, uses, modern identity, much more. Unsurpassed encyclopedic resource for scholars, botanists, nature lovers, students of Bible. Bibliography. Indexes. 123 black-and-white illustrations. 384pp. 6 × 9. 25069-5 Pa. $8.95

FAMOUS AMERICAN WOMEN: A Biographical Dictionary from Colonial Times to the Present, Robert McHenry, ed. From Pocahontas to Rosa Parks, 1,035 distinguished American women documented in separate biographical entries. Accurate, up-to-date data, numerous categories, spans 400 years. Indices. 493pp. 6½ × 9¼. 24523-3 Pa. $9.95

THE FABULOUS INTERIORS OF THE GREAT OCEAN LINERS IN HISTORIC PHOTOGRAPHS, William H. Miller, Jr. Some 200 superb photographs capture exquisite interiors of world's great "floating palaces"—1890's to 1980's: *Titanic, Ile de France, Queen Elizabeth, United States, Europa,* more. Approx. 200 black-and-white photographs. Captions. Text. Introduction. 160pp. 8⅞ × 11¾. 24756-2 Pa. $9.95

THE GREAT LUXURY LINERS, 1927–1954: A Photographic Record, William H. Miller, Jr. Nostalgic tribute to heyday of ocean liners. 186 photos of Ile de France, Normandie, Leviathan, Queen Elizabeth, United States, many others. Interior and exterior views. Introduction. Captions. 160pp. 9 × 12. 24056-8 Pa. $9.95

A NATURAL HISTORY OF THE DUCKS, John Charles Phillips. Great landmark of ornithology offers complete detailed coverage of nearly 200 species and subspecies of ducks: gadwall, sheldrake, merganser, pintail, many more. 74 full-color plates, 102 black-and-white. Bibliography. Total of 1,920pp. 8⅝ × 11¼. 25141-1, 25142-X Cloth. Two-vol. set $100.00

THE SEAWEED HANDBOOK: An Illustrated Guide to Seaweeds from North Carolina to Canada, Thomas F. Lee. Concise reference covers 78 species. Scientific and common names, habitat, distribution, more. Finding keys for easy identification. 224pp. 5⅜ × 8½. 25215-9 Pa. $5.95

THE TEN BOOKS OF ARCHITECTURE: The 1755 Leoni Edition, Leon Battista Alberti. Rare classic helped introduce the glories of ancient architecture to the Renaissance. 68 black-and-white plates. 336pp. 8⅜ × 11¼. 25239-6 Pa. $14.95

MISS MACKENZIE, Anthony Trollope. Minor masterpieces by Victorian master unmasks many truths about life in 19th-century England. First inexpensive edition in years. 392pp. 5⅜ × 8½. 25201-9 Pa. $7.95

THE RIME OF THE ANCIENT MARINER, Gustave Doré, Samuel Taylor Coleridge. Dramatic engravings considered by many to be his greatest work. The terrifying space of the open sea, the storms and whirlpools of an unknown ocean, the ice of Antarctica, more—all rendered in a powerful, chilling manner. Full text. 38 plates. 77pp. 9¼ × 12. 22305-1 Pa. $4.95

THE EXPEDITIONS OF ZEBULON MONTGOMERY PIKE, Zebulon Montgomery Pike. Fascinating first-hand accounts (1805–6) of exploration of Mississippi River, Indian wars, capture by Spanish dragoons, much more. 1,088pp. 5⅜ × 8½. 25254-X, 25255-8 Pa. Two-vol. set $23.90

A CONCISE HISTORY OF PHOTOGRAPHY: Third Revised Edition, Helmut Gernsheim. Best one-volume history—camera obscura, photochemistry, daguerreotypes, evolution of cameras, film, more. Also artistic aspects—landscape, portraits, fine art, etc. 281 black-and-white photographs. 26 in color. 176pp. 8⅜ × 11¼. 25128-4 Pa. $12.95

THE DORÉ BIBLE ILLUSTRATIONS, Gustave Doré. 241 detailed plates from the Bible: the Creation scenes, Adam and Eve, Flood, Babylon, battle sequences, life of Jesus, etc. Each plate is accompanied by the verses from the King James version of the Bible. 241pp. 9 × 12. 23004-X Pa. $8.95

HUGGER-MUGGER IN THE LOUVRE, Elliot Paul. Second Homer Evans mystery-comedy. Theft at the Louvre involves sleuth in hilarious, madcap caper. "A knockout."—Books. 336pp. 5⅜ × 8½. 25185-3 Pa. $5.95

FLATLAND, E. A. Abbott. Intriguing and enormously popular science-fiction classic explores the complexities of trying to survive as a two-dimensional being in a three-dimensional world. Amusingly illustrated by the author. 16 illustrations. 103pp. 5⅜ × 8½. 20001-9 Pa. $2.25

THE HISTORY OF THE LEWIS AND CLARK EXPEDITION, Meriwether Lewis and William Clark, edited by Elliott Coues. Classic edition of Lewis and Clark's day-by-day journals that later became the basis for U.S. claims to Oregon and the West. Accurate and invaluable geographical, botanical, biological, meteorological and anthropological material. Total of 1,508pp. 5⅜ × 8½. 21268-8, 21269-6, 21270-X Pa. Three-vol. set $25.50

LANGUAGE, TRUTH AND LOGIC, Alfred J. Ayer. Famous, clear introduction to Vienna, Cambridge schools of Logical Positivism. Role of philosophy, elimination of metaphysics, nature of analysis, etc. 160pp. 5⅜ × 8½. (Available in U.S. and Canada only) 20010-8 Pa. $2.95

MATHEMATICS FOR THE NONMATHEMATICIAN, Morris Kline. Detailed, college-level treatment of mathematics in cultural and historical context, with numerous exercises. For liberal arts students. Preface. Recommended Reading Lists. Tables. Index. Numerous black-and-white figures. xvi + 641pp. 5⅜ × 8½. 24823-2 Pa. $11.95

28 SCIENCE FICTION STORIES, H. G. Wells. Novels, *Star Begotten* and *Men Like Gods*, plus 26 short stories: "Empire of the Ants," "A Story of the Stone Age," "The Stolen Bacillus," "In the Abyss," etc. 915pp. 5⅜ × 8½. (Available in U.S. only) 20265-8 Cloth. $10.95

HANDBOOK OF PICTORIAL SYMBOLS, Rudolph Modley. 3,250 signs and symbols, many systems in full; official or heavy commercial use. Arranged by subject. Most in Pictorial Archive series. 143pp. 8⅛ × 11. 23357-X Pa. $5.95

INCIDENTS OF TRAVEL IN YUCATAN, John L. Stephens. Classic (1843) exploration of jungles of Yucatan, looking for evidences of Maya civilization. Travel adventures, Mexican and Indian culture, etc. Total of 669pp. 5⅜ × 8½. 20926-1, 20927-X Pa., Two-vol. set $9.90

CATALOG OF DOVER BOOKS

DEGAS: An Intimate Portrait, Ambroise Vollard. Charming, anecdotal memoir by famous art dealer of one of the greatest 19th-century French painters. 14 black-and-white illustrations. Introduction by Harold L. Van Doren. 96pp. 5⅜ × 8½.
25131-4 Pa. $3.95

PERSONAL NARRATIVE OF A PILGRIMAGE TO ALMANDINAH AND MECCAH, Richard Burton. Great travel classic by remarkably colorful personality. Burton, disguised as a Moroccan, visited sacred shrines of Islam, narrowly escaping death. 47 illustrations. 959pp. 5⅜ × 8½. 21217-3, 21218-1 Pa., Two-vol. set $17.90

PHRASE AND WORD ORIGINS, A. H. Holt. Entertaining, reliable, modern study of more than 1,200 colorful words, phrases, origins and histories. Much unexpected information. 254pp. 5⅜ × 8½. 20758-7 Pa. $4.95

THE RED THUMB MARK, R. Austin Freeman. In this first Dr. Thorndyke case, the great scientific detective draws fascinating conclusions from the nature of a single fingerprint. Exciting story, authentic science. 320pp. 5⅜ × 8½. (Available in U.S. only) 25210-8 Pa. $5.95

AN EGYPTIAN HIEROGLYPHIC DICTIONARY, E. A. Wallis Budge. Monumental work containing about 25,000 words or terms that occur in texts ranging from 3000 B.C. to 600 A.D. Each entry consists of a transliteration of the word, the word in hieroglyphs, and the meaning in English. 1,314pp. 6⅝ × 10.
23615-3, 23616-1 Pa., Two-vol. set $27.90

THE COMPLEAT STRATEGYST: Being a Primer on the Theory of Games of Strategy, J. D. Williams. Highly entertaining classic describes, with many illustrated examples, how to select best strategies in conflict situations. Prefaces. Appendices. xvi + 268pp. 5⅜ × 8½. 25101-2 Pa. $5.95

THE ROAD TO OZ, L. Frank Baum. Dorothy meets the Shaggy Man, little Button-Bright and the Rainbow's beautiful daughter in this delightful trip to the magical Land of Oz. 272pp. 5⅜ × 8. 25208-6 Pa. $4.95

POINT AND LINE TO PLANE, Wassily Kandinsky. Seminal exposition of role of point, line, other elements in non-objective painting. Essential to understanding 20th-century art. 127 illustrations. 192pp. 6½ × 9¼. 23808-3 Pa. $4.50

LADY ANNA, Anthony Trollope. Moving chronicle of Countess Lovel's bitter struggle to win for herself and daughter Anna their rightful rank and fortune—perhaps at cost of sanity itself. 384pp. 5⅜ × 8½. 24669-8 Pa. $6.95

EGYPTIAN MAGIC, E. A. Wallis Budge. Sums up all that is known about magic in Ancient Egypt: the role of magic in controlling the gods, powerful amulets that warded off evil spirits, scarabs of immortality, use of wax images, formulas and spells, the secret name, much more. 253pp. 5⅜ × 8½. 22681-6 Pa. $4.00

THE DANCE OF SIVA, Ananda Coomaraswamy. Preeminent authority unfolds the vast metaphysic of India: the revelation of her art, conception of the universe, social organization, etc. 27 reproductions of art masterpieces. 192pp. 5⅜ × 8½.
24817-8 Pa. $5.95

CHRISTMAS CUSTOMS AND TRADITIONS, Clement A. Miles. Origin, evolution, significance of religious, secular practices. Caroling, gifts, yule logs, much more. Full, scholarly yet fascinating; non-sectarian. 400pp. 5⅜ × 8½.
23354-5 Pa. $6.50

THE HUMAN FIGURE IN MOTION, Eadweard Muybridge. More than 4,500 stopped-action photos, in action series, showing undraped men, women, children jumping, lying down, throwing, sitting, wrestling, carrying, etc. 390pp. 7⅞ × 10⅝.
20204-6 Cloth. $19.95

THE MAN WHO WAS THURSDAY, Gilbert Keith Chesterton. Witty, fast-paced novel about a club of anarchists in turn-of-the-century London. Brilliant social, religious, philosophical speculations. 128pp. 5⅜ × 8½.
25121-7 Pa. $3.95

A CEZANNE SKETCHBOOK: Figures, Portraits, Landscapes and Still Lifes, Paul Cezanne. Great artist experiments with tonal effects, light, mass, other qualities in over 100 drawings. A revealing view of developing master painter, precursor of Cubism. 102 black-and-white illustrations. 144pp. 8¾ × 6⅜.
24790-2 Pa. $5.95

AN ENCYCLOPEDIA OF BATTLES: Accounts of Over 1,560 Battles from 1479 B.C. to the Present, David Eggenberger. Presents essential details of every major battle in recorded history, from the first battle of Megiddo in 1479 B.C. to Grenada in 1984. List of Battle Maps. New Appendix covering the years 1967–1984. Index. 99 illustrations. 544pp. 6½ × 9¼.
24913-1 Pa. $14.95

AN ETYMOLOGICAL DICTIONARY OF MODERN ENGLISH, Ernest Weekley. Richest, fullest work, by foremost British lexicographer. Detailed word histories. Inexhaustible. Total of 856pp. 6½ × 9¼.
21873-2, 21874-0 Pa., Two-vol. set $17.00

WEBSTER'S AMERICAN MILITARY BIOGRAPHIES, edited by Robert McHenry. Over 1,000 figures who shaped 3 centuries of American military history. Detailed biographies of Nathan Hale, Douglas MacArthur, Mary Hallaren, others. Chronologies of engagements, more. Introduction. Addenda. 1,033 entries in alphabetical order. xi + 548pp. 6½ × 9¼. (Available in U.S. only)
24758-9 Pa. $11.95

LIFE IN ANCIENT EGYPT, Adolf Erman. Detailed older account, with much not in more recent books: domestic life, religion, magic, medicine, commerce, and whatever else needed for complete picture. Many illustrations. 597pp. 5⅜ × 8½.
22632-8 Pa. $8.50

HISTORIC COSTUME IN PICTURES, Braun & Schneider. Over 1,450 costumed figures shown, covering a wide variety of peoples: kings, emperors, nobles, priests, servants, soldiers, scholars, townsfolk, peasants, merchants, courtiers, cavaliers, and more. 256pp. 8⅜ × 11¼.
23150-X Pa. $7.95

THE NOTEBOOKS OF LEONARDO DA VINCI, edited by J. P. Richter. Extracts from manuscripts reveal great genius; on painting, sculpture, anatomy, sciences, geography, etc. Both Italian and English. 186 ms. pages reproduced, plus 500 additional drawings, including studies for Last Supper, Sforza monument, etc. 860pp. 7⅞ × 10¾. (Available in U.S. only) 22572-0, 22573-9 Pa., Two-vol. set $25.90

THE ART NOUVEAU STYLE BOOK OF ALPHONSE MUCHA: All 72 Plates from "Documents Decoratifs" in Original Color, Alphonse Mucha. Rare copyright-free design portfolio by high priest of Art Nouveau. Jewelry, wallpaper, stained glass, furniture, figure studies, plant and animal motifs, etc. Only complete one-volume edition. 80pp. 9⅜ × 12¼. 24044-4 Pa. $8.95

ANIMALS: 1,419 COPYRIGHT-FREE ILLUSTRATIONS OF MAMMALS, BIRDS, FISH, INSECTS, ETC., edited by Jim Harter. Clear wood engravings present, in extremely lifelike poses, over 1,000 species of animals. One of the most extensive pictorial sourcebooks of its kind. Captions. Index. 284pp. 9 × 12.
23766-4 Pa. $9.95

OBELISTS FLY HIGH, C. Daly King. Masterpiece of American detective fiction, long out of print, involves murder on a 1935 transcontinental flight—"a very thrilling story"—NY Times. Unabridged and unaltered republication of the edition published by William Collins Sons & Co. Ltd., London, 1935. 288pp. 5⅜ × 8½. (Available in U.S. only) 25036-9 Pa. $4.95

VICTORIAN AND EDWARDIAN FASHION: A Photographic Survey, Alison Gernsheim. First fashion history completely illustrated by contemporary photographs. Full text plus 235 photos, 1840–1914, in which many celebrities appear. 240pp. 6½ × 9¼. 24205-6 Pa. $6.00

THE ART OF THE FRENCH ILLUSTRATED BOOK, 1700–1914, Gordon N. Ray. Over 630 superb book illustrations by Fragonard, Delacroix, Daumier, Doré, Grandville, Manet, Mucha, Steinlen, Toulouse-Lautrec and many others. Preface. Introduction. 633 halftones. Indices of artists, authors & titles, binders and provenances. Appendices. Bibliography. 608pp. 8⅜ × 11¼. 25086-5 Pa. $24.95

THE WONDERFUL WIZARD OF OZ, L. Frank Baum. Facsimile in full color of America's finest children's classic. 143 illustrations by W. W. Denslow. 267pp. 5⅜ × 8½. 20691-2 Pa. $5.95

FRONTIERS OF MODERN PHYSICS: New Perspectives on Cosmology, Relativity, Black Holes and Extraterrestrial Intelligence, Tony Rothman, et al. For the intelligent layman. Subjects include: cosmological models of the universe; black holes; the neutrino; the search for extraterrestrial intelligence. Introduction. 46 black-and-white illustrations. 192pp. 5⅜ × 8½. 24587-X Pa. $6.95

THE FRIENDLY STARS, Martha Evans Martin & Donald Howard Menzel. Classic text marshalls the stars together in an engaging, non-technical survey, presenting them as sources of beauty in night sky. 23 illustrations. Foreword. 2 star charts. Index. 147pp. 5⅜ × 8½. 21099-5 Pa. $3.50

FADS AND FALLACIES IN THE NAME OF SCIENCE, Martin Gardner. Fair, witty appraisal of cranks, quacks, and quackeries of science and pseudoscience: hollow earth, Velikovsky, orgone energy, Dianetics, flying saucers, Bridey Murphy, food and medical fads, etc. Revised, expanded In the Name of Science. "A very able and even-tempered presentation."—The New Yorker. 363pp. 5⅜ × 8.
20394-8 Pa. $6.50

ANCIENT EGYPT: ITS CULTURE AND HISTORY, J. E Manchip White. From pre-dynastics through Ptolemies: society, history, political structure, religion, daily life, literature, cultural heritage. 48 plates. 217pp. 5⅜ × 8½. 22548-8 Pa. $4.95

SIR HARRY HOTSPUR OF HUMBLETHWAITE, Anthony Trollope. Incisive, unconventional psychological study of a conflict between a wealthy baronet, his idealistic daughter, and their scapegrace cousin. The 1870 novel in its first inexpensive edition in years. 250pp. 5⅜ × 8½. 24953-0 Pa. $4.95

LASERS AND HOLOGRAPHY, Winston E. Kock. Sound introduction to burgeoning field, expanded (1981) for second edition. Wave patterns, coherence, lasers, diffraction, zone plates, properties of holograms, recent advances. 84 illustrations. 160pp. 5⅜ × 8¼. (Except in United Kingdom) 24041-X Pa. $3.50

INTRODUCTION TO ARTIFICIAL INTELLIGENCE: SECOND, ENLARGED EDITION, Philip C. Jackson, Jr. Comprehensive survey of artificial intelligence—the study of how machines (computers) can be made to act intelligently. Includes introductory and advanced material. Extensive notes updating the main text. 132 black-and-white illustrations. 512pp. 5⅜ × 8½. 24864-X Pa. $8.95

HISTORY OF INDIAN AND INDONESIAN ART, Ananda K. Coomaraswamy. Over 400 illustrations illuminate classic study of Indian art from earliest Harappa finds to early 20th century. Provides philosophical, religious and social insights. 304pp. 6⅜ × 9⅜. 25005-9 Pa. $8.95

THE GOLEM, Gustav Meyrink. Most famous supernatural novel in modern European literature, set in Ghetto of Old Prague around 1890. Compelling story of mystical experiences, strange transformations, profound terror. 13 black-and-white illustrations. 224pp. 5⅜ × 8½. (Available in U.S. only) 25025-3 Pa. $5.95

ARMADALE, Wilkie Collins. Third great mystery novel by the author of *The Woman in White* and *The Moonstone*. Original magazine version with 40 illustrations. 597pp. 5⅜ × 8½. 23429-0 Pa. $7.95

PICTORIAL ENCYCLOPEDIA OF HISTORIC ARCHITECTURAL PLANS, DETAILS AND ELEMENTS: With 1,880 Line Drawings of Arches, Domes, Doorways, Facades, Gables, Windows, etc., John Theodore Haneman. Sourcebook of inspiration for architects, designers, others. Bibliography. Captions. 141pp. 9 × 12. 24605-1 Pa. $6.95

BENCHLEY LOST AND FOUND, Robert Benchley. Finest humor from early 30's, about pet peeves, child psychologists, post office and others. Mostly unavailable elsewhere. 73 illustrations by Peter Arno and others. 183pp. 5⅜ × 8½. 22410-4 Pa. $3.95

ERTÉ GRAPHICS, Erté. Collection of striking color graphics: *Seasons, Alphabet, Numerals, Aces* and *Precious Stones*. 50 plates, including 4 on covers. 48pp. 9⅜ × 12¼. 23580-7 Pa. $6.95

THE JOURNAL OF HENRY D. THOREAU, edited by Bradford Torrey, F. H. Allen. Complete reprinting of 14 volumes, 1837–61, over two million words; the sourcebooks for *Walden*, etc. Definitive. All original sketches, plus 75 photographs. 1,804pp. 8½ × 12¼. 20312-3, 20313-1 Cloth., Two-vol. set $80.00

CASTLES: THEIR CONSTRUCTION AND HISTORY, Sidney Toy. Traces castle development from ancient roots. Nearly 200 photographs and drawings illustrate moats, keeps, baileys, many other features. Caernarvon, Dover Castles, Hadrian's Wall, Tower of London, dozens more. 256pp. 5⅜ × 8¼. 24898-4 Pa. $5.95

AMERICAN CLIPPER SHIPS: 1833–1858, Octavius T. Howe & Frederick C. Matthews. Fully-illustrated, encyclopedic review of 352 clipper ships from the period of America's greatest maritime supremacy. Introduction. 109 halftones. 5 black-and-white line illustrations. Index. Total of 928pp. 5⅜ × 8½.
25115-2, 25116-0 Pa., Two-vol. set $17.90

TOWARDS A NEW ARCHITECTURE, Le Corbusier. Pioneering manifesto by great architect, near legendary founder of "International School." Technical and aesthetic theories, views on industry, economics, relation of form to function, "mass-production spirit," much more. Profusely illustrated. Unabridged translation of 13th French edition. Introduction by Frederick Etchells. 320pp. 6⅛ × 9¼. (Available in U.S. only)
25023-7 Pa. $8.95

THE BOOK OF KELLS, edited by Blanche Cirker. Inexpensive collection of 32 full-color, full-page plates from the greatest illuminated manuscript of the Middle Ages, painstakingly reproduced from rare facsimile edition. Publisher's Note. Captions. 32pp. 9⅜ × 12¼.
24345-1 Pa. $4.95

BEST SCIENCE FICTION STORIES OF H. G. WELLS, H. G. Wells. Full novel *The Invisible Man*, plus 17 short stories: "The Crystal Egg," "Aepyornis Island," "The Strange Orchid," etc. 303pp. 5⅜ × 8½. (Available in U.S. only)
21531-8 Pa. $4.95

AMERICAN SAILING SHIPS: Their Plans and History, Charles G. Davis. Photos, construction details of schooners, frigates, clippers, other sailcraft of 18th to early 20th centuries—plus entertaining discourse on design, rigging, nautical lore, much more. 137 black-and-white illustrations. 240pp. 6⅛ × 9¼.
24658-2 Pa. $5.95

ENTERTAINING MATHEMATICAL PUZZLES, Martin Gardner. Selection of author's favorite conundrums involving arithmetic, money, speed, etc., with lively commentary. Complete solutions. 112pp. 5⅜ × 8½.
25211-6 Pa. $2.95

THE WILL TO BELIEVE, HUMAN IMMORTALITY, William James. Two books bound together. Effect of irrational on logical, and arguments for human immortality. 402pp. 5⅜ × 8½.
20291-7 Pa. $7.50

THE HAUNTED MONASTERY and THE CHINESE MAZE MURDERS, Robert Van Gulik. 2 full novels by Van Gulik continue adventures of Judge Dee and his companions. An evil Taoist monastery, seemingly supernatural events; overgrown topiary maze that hides strange crimes. Set in 7th-century China. 27 illustrations. 328pp. 5⅜ × 8½.
23502-5 Pa. $5.95

CELEBRATED CASES OF JUDGE DEE (DEE GOONG AN), translated by Robert Van Gulik. Authentic 18th-century Chinese detective novel; Dee and associates solve three interlocked cases. Led to Van Gulik's own stories with same characters. Extensive introduction. 9 illustrations. 237pp. 5⅜ × 8½.
23337-5 Pa. $4.95

Prices subject to change without notice.
Available at your book dealer or write for free catalog to Dept. GI, Dover Publications, Inc., 31 East 2nd St., Mineola, N.Y. 11501. Dover publishes more than 175 books each year on science, elementary and advanced mathematics, biology, music, art, literary history, social sciences and other areas.